CLIMATE CHANGE, NEW SECURITY CHALLENGES AND THE UNITED NATIONS

CLIMATE CHANGE, NEW SECURITY CHALLENGES AND THE UNITED NATIONS

Dr Sabita Mohapatra

Routledge
Taylor & Francis Group

LONDON AND NEW YORK

First published 2018
by Routledge
4 Park Square, Milton Park, Abingdon, Oxon OX14 4RN
605 Third Avenue, New York, NY 10017

First issued in paperback 2023

Routledge is an imprint of the Taylor & Francis Group, an informa business

© 2018 Sabita Mohapatra and KW Publishers Pvt Ltd

The right of Sabita Mohapatra to be identified as author of this work has been asserted by her in accordance with sections 77 and 78 of the Copyright, Designs and Patents Act 1988.

Publisher's Note
The publisher has gone to great lengths to ensure the quality of this reprint but points out that some imperfections in the original copies may be apparent.

Print edition not for sale in South Asia (India, Sri Lanka, Nepal, Bangladesh, Afghanistan, Pakistan or Bhutan)

British Library Cataloguing-in-Publication Data
A catalogue record for this book is available from the British Library

Library of Congress Cataloging-in-Publication Data
A catalog record for this book has been requested

ISBN-13: 978-1-138-24379-8 (hbk)
ISBN-13: 978-1-03-265245-0 (pbk)
ISBN-13: 978-1-315-27208-5 (ebk)

DOI: 10.4324/9781315272085

Typeset in Times New Roman
by KW Publishers

KNOWLEDGE WORLD

CONTENTS

ACKNOWLEDGEMENTS

The doors of knowledge are always anchored in the attitude of gratitude that I am to express to those and sources I always remain indebted to. Acknowledgement is one way of such expression. This kind of humility is not belittling one's knowledge rather opening one's door for divergent ideas to come in for strengthening and refurbishing your ideas and thinking. It means demystifying ethnocentrism of knowledge. Ideas remain abstruse and have the chance of being eroded into oblivion unless tried to be put in black and white. Participation in national seminar sponsored by UGC as keynote speaker on climate change propelled me to think of writing a book on this, which ultimately finds its reification. This long intellectual stride commenced from the day I conceived of inking a book on this taking not only my own ideas but also ideas and facts from various conferences and seminars abroad and in India I attended, imbibing new thinking, approaches and views from scholars I met, and from books I rummaged through, getting articles, and papers published in various journals of international repute, and reached its logical conclusion when all these variant strands of ideas seeped through my in depth analytical acumen and rigour of scientific scrutiny to be blended nicely into a synthesised and pithy tapestry. If the collections from others are likened to bricks, mine is the architectural design, adding blood and flesh to the structure and finally shaping the ultimate attic in the form of this book. But for the help and cooperation from others it could not have been possible to bring out this book in the present form. My unbounded thanks and gratitude flow from the core of my heart to a litany of scholars, intellectuals and critics who have indirectly or directly helped me in this monumental task. From the fathomless depth of my heart I always convey my sense of gratification to a host of erudite scholars for the materials, articles and books they have provided me without exhibiting even a shred of evidence of their being hesitant.

Nothing happens in this universe even a blade of grass does not grow without God's will. What I did here is due to His will and bliss. I proffer me heartfelt love and prayer to God and Gurus for showering on me their choicest blessings.

I am very much grateful to Prof. Dr Narottam Gaan, Prof. Dr Brahmananda Satpathy of Deptt. of Political Science, Utkal University, Bhubaneswar, Prof. Dr Prafulla Kumar Mishra, Vice Chancellor, Uttar Odisha University, Baripada, Dr. Ajit Kumar Tripathy, Ex-Chief Secretary and Development Commissioner, Ex-State Election Commissioner, Odisha, Bhubaneswar, Dr Prafulla Chandra Tripathy, an eminent Historian and educationist, Sri Niranjan Rath, an educationist and General Secretary, Loksevak Mandal, New Delhi, Sri Kaliprasad Samantray, an educationist, Bhubaneswar and Smt. Minati Mohapatra (my mother) for their valuable advice and guidance.

Dr **Sabita Mohapatra**

PREFACE

In the conventional paradigm the concept of security is essentially defined and articulated in terms of and around state and its corresponding military apparatus. In the state centric conceptualisation of security, the sources of security threats to the other remain anchored in the state itself. In international relations the security threat is always configured in terms of the other/outside/external. In the internal domain the sources of providing security to the insiders remain with the state. In its strive for perpetuating its survival and identity as a moral political community the state secures its own citizens. In this way the state remains predominantly the main referent object of security. This kind of conventionally established thinking and conceptualisation about security hovering on state continues to persist as the foundational mainstream of the foreign policy of major powers. Since the anarchical international system has been the guiding metaphor for states to position themselves in a self-help situation, the states taking on to them the onus of providing security to its insiders, ensuring survival and identity remain busy increasing competitively the military capability to overshadow the other in their bids to obviate security dilemma. It becomes their primary goal to look more invulnerable and impregnable militarily. This kind of understanding and conceptualising security centring on state precludes the policy makers from unveiling the other sources of security threats which are not necessarily state centric. Nowhere this was seen more glaringly than in the foreign policy of the two super powers during the cold war and other powers following them. When the two superpowers, former Soviet Union and the United States were engaged in a military showdown in Cuban Missile crisis in 1962 virtually on the brink of a nuclear Armageddon, the world could little notice the explosive significance of the theme that Rachel Carson points to in her famous book 'The Silent Spring' published in the same year contesting the traditional state centric security. It is not the weapons that pose security threats but the use of pesticides, insecticides, fertilizers and chemicals in agriculture that can kill human beings. What she raised in this book is worth taking note of. Can the environmental crisis be not a security threat if the loss of human beings is the criterion? It matters little whether threats are military or non-military. Such an important issue raised in 1960s by Carson could hardly find any place in policy making and could not fit rightly into the procrustean bed of realism and neo-realism as to be securitised.

In the cacophony of sabre rattling that was taking place in the backyards of the arsenals of major powers and others was not heard the alarming note that Carson was tendering to the world community not until 1980s when there was an intellectual upsurge in the west echoing the former and contesting the adequacy of security under the panoply of realism and neo-realism. Amongst the proliferation of non-traditional sources of security threats in recent years what stands prominent is the environmental crisis and climate change which was due to the relentless interference of human beings into the functioning of nature in the name of development. The consequences of climate change in terms of global warming, sea level rise, storms and cyclones, spread of diseases and others could not even be addressed by the nuclear weapons. The question started nagging what is the use of spending frantically so much on weapons when these utterly fail to meet the devastating challenges of climate change. During the 1990s and more particularly after the end of the cold war, the world started thinking very seriously and accordingly taking action to address these impending dangers. What was neglected and could not find any place in the security agenda of policy makers has now come to the forefront of policy making of all countries of the world. Apart from this, many international institutions, NGOs and over and above all the United Nations have played a pioneering role in putting pressures on all states to take certain positive steps in not allowing the temperature of the earth increase beyond 2 degree Celsius as warned by IPCC.

The landscape of international politics was dotted by specks of conflicts and war between and among nations. The issues were understood in realist and neo-realist idioms and tried to be solved by military means. The United Nations was the reflection of super power politics. The super powers' national interest, strategic goals and global outlook became the guiding principle of the UN which ultimately turned out to be a cold war in miniature. With climate change making the extinction of humanity a near certainty the United Nations has been playing an active role in making the conscience of all the nations of the world move towards addressing the issue collectively. Under the aegis of United Nations Framework Convention on Climate Change(UNFCC) the leaders of the nations have met at Rio, Kyoto, Copenhagen, Durban and now at Paris to chalk out certain and concrete measures for addressing this non-traditional security challenge from climate change.

Changing climate patterns affect both inter-state relations and international and national security in a narrow, geo-strategic sense as well as the well-being and survival of human beings. Climate change increasingly undermines human security and will increasingly do so in the future, by reducing access to, and the quality of, natural resources that are important to sustain livelihoods. Climate change is also likely to undermine the capacity of states to provide the opportunities and services that help people to sustain their livelihoods. Climate change may increase the risk of violent conflict.

Environmental changes can threaten global, national and human security. Environmental issues include land degradation, climate change, water quality and quantity, and the management and distribution of natural resources such as oil, forests and minerals. Climate change will trigger enormous physical and social changes like water shortages, natural disasters, decreased agricultural productivity, increased rates and scope of infectious diseases, and shifts in human migration. These could impact significantly on international security which may create humanitarian crises.

The Intergovernmental Panel on Climate Change (IPCC) of its Fourth Assessment Report in 2007 showed the poor planet is most likely to suffer the worst effects of climate change. The 'poor' includes the frequently invisible group that is, women. 70 per cent living in poverty line are women for whom climate change represents very specific threats to security. Women are the primary managers of family, food, water and health. Human security takes the basic needs of food, water, health and livelihood - the issues addressed in the Millennium Developments Goals.

The development of the human security framework by the global 1994 Human Development Report (HDR) of the United Nations Development Programme (UNDP) was a pioneering step. The report shifted the focus of security from the protection of the state and its borders by military means to the protection of individuals from a wider range of threats to their well-being and security, and by a wider range of measures and policies, from the local and community levels to the national and international.

There are three different levels of food security (global, national and household food security) to describe the effects of climate change. Climate change will affect all four dimensions of food security: food availability, food accessibility, food utilisation and food systems stability. Agriculture based livelihood systems that are already vulnerable to food insecurity face immediate risk of increased crop failure, new patterns of pests and diseases, lack of appropriate seeds and loss of livestock. Low-income people in urban areas will be at risk of food insecurity in developing and developed countries. Food systems will also be affected through internal and international migration, resource-based conflicts and civil unrest triggered by climate change and its impacts. Agriculture, forestry and fisheries will not only be affected by climate change, but also contribute through emitting greenhouse gases.

The key concept in the United Nations to measure the food and nutrition situation of people and groups is the term "food security". UNICEF and other nutritionists' emphasise the fact that half of the children dying from hunger are not dying because of missing food supply, but because of undigested food due to sickness. The impact of climate change will be particularly on smallholder and subsistence agriculture. The farming system will be directly affected by changing weather patterns, sea level rise, and the increase in frequency and intensity of extreme events. The nations of the world are being able to meet the two degree Celsius target of maximum average temperature rise set by the UNFCCC negotiations in Cancun. Agricultural activities include indirect effects through deforestation and forms of land conservation account for about one-third of total global warming potential from GHG emissions today, so reducing the direct and indirect emissions from agriculture is an essential part of the larger effort to slow the pace of climate change.

Adaptation is reducing and responding to the risks, climate change poses to people's lives and livelihoods. Adaptation management can particularly be a valuable tool for devising risks to which different ecosystems and livelihood groups are exposed. Mitigating climate change means reducing greenhouse gas emissions and storing carbon in the short term and curbing emissions in the long-term.

Several funds within the United Nations system finance specific activities aimed at reducing greenhouse gas emissions and increasing resilience to the negative impacts of climate change. Because many mitigation actions that would have high payoffs also represent good options for adaptation within the food and agriculture sectors of low-income developing countries to obtain additional resources from bilateral and multilateral aid agencies. The ultimate goal of FAO's climate change work is to inform and promote local dialogue about the impacts of climate change for reducing vulnerability and to provide local communities the specific solutions.

One recommendation to the global community of policy makers, was collectively to create an enabling environment for a sustainable global food system. Climate change will impact on both irrigated and rain-fed agriculture across the globe. Both the livelihoods of rural communities and the food security of a predominantly urban population are at risk from water related impacts linked primarily to climate variability. Various adaptation measures that deal with climate variability and water management practices have the potential to create resilience to climate change and to enhance water security. A focus on water security is a sound early adaptation strategy; delivering immediate benefits to vulnerable and underserved populations, thus advancing the millennium development goals, while strengthening systems and capacity for longer-term climate risk management. A water secure world will need investment in the three I's: better and more accessible 'Information', stronger and more adaptable 'Institution', and natural and man-made 'Infrastructure' to store, transport and treat water.

Climate change affects human health in all sectors of society, both domestically and globally. The sea-level rise, changes in precipitation resulting in flooding and drought, heat waves, more intense hurricanes and storms, and degraded air quality, will affect human health, both directly and indirectly. The effects of climate change on human health are challenging for both the surrounding environment and people those influence health. Certain adverse health effects can be avoided if decisions made prior in identification of vulnerable populations such as children and elderly ensured to preventive measures. For this reason, an adhoc Interagency Working Group on Climate Change and Health (IWGCCH) assembled to develop research on mitigation and adaptation strategies.

The gender differences in health risks are exacerbated by climate change and the adaptation and mitigation measures can help to protect and promote health. The aim is to provide a framework to strengthen World Health Organisation (WHO) support to member states in developing health risks assessments and climate policy interventions that are beneficial to both women and men. These research and science needs broadly include basic and applied science, technological innovations and capacities, public health infrastructure, and communication and education.

Climate change leads to natural disasters. The number of disasters linked to natural hazards on human lives, livelihoods, assets and economies. Disaster impacts undermine livelihoods and progress towards poverty reduction and the Millennium Development Goals. Disaster events are hydrological, meteorological or climatological in nature. The 2004 Indian Ocean tsunami triggered in promoting ecosystem management approaches for reducing disaster risk.

In 2005, the Hyogo Framework for Action, the first global agreement on disaster reduction, recognised the importance of sustainable ecosystems and environmental management in reducing disaster risk. The UNFCCC negotiations for a global climate agreement and in particular since the conference of parties (COP) in Copenhagen in 2009, ecosystem-based approaches have been recognised as a key climate change adaptation strategy. Therefore, sustainable ecosystems management is an effective approach for achieving both disaster risk reduction and climate change adaptation priorities. Ecosystem-based approaches are cost-effective, but also locally accessible and applicable and World Bank plays great role in disaster reduction strategies and ecosystem-based approach.

National societies can make a major contribution to global efforts in all four areas of the federation strategy 2010 - disaster preparedness, disaster response, health and care in the community, and principles and humanitarian values - are critical elements of the response to weather and climate related disasters. Climate change and disaster risk reduction are closely linked. More extreme weather events in future are likely to increase the number and scale of disasters, while at the same time, the existing methods and tools of disaster risk reduction provide powerful capacities for adaptation to climate change. For example, recently on October 10, 2013 in Odisha, 99.5 per cent people were saved due to early warning and alertness of the local government from severe Phailin Cyclone.

The United Nations International Strategy for Disaster Reduction (UNISDR) outlines the nature and significance of climate change for disaster risk, as well as the perspectives and approaches of disaster risk reduction and adaptation strategies. In 1990, the Intergovernmental Panel on Climate Change (IPCC) noted that the greatest impact of climate change could be on human migration with millions of people displaced due to natural disasters such as

shoreline erosion, coastal flooding and agricultural disruption. Displacement due to natural disasters of climate change can be divided into two ways - (i) climate processes, such as sea-level rise, salinisation of agricultural land, desertification and growing water scarcity, and (ii) climate events, such as flooding, storms and glacial lake outburst floods.

Temporary migration as an adaptive response to climate stress is already apparent in many areas. However, there has been a collective and successful attempt to ignore the scale of problem. Forced climate migrants include international refugee and immigration policy and political refugees to incorporate climate "refugee". There is no "home" for climate migrants in the international community, both literally and figuratively. At COP 16, parties developed the Cancun Adaptation Framework, which helped identify areas of activities that qualified as "adaptation" and which later may be considered for climate finance support. On November 15, 2011, the Global Migration Group adopted a joint statement on the impact of climate change on migration-to be adopted at the level of United Nations.

The adverse impacts of climate change, an analysis of their security implications, and policy recommendations for strengthening the 'United Nations' capacity to respond to climate related security threats are the vital functions of the UNFCCC. In 2007, the UN Security Council held its first debate on climate change and its implications and emergency appeals for humanitarian aid for international security. Unmitigated climate change beyond 2 degree Celsius will lead to unprecedented security scenarios. Investment in mitigation to avoid such scenarios will address the international security threats created by climate change. It is viewed as a treat multiplier which exacerbates existing trends, tensions and instability.

COPs have adopted a number of decisions linking disaster risk reduction to climate change adaptation. These decisions include the "Adaptation Committee" and the "Loss And Damage" work programme that were detailed in COP 17, in Durban, South Africa in 2011 as part of the actions envisioned in the "Cancun Adaptation Framework" adopted, in Cancun, Mexico 2010 at COP 16. "The Bali Action Plan" under the Ad-hoc Working Group on Long-Term Cooperative Action under the convention (AWG-LCA), Strengthening Disaster Risk Reduction in climate change adaptation related agendas at COP 18, Doha, Qatar, and "Nairobi Work Programme" on impacts, vulnerability and adaptation to climate change have considered and supported stronger efforts to reduce the risks of disasters.

Security is fundamental to people's livelihoods, reducing poverty and achieving the Millennium Development Goals. It relates to personal and state safety, access to social services and political processes. It is a core government responsibility, necessary for economic and social development and vital for the protection of human rights. Organisation for Economic Co-operation and Development (OECD), governments and their development actors aim to help partner countries establish appropriate structures and mechanisms to manage change and resolve disputes through democratic and peaceful means. The Security System Reform (SSR) is a key component of the broader "human security" agenda, developed with leadership from the United Nations Development Programme (UNDP) and described in Human Security Now, the report of the UN commission on Human Security.

The Development Assistance Committee (DAC) helps their own governments, organisations, developing countries and international organisations to reinforce work on SSR. This requires strategic planning for improved policies, practices and partnerships amongst all actors. The DAC commitment to work on the security and development nexus agreed in the DAC Guidelines and Policy Statement: "Helping Prevent Violent Conflict".

The present book is an attempt to unravel the new non-traditional challenges that the United Nations will face in coming decades if the failure of the states to keep the climate change fixed at the threshold i.e. 2 degree Celsius as agreed to COP 21 conference in Paris by all the states.

1. INTRODUCTION

Climate change and its adverse impact and an analysis of their security implications and policy recommendations for strengthening United Nations' capacity to respond to climate related security threats have become the main theme of this chapter.

The global scientific consensus says that the earth's atmosphere is warming rapidly on an unprecedented scale due to human activity. The Intergovernmental Panel on Climate Change (IPCC), the multi-lateral body predicts that global warming will trigger enormous physical and social changes. The implications of climate change, they identified are the following physical and socio-economic effects:

The physical effects of climate change include,

- Higher average surface and ocean temperatures.
- More rainfall globally from increased evaporation.
- More variability in rainfall and temperature with more frequent and severe floods and droughts.
- Rising sea levels from warming water, expanded further by runoff from melting continental ice fields.
- Increased frequency and intensity of extreme weather events e.g. hurricanes and tornadoes.
- Extended ranges and seasons for mosquitoes and other tropical disease carriers.

All socio-economic effects of climate change will not be negative but number of adverse socio-economic impacts are anticipated. These are:

- Shortfalls in water for drinking and irrigation, with concomitant risks of thirst and famine.
- Changes and possible decline in agricultural productivity from temperature, rainfall or pest patterns.
- Increased rates and geographic scope of malaria and other diseases.
- Associated shifts in economic output and trade patterns.
- Changes and possibly large shifts in human migration patterns.

Large economic and human losses are attributable to extreme weather events, such as hurricanes. The security implication of these physical and socio economic changes is significant. What kinds are the nature of threats? And where those threats are more likely to occur. There are many kinds of threats. These are:

Violence and armed conflict
Some researchers have speculated that these changes could be caused due to armed conflict and violence. The general link between the environment and armed conflict is well established. Competition for natural resources (e.g. diamonds, timber, oil and water) has motivated violence in the places as Kuwait, Colombia and Afghanistan. Natural resources will be exploited for quick financial reward when climate change and violence will be strong. Climate change happens gradually. Global warming is to be the primary cause of any particular armed conflict. Regional climate changes and environmental degradation could make armed conflict.

Natural disasters and humanitarian crises
A warmer world will generate more natural disasters and more humanitarian crises. Natural disasters have already been a major security threats between 1990 and 1999; an estimated 188 million people per year were affected by natural disasters, 6 times more than the 31 million people annually affected by armed conflict. Many people affected by natural disasters become refugees or Internally Displaced Persons (IDPs). Both refugees and IDPs are vulnerable not only to the physical and socio economic effects and diseases, malnutrition and loss of income, but they can also become personally insecure and subject to crime and violence. Natural disasters become wider security challenges when a country lacks capability to help affected populations on determining the government's legitimacy and increasing popular grievance.

Destabilising forces
In a weak state drought, disease and economic stagnation may reach a critical level. The global HIV/AIDS has renewed international concern that widespread death from infectious diseases could destabilise vulnerable nations. The vulnerable nations fail to provide effective security, education and health care. A recent study from the World Health Organisation (WHO) and London school of hygiene and tropical medicine estimates more than 160,000 people die annually from ancillary effects of global warming such as malaria and malnutrition. It will be doubled by 2020.

There are three types of vulnerable nations to the security risks of climate change. They are:

Least developed nations
Poor developing countries suffer from the effects of climate change. These states lack the economic, governance or technical capabilities to adapt. They lack the capacity to prevent the humanitarian disasters e.g. weather events, drought, famine and disease.

Weak states
Failed and failing states have a weak government, poor border control, repressed population or marginal economies. Weak states have no capacity to respond to climate change and to prevent humanitarian disasters. In 1990 in Somalia 10,000 people died because of drought, crop failure and also, state failure.

Undemocratic states
20 years ago economists Amartya Sen noted that democratic leaders had to be responsive to people who can vote them out of power, the leaders don't produce famine. But undemocratic states fail to protect populations at a risk of drought, floods and other weather related phenomena. While modern India has never suffered a famine, tens of millions died in China under Mao. North Korea is able to produce nuclear weapons but remains unable to meet its people's basic nutritional needs. Thus populations in undemocratic states will be vulnerable to humanitarian crises induced by the climate change.

The United Nations' strategy for addressing climate change is to facilitate agreements among nations.
- To mitigate those nations' greenhouse gas emissions and concentrate these gases at a safe level and
- Help vulnerable nations to adapt to the adverse consequences of global warming.

While these goals are the right ones, the UN system is not acting

with sufficient ambition or effectiveness to deal with the security risks posed by climate change.

Global warming will continue until concentration of greenhouse gases in the atmosphere, which will only occur after net global annual emissions of these gases decline to zero. Global emissions are still rising rapidly in the majority of nations. A major focus of the UN's climate change security strategy must be to facilitate emission mitigations in both developed and developing nations. Global efforts to arrest climate change have been carried out largely in the context of the 1992 UN Framework Convention on Climate Change and its 1997 Kyoto Protocol. Today, those efforts have produced very modest results. Developed nations ignored the political commitment. They made under convention to return their emissions to 1990 levels by 2000. If the Kyoto Treaty goes into force, it will cover only 25 per cent of global emissions. By 2012, Kyoto will have reduced emissions in participating industrialised countries by only less than 3 per cent below 1990 levels.

The climate architecture associated with the Kyoto Protocol has become increasingly divisive, not only among advanced industrialised countries, but also within the North-South dialogue. Since 1992 earth summit, the environment has lost ground politically submerged under broader sustainable agenda. The Secretary General must raise the visibility of climate change and play a more active role to run speedy mitigation efforts. One complication is that while developed nations should take the lead in reducing emissions, but in developing nations emission mitigation could be more cost effective. Until the international community develops the political will necessary for public private financing of emission reduction.

Security challenges are not met properly as for example Iraq's people remain vulnerable to long term security threats due to climate change and environmental crisis. One option would be for Secretary General to advocate the creation of UN High Commissioner for the environment. The high commissioner's mandate would be to raise global awareness about environmental degradation including climate change. Climate would be only part of agenda. This official should also have a role in building political will to meet other international environmental goals e.g. providing safe drinking water and sanitation for all. Locating the office in Geneva would help integrate environmental concerns and climate change into the UN system in a way that UN environmental programme Nairobi has been unable to accomplish.

A two part of strategy is needed to deal with the adverse effects of climate change. First, the UN should strengthen those programmes that handle disaster and humanitarian crises and that are already beginning to take climate change into account. Second, the UN should create new effort focussed on predicting, preventing and handling climate change related disasters in weak states and those with repressive governments.

Strengthening ongoing disaster work

Shift priority from relief to prevention
Very little money is spent on disaster risk reduction. Even among countries with responsive decision makers, there is too little awareness of the priority of disaster risk reduction. One strategy would be the decision to dedicate at least 5 per cent to 10 per cent of humanitarian relief funds to disaster risk reduction. While the precise target should be resolved by member states, the Secretary General should take the lead in proposing the establishment of such a principle. The UN's Inter Agency Task Force on Disaster Reduction (IATF/DR) and the Inter Agency Secretariat of the International Strategy for Disaster Reduction (UN/ISDR) are existing frameworks which give the early warning systems vulnerability assessments. In January 2005, the second world conference on disaster reduction took place in Kobe, Japan. The parties reviewed the 1994 Yokohama strategy on natural disasters and established the disaster reduction action plan for the next decade. These processes provide opportunities more prominently on prevention.

Integrate disaster and climate planning
The UN system needs to integrate about the topic of climate change, its security, natural disaster prevention and humanitarian response activities. In 2003, UN/ ISDR launched a project to do on disaster reduction and its progress seems promising. IATF/DR created a new working group in May 2004 on climate adaptation and disaster reduction and the UN/ ISDR secretariat is coordinating an expert dialogue among disaster relief, climate and development communities (UN/ISDR-2004). Climate change bodies are in danger of reinventing the wheel on disaster prevention and response. The disaster experts should more fully integrate to avoid the potential problem.

New strategy needed for vulnerable states
These systems would not be adequate to face dangerous security challenges, massive migration, armed conflict and state collapse in undemocratic and weak states. New multipart strategy is needed.

Improve early warning systems and vulnerability indices
The UN system needs for predicting which states and regions are most vulnerable to severe security threats related to climate change. Early warning systems are necessary to deal with different challenges in international community. The UN's Humanitarian Early Warning Systems

(UN/HEWS) is an internal tool to identify countries in pre-crisis situations. Office for the Coordination of Humanitarian Affairs (OCHA) is an external system for natural disasters and complex emergencies. At the regional and country levels, OCHA has an Integrated Regional Information Network. In agriculture, FAO has the global information and Early Warning System on Food and Agriculture. In terms of weather related warning systems UN Development Programme and UN Environment Programme have developed a Disaster Risk Index. World Bank and Columbia University have completed the project on Global Disaster Risk.

Much of this work is positive. But the emerging early warning systems in the disaster reduction community must take political indicators of vulnerability, such as the repressive nature of political regimes and other governance factors.

Preventive diplomacy
UN has identified high risk countries; it should develop contingency plans for climate change. The involvement of UN in systematic forward planning, is not clear. Contingency plan includes plans for providing shelter, nutrition, medicines and policing. The local UN staff should open discreet channels of communication with the decision makers in high risk countries to discuss and encourage risk reduction strategies. UN officials should also share information concerning disaster prevention with relief agencies e.g. the UN High Commissioner for Refugees (UN/HCR), the International Red Cross and NGO relief community.

Conflict and post conflict engagement legitimacy and force
Sometimes diplomatic preparedness will not reduce humanitarian catastrophe. The world will face using force to prevent mass starvation. The security risk posed by climate change needs humanitarian intervention force. The international community needs to revisit the institutional arrangement concerning the use of force in response to disasters just as it is doing with respect to terrorism and weapons of mass destruction. The UN should be facilitating this dialogue including potential climate induced catastrophe for post conflict reconstruction.

The IPCC has summarised that the climate change is and will increasingly have dramatic impacts on ecological and social systems. Climate change will effect some major environmental changes which create development problems and security problems for some individuals, social groups and countries. Climate change means it is global environmental change having impact on social and ecological systems. Environmental change is being understood as a security issue. It may undermine human security reducing the quality of natural resources to sustain livelihoods, undermine the capacity of states to provide services that help sustain livelihoods and creates violence.

Climate change can occur due to both natural and human causes. Human causes induced climate change springs primary from the burning of fossil fuels and changes in land use. These activities release greenhouse gases (e.g. carbon dioxide and methane) with long wave radiation in the climate system and warm the planet. Natural causes of climate change include volcanic eruptions, changes in the sun's activity and changes in ocean circulation. In the last 100 years, number of changes have been observed in the global climate including an increase in global surface temperature, increased precipitation in the eastern parts of North and South America, northern Europe and Central Asia, an increase in the destructiveness of tropical cyclones; a significant decrease in snow cover especially in spring: a reduction in arctic sea ice coverage and a rise in the global mean sea level. In the latter part of the 20th century, climate system has attributed many of these changes to human produced greenhouse gases.

Climate change in the long distant past reveals that the climate system can abruptly change over relatively short time periods. These changes occur when the climate system is forced to cross a threshold and changes dramatically beyond the level from initial forcing. In other words the climate system can be forced to a point where it suddenly flips into another state. These points are collectively known as tipping points which include the shutdown of the Gulf Stream in the Atlantic, the die-back of the Amazon tropical rain forest, stronger and prolonged ENSO events and the collapse of ice sheets leading to large sea level rises. These climate surprises signify massive demographic change, a loss of control of change of the climate system by humans, increase greenhouse gases, large and widespread changes in the atmosphere. At present the IPCC (Intergovernmental Panel on Climate Change) considers that currently tipping points in the climate system occur in the 21st century a project to 2050 that the world may have time to adjust to their potential impact. However, it is recognised that many of the climate feedbacks are responsible for causing abrupt changes in the climate after the initial forcing.

The aim of the UN Framework Convention on Climate Change is to avoid 'dangerous' interference in the climate system and such impacts have been defined as a threat to security.[1] Climate change may increase the risk of violent conflict.[2] The majority of interpretations of environmental security focus on the way environmental change may interact with the same national security concerns that dominated policy throughout the 20th century in particular the way, environmental change may trigger violent conflict.[3]

Security has given stress on vulnerability and attached importance to the most vulnerable entities, for example, the nation (national security), basic needs (human security), income (financial security) and property (home security). In a general sense, security is the condition of being protected from or not exposed to danger. Waever (1995) said that the process of discursively 'securitising' vulnerable referent objects and defining particular risk is a political one. Soroos (1997) defines security as, ''the assurance people have that they will continue to enjoy those things that are most important to their survival and well-being" (p.236). 'Who is to be secured and how environmental change threatens them, environmental change can be considered as a security issue. A question as which environmental problems can be considered security issues'? [4]

From the above statement, it can be argued that climate change is a security issue for some nation states, communities and individuals. In the case of countries like Tuvalu or Kiribati, there is widespread agreement that climate change and associated sea level rise threatens the long term ability of people to remain living on their islands.[5] So, the countries face most serious form of environmental change and security problem. In the Arctic circle where snow cover is less predictable and thinner ice sheets restrict hunting. In Bangladesh families are living in low laying deltas due to frequent floods and the people are suffering diseases spread by mosquitoes due to changing temperature and rain fall in the highlands of Papua new Guinea. So, climate change poses cultural, health and life threatening risks. Climate change is a security issue for certain communities, cultures and countries. 'The UNFCCC is an important security treaty making certain negotiating groups such as the Alliance of Small Island States security coalition'.[6]

The following (figure-1) an heuristic guide to environment and security, the linkages between the particular problem of climate change and security:

Figure I: A Guide to Environment - Security Linkages.

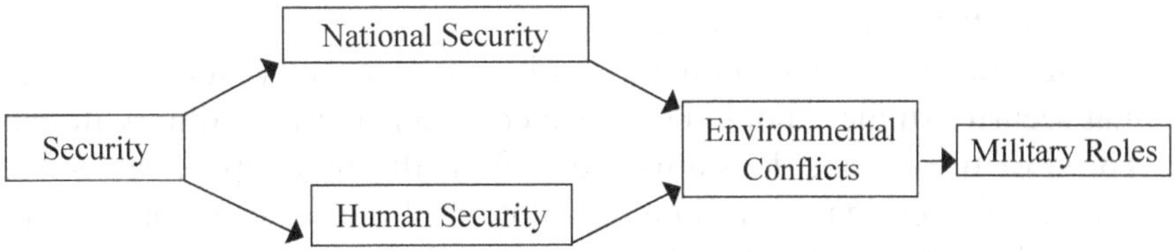

In 1977 Lester Brown explored the links between environmental degradation- including climate change and security. Brown said the deterioration of biophysical systems on national security, identifying four systems under stress: - fisheries, grasslands, forests and croplands. Brown argued that armed forces are incapable of meeting the challenges posed by climate change.

According to Smil (1977), environmental security has replaced the threat of global nuclear warfare. The Toronto conference in 1988 - The first international meeting of scientists and national policy makers to highlight the dangers of climate change was called, "The Changing Atmosphere, Implications for global security". The conference concluded that "humanity is conducting an unintended, uncontrolled, globally pervasive experiment whose ultimate consequences could be second only to be global nuclear war". Another melting of the West Antarctic Ice sheet would cause sea level to rise by some 6 meters.[7] The past two decades have increased the new security threats and risks from environmental issues to terrorism and economic instability.

The traditional conception of security has traditionally restricted its remit to the discussion of threats to a state's security posed by the military activities of other states or terrorist groups. Security has been characterised as, "the effort to protect a population and territory against organised force while advancing state interests through competitive behaviour".[8] The traditional account of security was cold war period. At that time virtually all sectors of security community held that the threat to national security was the capacity of another state to mount a decisive nuclear attack. This reflected the fact that cold war security theorists were pre-occupied with the possibility of decisive strike against the people and institutions of the state. The consequence of this preoccupation was that the politics of nuclear confrontation and nuclear weapons proliferation became the key focus of security scholarship during this period and further that threats to national security that did not have immediacy and decisiveness of nuclear attack were neglected.[9] The view was that threats arising from highly complex and unpredictable biological and physical systems associated with environmental change, could not threaten a state's security.

Several developments have combined in recent years in the view that security studies should be restricted to considering military threats. Three of these considerations are: - first, the breakup of the Soviet Union was a critical event in the evolution of the security concept, this reduced the threat of full scale nuclear confrontation between the super powers and created space for the discussion of other potential threats to national security. Second, the growing awareness of economic relations throughout the world, as manifested in a series of global economic and financial crises. Third, certain environmental problems could endanger the existence of whole communities as well as social problems e.g. poverty, mortality, morbidity, over population and so on. On future extreme will be impacts of environmental degradation and climate change on coastal zones, Small Island States and the developing world in broader terms.[10]

National security and climate change
The impacts of climate change have been considered as a national security. Sea level rise may undermine national security in serious ways, for example: - 45cm rise in sea level will potentially result in a loss of 10-9 per cent of Bangladeshi's territory forcing some 5.5 million people to relocate.[11] There is loss of habitable territory in island due to rising seas and increasing climate variability. The socio- economic impacts of global warming on Islands may be so profound that they dwarf any strategic issue currently confronting a major peace time economy.[12]

National security also has an internal dimension that it is partly a function of state legitimacy. For Bangladesh, 5.5 million refugees pose potentially serious problems for state legitimacy and internal harmony. Climate change means more exogenous shocks to all countries due to increasingly frequent and severe hazardous events. Climate change may have many other indirect negative effects that can undermine legitimacy. They are undermined individuals' collective economic livelihood, affects human health due to lack of fresh water and food, exposing people to new diseases undermining state wealth and military capability and exacerbating inequalities between people.

The impact of climate change will have financial cost and it is justified that climate change is a security issue. Measures implemented to reduce greenhouse gas emissions will impose costs to national economies. It is their assumptions about the cost of reducing emissions and about lost comparative advantage that apparently underlie US reticence on the Kyoto Protocol. Similarly, the Australian government has argued that Kyoto Protocol would adversely affect Australia's economy and so the country's final Kyoto target is a 108 per cent change above 1990 levels of emissions. But it is the oil exporting economies that are arguably the most at risk from an implemented Kyoto Protocol. Most models suggest that policies to implement the Kyoto Protocol will increase oil prices and reduce demand in developed countries which account for 60 per cent of world oil consumption, thereby driving down global oil demand and prices and projected revenues for oil exporters. For example, 0.45 per cent decline in projected GDP in OPEC countries for 2010.[13]

Another point, the national security in military terms defined by Richard Ullman in two key reasons: - first, it ignores the salience of other variables, such as environmental degradation that can be more harmful to the security interests of a state than intra-state or inter-state, military conflict. Second, he outlines the normative objection that to retain a traditional, militarised view of security, 'contributes to a pervasive militarisation of international relations that in the long run can only increase global insecurity'.

Ullman's argument appears to have five steps:

- States commonly devote a great deal of resources to military defence in the belief that this increases national security.
- This investment is justified in terms of the omnipresent risk of military conflict with other states.
- Much research by natural and social, scientists demonstrates that developed and developing states are far more vulnerable to non-military threats (environmental degradation) than they are to traditional armed confrontations with other states.
- In order to protect national security, more attention and resources should be focussed on environmental threats.
- It should be recognised by theorists and policy makers that the concept of national security needs to be broadened to incorporate a sensitive to non- military threats.

It has two advantages; first, it can be easily accommodated within the mainstream of traditional thinking on security. Second, it is compatible with, but not dependent on, the existence of the environmental degradation or violence. Ullman does not reject violence as a trigger of insecurity but rather supplements it with additional triggers.

One problem is that instances of organised violence and environmental degradation often pose different kinds of threats to human communities. Environmental problems such as climate change, ozone depletion and deforestation undermine the security of the states where they hit quite hard in the same way as wars and other violent conflicts seem problematic.

Climate change and violent conflict

Environmental change is a factor leading to violent conflict, environmental change such as degradation of grass lands and boreal and tropical forests, desertification of water resource, stress and coral leaching, climate change may be an exacerbating factor in violent conflict in the future.[14] A common assumption amongst security scholars, policy makers and the media, is that environmental degradation is persistent and growing, cause of violent conflict within and between the states is defined as a crisis between groups leading to deaths, a category short of war which is defined as crisis leading to more than 1000 deaths.[15]

It is long standing snags in finding meaningful evidence of the determinants of violent conflict and war. General findings for which there is some evidence are that: Major Powers are more likely to be involved in war. High level of rapid growth in military spending tend to be associated with war, poverty and inequality area prevalent in many cases of sub national conflict, recent violence is a good predictor of future violence, 'strong states' with an ability to monopolise the use of force and manage collective actor problems tend to be less prone to internal conflict: democracies tend to be less prone to internal violent conflicts and war; The most important disputed issue in past violent conflicts has been territory and violence is more likely to happen between neighbour groups and countries.[16] It is necessary to be cautious about the links between climate change and conflict. There are few studies that explain in details, ways in which human insecurity increases the risk of violent conflict.[17]

Table-I explains the key point of climate change which may undermine human security and increase the risk of violent conflict. It explores the connection between human insecurity and the risk of violent conflict, which is given below:

Table 1

Factors affecting violent conflicts	Process which climate change could affect exacerbate
Vulnerable livelihoods	Climate change is likely to cause widespread on water availability, coastal regions, agriculture, extreme events and diseases. These affect livelihoods by exposing people to risk, thereby increasing their vulnerability, the impacts will be more significant in sectors of populations with high resource dependency and located in more environmentally and socially marginalized areas. Some of these climate driven out comes are long term and chronic (e.g. declining productivity of agricultural land). While others are episodic (such as floods).
Poverty (relative, chronic, transitory)	Poverty (and particularly relative deprivation) is affected by the spatial differentiation of climate impacts and the Sensitivity of place to them. Climate changes may directly increase absolute, relative and transient poverty by undermining access to natural capital. It may indirectly increase poverty through its effects on resource sectors and the state. Stresses from climate change will differently affect those vulnerable by present political economic process.
Weak states	The impacts of climate change are likely to increase the costs of providing public infrastructure such as water resources and services such as education and may decrease the state revenues. So, climate change may decrease the state's ability to create opportunities and provide important freedom for people, as well as decrease the state's capacity to adapt and respond <u>to climate change itself.</u> Migration may be one response of people whose livelihoods are undermined by climate change. However, climate is likely to be the sole, or even the most important 'push' factor in migration decisions.
Migration	Yet large scale movements of people may increase the risk of conflict in host <u>communities.</u>

In the summery of factors above in the table-1, it is important to stress that climate change will not undermine human security or increase the risk of violent conflict in isolation from other important social factors. In the table discussed above, it builds what is known about the vulnerability of individuals and groups to climate change and it should be read as a simple statement of the ways climate change can be a security problem, nor as a blueprint for reductionist research. So, climate change factors do not cause violent conflict but rather merely affect the parameters that are sometimes important in generating violent conflict.

The changes in environmental variables e.g. those affecting access to food and fresh water; trigger social-political effects e.g. increasing competition for scarce resources which in turn trigger insecurity enhancing violent conflicts amongst those affected.[18] If the ensuing violence is sufficiently intense and widespread, both national and international security interests are involved.[19] According to Elliot , environmental change gives rise to violent conflict, it does this by interacting with other social, economic, political and cultural drivers which reduce stability in a given domain'. The impact of growing environmental stress on demographic variables, e.g. mass migrations and refugee crises; economic variables, e.g. employment and competitiveness; inequalities and socio-political cleavages. Homer-Dixon's strategy was to supervise a targeted set of empirically focussed studies of the way in which environmental stress gives rise either directly or indirectly to violent conflict in various parts of the world. The author identified several case studies, a link between environmental change and violent conflict.

The findings of the research are outlined in a much criticised series of journal articles, books and opinion pieces by Homer-Dixon and other research group members and four key claims are outlined in this research.[20] First, the most important environmental resources as far as environmental causes of violent conflict are concerned are the land forests, water and fish. Second, there are three main sources of environmental stress: environmental change, population growth and social inequality. Third, these three sources of environmental resource scarcity are mutually reinforcing. Fourth, societies that are able to adapt to environmental stress are more likely to avoid significant turmoil and conflict than those that are not. Putting these findings together, Homer-Dixon, found that there was 'substantial evidence' for the claim that environmental scarcity causes violent conflict and environmental pollution plays a minor role in violent conflict.

Homer-Dixon (2007) summarises that 'within our children's lifetime, severe droughts, storms and heat waves are caused by climate change to international security just as dangerous and more intractable than the arms race between the United States and the Soviet Union during the cold war or the proliferation of nuclear weapons among rogue states today'.

Echoing Homer-Dixon's analysis, a series of high profile scientists, commentators and security agencies have warned of national security threats posed by environmental problems such as climate change. David King, former Chief Scientific Advisor to the UK government agreed in 2004 that 'climate change is the most severe problem that we are facing today - more serious than the threat of terrorism. Ban Ki Moon (2007) attributed past and potentially future conflicts in Africa to climatic changes in the region. Climate change as a 'threat multiplier' exacerbates pre-existing trends, tension and instabilities underlying intra-state and inter-state violent conflict such as competition for energy resources, migration, water and food shortages and environmental disasters.[21]

Bachelor argues there is a need for more, "elaborate case studies which are linked with other studies of conflict that deal with interacting crucial issues e.g. poverty, ethnicity and state". Three criteria can be used to frame and scale such a research programme: political scale, the nature of governance and the nature of the environmental changes affected by climate change.

- Political scale: According to Homer-Dixon, environmental factors are the only or even important factors leading to conflict. Other factors e.g. poverty, and inequities between groups, the availability of weapons, ethnic tension, external indebtedness, institutional resilience, state legitimacy and its capacity and willingness to intervene, seem to matter as much if not more than environmental change.[22] Environmental factors do not and nor are they likely to, trigger open conflict between nation states, even in transnational water catchments.[23] Climate change may be a factor of international violence, climate change mitigation where it seems extremely that violence will erupt between states over disagreements about greenhouse gas emission reductions, although changes in the political economy may lead to new rivalries between states.
- The Nature of Governance: Industrialised economies take a global division of labour and resources which affects a global division of environmental degradation to developing countries. The political and economic structure of the state is critical in preventing environmental conflicts. The levels of wealth in the industrialised world allow for institutions that provide stability and resilience to environmental change, well-financed government, the insurance industry, transport and communication infrastructure, a degree of democratic participation and a base

level of personal affluence all seem to help hedge against turmoil in the face of environmental stress.[24]

- 'Strong states' tend to be less prone to internal conflicts. Strong states have capacity to take collective action and identify mitigations against conflicts among heterogeneous groups. They effect administrative hierarchies and they control the legitimate use of force, which helps manage potential internal challengers. They have the capacity to mediate impending conflicts before they turn violent. Both democracies and strongly authoritarian regimes appear to experience relatively less inter-state conflicts.[25] Weak states and states undergoing transition to alternative governance structures are relatively more prone to internal violent conflict. Importantly, inequities within states has been proven to be a factor in many violent conflicts, as well as being a factor in relatively greater levels of environmental damage.[26]

  Hauge and Ellingsen (2001) find that there is a correlation between environmental degradation, state regime type and internal conflict. At the intra-state level, a climate change conflict research agenda would focus on those transition economies and transition democracies where income inequalities are high.

- The Nature of Environmental Change: Discussions of violence and environmental conflicts are directly and indirectly informed by a more long standing concern of conflict studies with resource conflicts. What is an environmental conflict? Is it an important question? Libiszewski's answer is that environmental conflicts are characterised by the degradation of one or more of renewable resources, environmental sinks and living spaces. Fresh water, forests, fisheries and soil can all be seen as providing resources as well as ecological services and all are at risk of reduction and degradation from climate change.

The IPCC suggests that the most sensitive natural systems to climate change, are coral reefs, mangrooves, boreal and tropical forests, polar and alpine ecosystems, prairie wetlands and remnant native grass-lands. Climate change may affect scarcities of renewable environmental resources in these regions. Human systems that are most sensitive to climate change include; water supply systems, forestry activities, agricultural systems and coastal zones and fisheries.[27]

The IPCC sees, Africa is highly vulnerable to climate change, particularly due to decreased water availability, enhanced food insecurity, impacts on human health and increased desertification.[28] Asia is likely to have problems with food security and flooding, but less vulnerable than Africa. Latin America is also, less vulnerable than Africa, but is likely to be severe and possible increasingly frequent climatic variations largely due to changes in the EI Nino Southern Oscillation (ENSO), as well as decreasing biodiversity and reduced crop yields. Small Island States that are most vulnerable to climate change through sea surface warming and coral bleaching, droughts and floods and changes in ENSO. Europe, Australia, New Zealand and North America are relatively much less vulnerable to climate change than developing regions largely by virtue of their considerably greater adaptive capacity.

Impacts of climate change and migration

One sixth of world population lives in glacier or snowmelt fed river basins, a decline in water volumes stored in glaciers and snow cover in the future due to warming of the climate. The impact of this will be increased flows and flooding due to melting in the short term and permanently reduced summer and autumn.

These changes will probably increase the frequency and severity of floods and droughts. The number of people living in severely stressed river basins is projected to increase from 1.4 - 1.6 billion in 1995 to 4.3 - 6.9 billion in 2050. Fresh water availability will also be reduced in coastal areas as sea level will rise. Ecosystems, human health and water system will be impacted by water pollution resulting from higher water temperatures, increased precipitation for longer periods of low flow.

Human societies rely on ecosystems, not only for food and resources but also for free services e.g. water purification and defence against natural hazard. The natural ability of many ecosystems to adapt both the rate of climate change as well as a combination of changes in climate variables e.g. increased flooding, drought conditions, wild fires and ocean acidification. The most vulnerable ecosystems to climate change impacts are those of Tundra, Boreal forest, and mountain and Mediterranean ecosystems. Ocean acidification is brought about by increased atmospheric carbon dioxide concentrations.

Production of food, fiber and forest, the carbon dioxide in the atmosphere will initially increase plant in locations where water and nutrients are not limited. In mid to high latitude regions, moderate warming will benefit cereal crops and pasture yields, however decreases in yields are expected in seasonally dry and tropical regions even for a slight warming. If the temperature increases above 3^0C, global food production is very likely to decrease with yield throughout the world. Due to more intense coastal storms and sea level rise, the coasts and low-lying areas will be exposed like Asian mega deltas e.g. the Ganges-Brahmaputra in Bangladesh and West Bengal, low-lying coastal urban areas and tropical storm landfall e.g. new Orleans, Shanghai, small islands and the low-lying Maldives.

The point is that environmental change is a factor in migration decisions. 'Most migration is not international but rather occurs within individual countries and most international migration occurs between developing countries'.[29] The migration is mostly seasonal and cyclical rather than permanent. People rarely migrate for environmental reasons. People also migrate for economic opportunity and operate in unison as a consequence of the economic and cultural effects of globalisation. The Climate change also impacts on social and ecological systems.

Large migrations have at times led to conflict and are likely as consequence of climate change. If they are to occur at all, climate induced conflicts are most likely as a result of migration.[30] It will be climatic extremes and increasing climate variability that will enhance migration as soils are degraded, water supplies contaminated and depleted, housing-livestock and infrastructure damaged, insurance costs' rise and lives are lost. Sea level rise is very likely to induce large scale migration in the long term. By 2080, the flood risk for people living in islands will be 200 times greater than in a situation where there was no global warming.[31]

In developing countries, planning for enhanced internal migration and internal immigration is required given that they are more vulnerable to the impact of climate change and most migration is within and between developing countries. For example, most of the 5.5 million people living on the Ganges delta in Bangladesh will be forced to relocate with a 4.5 cm rise in sea-level may seek to move to neighbouring India and Pakistan and previous migration of this kind has been a factor in violence in the region.[32] 'Environmental refugees' may also be indicative of the places from areas already under environmental stress and possibly under increasing stress due to climate change.

According to United Nations estimates, the world's population will rise between 7.8 and 11.9 billion people by 2050 with a 'medium variant' estimate of 9.2 billion. The percentage of international migrants in the world continues to rise at the same rate as in the last decades of the 20th century. Stern's observation that mass migration due to cause of climate change will lead to hundreds of millions of more people without sufficient water or food to survive or threatened by dangerous floods and increased diseases.[33] The 4th Assessment report of the IPCC describes the estimates of number of environmental migrants as 'at best', 'guess work' because of a host of intervening factors that influence both climate change impacts and migration patterns suggesting the need for extreme caution.

Extreme weather events and migrations are concerned with the effects of drought or changing rainfall patterns on migratory behaviour. A study of Southwest Mexico found a correlation between declining rainfalls and rising migration to the US, many rural communities depend on rain-fed agriculture. It is estimated that 2.5 million people were displaced by drought and dust storms in the mid-west of US in the 1930s, moving to neighbouring states, but 3,00,000 moving to California as some of the world's first 'eco migrants'.

The effects of declining rainfall and drought on migration depend on the socio-economic situation of the people concerned. In Northern Sudan, Northern Ethiopia, where a survey of more than 100 peasant farmers concluded that people in marginal regions have developed a great variety of adaptation mechanisms which strengthen their ability to cope with both slow climatic changes and extreme climatic events. Migration out of the Mid West had already begun before the periods of drought, reflecting over capitalisation of agriculture in the region and the overall depression of the US economy from 1929. Most prominent example is of the migration linked to tropical cyclones and hurricanes, which hit parts of the US states of Albania, Mississippi and Louisiana and destroyed the city of New Orleans in 2005.

Overall, it would seem that empirical studies into the relationship between climates related environmental events and migration are few and their results are not conclusive. Climate change and climate variability is likely to be associated with migration in the medium term. As discussed above, there are two ways of thinking about this interaction - (a) to identify the major types of socio-economic impacts that have been identified in climate science and consider ways in which these might lead to increased migration, (b) to identify the major patterns and drivers of existing migration and consider the ways in which these might be sensitive to climate change impacts.

The weather systems severely undermine rural livelihoods and could play a key role in decision making on migration and other possible livelihoods- particularly for the poor. At drought risk areas, it is clear that migration is already a significant response to the risk of drought in many dry land areas, as part of a livelihood diversification strategy that seeks to minimise vulnerability. The drought events are high throughout parts of dry land Africa. In Ethiopia, research documented by the UNDP found that half of all Ethiopians had experienced at least one major drought during the brief period from 1999- 2004. Recurrent drought increases societal vulnerability to future drought with the destruction of key social insurance mechanisms e.g. livestock, poor terms of trade and preventing households from replenishing stocks. Increased drought risk may be expected more than the migration, particularly in the Sahel zone of America which have the world's highest fertility rates and so a rising number of potentially affected people. It is also clear that international migration is an expensive endeavour with significant resources required both to undertake the journey from Africa to Europe or North America and especially to cross international borders.

The IPCC observes that in Latin America, already dry regions e.g. Southern Chile, South-west Argentina, Southern Peru and Western Central America have become drier and the other areas e.g. parts of Bolivia, increases in rain fall have affected land use and crop yields. Most current migration to the US takes place from Central America with about 1, 64,000 Mexican migrants crossing the border to the US every year.

Turning to flood risk, increased flooding of coastal and low-laying areas is also a major livelihood threats especially in vulnerable societies that do not passes the economic and technical means to cope. 16 of the 22 largest cities in the world projected from 2015, representing more than 260 million people are port cities, many of them in developing countries. The Asian Tsunami in 2004, killed more than 2, 00,000 people and displaced over 4, 00,000, but most were received locally, and most have returned. The fact that there was a swift and internationally supported humanitarian response, but it is not the only reason why people did not move to richer countries. For example— in Bangladesh, millions of people are left homeless each year due to flooding, although most of them travel very short distances and try to return and rebuild their houses after the disaster. The majority of Bangladesh migrants in the UK- have very low levels of flood risk e.g. poverty, landlessness and lack of economic opportunities all play into the decision-making process of migrants. The drought related migration, it is reasonable to expect much greater rural-urban, rural-rural and distress migration as a result of increased flood hazard in Bangladesh. This will impact on flood of migration to developed countries e.g. UK.

The coast of the Gulf of Mexico is another example of region in which frequent flooding occurs. The most recent example forced many to seek refuge in the neighbouring state of Veracruz. However, whether more devastating hurricanes and more frequent flooding in the region will cause only temporary displacement or whether people will migrate more permanently is unclear.

Both the drought risk and flood risk on migration are likely to promote an increase in internal rural-urban migration but those increased populations in coastal cities are themselves likely to be more vulnerable to flooding. In Africa, UN projections suggest 123 million people or over 10 per cent of the continents' population will already be living in some 45 cities of million people or more by 2015. 5.3 million people will live in 18 coastal cities vulnerable to flooding.

One problem is that over the last half of the 20th century, a number of countries in Southern Europe and East Asia have shifted from being net emigration to net immigration countries as a result of dynamic economic growth and it can be expected to continue into the future with countries e.g. Malaysia, Thailand and Turkey to undergo a migration transition in the coming decades yet such trends are not built into UN Projections, for example- Turkey is projected to have net positive migration to 2015 whilst Thailand is never considered to have had net negative migration.

UN projections estimated international migration to more developed countries from 2000-2005 with million people added to their population during that period. So, UN migration projections are based on 'past international migration estimates and consideration of the policy stance to each country with regard to future international migration', these projections tell the balance between immigration and voluntary. A first attempt to generate a global origin destination international migration database, based partly on censuses and partly on UN global migration stock data has been made by the Development Research Centre on migration, globalisation and poverty.[34]

The global migrant origin database provides some evidence on key migration systems that are significant in different world regions. It signifies south-south migration and intra-regional migration. Around half of all migrants from developing countries live in other developing countries.[35]

Whilst intra-regional moves account for 2/3rd of all migration within sub-Saharan Africa, as well as 2/3rd of moves within Europe and Central Asia.[36]

The military and climate change

For the national security, the word's militaries will increasingly have to adapt and to command a large share of public resources for that purpose of challenges of climate change. Militaries are major emitters of greenhouse gases. The share of a country's GAP spent on its military as representative of the military's share of that country's overall greenhouse gas emissions. Military expenditure was 11.7 per cent of 1995 GNP in the Russian federation, so the Russian armed forces emit roughly 185 million metric tons of CO_2; military expenditure was 3 per cent of 1995 GNP in the United Kingdom, so the UK armed forces emit some 17 million metric tons of CO_2 and military expenditure was 3.8 per cent of 1995 GNP in the US, so the US armed forces emit some 210 million metric tons of CO_2. So, worldwide military activity may be responsible for greener house gas emissions. In this respect, militaries are a problem rather than a solution to environmental insecurity.

Recognising the growing need for national governments to reduce greenhouse gas emissions, number of armed forces are voluntarily becoming involved in greenhouse gas reduction programs. The Australian department of defence has joined the Australian greenhouse challenge and is seeking to cut its emissions by 13 per cent by 2004 in February 2001, the United Nations Environmental Programme, the US environmental protection authority and the US Department of Defence hosted a conference on, 'The importance of Military organisations in stratospheric ozone protection and climate protection' which was attended by representatives from more than 35 countries and sough to share experiences of greenhouse gas reduction within the armed forces.

Climate change and human security

There is now widespread agreement that the changes underway in the earth's climate system have no precedent in the history of human civilisation.[37] As a macro- driver of many kinds of environmental changes e. g. coastal erosion, declining precipitation and soil moisture, increased storm intensity, and species migration, climate change poses risks to human security.[38] In most parts of the world the impacts of climate change on social-ecological systems e.g. temperature, sea level and annual precipitation over long time scales and also increases the frequency of floods, droughts, storms, and cyclones, fires, heat waves and epidemics which are projected to occur with high levels of events. These include melting of glaciers and permafrost which may add several meters to global sea levels, collapse of the thermohaline circulation which may cause regional climate changes in the northern hemisphere, and large scale shifts in the Asian monsoon and the El Nino southern oscillation phenomenon.[39]

In the past 10,000 years the rate of change is unprecedented and triggered large scale social disruptions. It impacts on human systems. According to Davis (2001), the El Nino events and famines killed tens of millions across the tropics in the late 19th century. Famine was triggered by drought, but caused by the way political and economic colonisation deprived people of their entitlements to natural resources. Famines now identify poverty, inequality, market failures and policy failures. David's arguments about the ways climatic variations have combined with stressed social-ecological systems to result in democratic social change.

The vulnerability of people to climate change depends on natural resources and ecosystem services. The more people are dependent on climate sensitive forms of natural capital, the less they rely on economic or social forms of capital and they are affected too much by climate change.

Climate change does undermine human security which varies across the world because the natural resources and social determinants of adaptive capacity are varied. For example- in contrast to many industrialised countries where agriculture represents 1-2 per cent of workforce, in subsistence farming so that 46 per cent of rural people live below the poverty line of US $0.55 per day.[40] The risks of climate change to social system's characteristics change the environmental systems.

In terms of environmental change, for example up stream users of water, distant atmospheric polluter, multinational logging and mining companies, regional-scale climatic process and a host of other distant actors and larger scale processes influence the security of individuals' entitlements to natural resources and services similarly in terms of the social determinants of vulnerability, warfare, corruption, trade dependency, macro-economic policies and a host of other larger scale processes associated with 'globalisation' shape, the social and economic entitlements that are necessary to reduce an individuals' vulnerability to environmental changes.[41]

They refer these large scale processes that shape people's entitlements to natural, economic and social capital which may themselves be vulnerable to climate change. Production sectors may be at risk, for example - it is not farmers whose livelihoods are at risk from climate change and those livelihoods who depend on agricultural production e.g. suppliers of inputs, people who work in transporting and processing agricultural commodities, people who work as extension officers and people who work in agricultural lending services. Some causes may occur, for example — rural decline can cause migration to urban areas, placing increasing demand on urban services and increasing political pressure on the state, which is varied entitlements e.g. education, healthcare, law and order, credit and protective security.

Climate change may be a national security issue.[42] The risk to national security may be both a cause and a consequence of human insecurity. So, human security is a function of multiple processes operating across space, overtime and at multiple scales. Climate change may affect human security a daunting task, which is not helped by the difficulty of ascertaining whether there are indeed any existing environmental changes that can be attributed to climate change.[43] Marginalised people are vulnerable to environmental change and it all helps to substantiate the argument that climate change poses significant risks to human security in many parts of the world. It is clear that human insecurity lead to violent conflict. This is important to consider that violent conflict is powerful cause of human insecurity and vulnerability to climate change.[44]

Myres expressed, 'in essence, though this is generally recognised by governments, security applies to most of the level of the individual citizen.

It amounts to human well-being; not only protection from harm and injury but access to other basic, requisites that are due of every person on earth'.[45] The demilitarised security can be found in the research being carried out under the Global Environmental Change and Human Security (GECHS) and Human Security Now (HSN) projects .

Human Security is achieved when and where individuals and communities have the options necessary to end, mitigate or adapt to threats to their human, environmental and social rights; and have the capacity and freedom to exercise these options and actively participate in attaining these options'.[46] Human security is concerned with reducing and -- when possible--removing the insecurities that plague human lives. It contrast with the notion of state security, which concentrates primarily on safeguarding the integrity and robustness of the people and thus has only an indirect connection with the security of the human beings who live in these states.[47]

GECHS and HSN raise a question that is central to the development of the concept of environmental security. They raise the question of the link between the concept of security and the value of security. The GECHS implies that democracy is a necessary condition of security and in so doing plays down the importance of geological, demographic, ethnic and historical factors that influence the political and socio-economic significance of environmental events. These factors, they relate to and seem more relevant to the issue of environmental insecurity than issues of democracy or autonomy.

The United Nations Development Programme (UNDP) has put forward the concept of human security to assist in the framing of development and equity issues, it is concerned with how people live and breathe in a society, how freely they exercise their many choices, how much access they have to market and social opportunities, and whether they live in conflict or peace. Human security is not a concern with weapons - it is a concern with human life and dignity.[48] Environmental insecurity is the double vulnerability of people that arises when under-development and impoverishment are compounded by human induced environmental change.[49] For example, Bangladeshis have a life expectancy 21 years less than Australians and a Bangladeshi woman is 90 times more likely to die when giving birth than a woman from Australia.[50] However, Australians produce 80 times more greenhouse gases than Bangladeshis and within 50 years up to 11 per cent of Bangladesh could be flooded due to sea level rise, whereas a much smaller amount of Australia's surface is likely to be flooded. The difference is that for most Australians, climate change is a problem of adaptation, but for the majority of Bangladeshis, is a matter of survival; they are insecure and this underlies the dialectical nature of human environmental insecurity.[49]

The state and human security
Without human security the state's role is redundant and meaningless. The state cannot be separated from human security, and can actively promote or repress rights to personal security, social services and economic opportunities. The role of the state is also central to understanding the causes of and solutions to violent conflict.[51] States play critical roles in creating the conditions whereby people can act in ways to pursue the lives they value.[52]

The states functions are:
- They can provide protective guarantees to assist people when their livelihoods suddenly contract through income support, food aid, or short term local employment programs.
- They can provide economic freedoms that are important for people to seek employment and to interact to seek mutually advantageous outcomes in terms of consumption and production.

- The states can provide political freedoms e.g. the freedom of speech, freedom of media, civil liberties and the freedom to vote for parties, leaders and policies.
- The state provides social opportunities such as education and healthcare which is another important state role.
- The states can provide transparency guarantees to ensure openness and accountability in transactions to mitigate against corruption and to maintain faith in market processes.
- The states function to give particular importance to mitigate against the generation of violent conflicts include the provision of health care and education, the protection of human rights, establishment and maintenance of a strong and independent judiciary, accountable and transparent police services and armed forces and the protection of democratic processes.[53] The state functions given above are interconnected, they 'supplement' and 'reinforce' each other.[52] When states cannot provide all these functions, the risk of violent conflict increases. Thus internal wars are more likely to increase in countries.

The characteristics of the states are: (i) States are legitimate, people have opportunities to develop and have less anxiety about the future, conflict resolution mechanisms tend to be effective and economies tend to grow and poverty levels tend to fall.[52] (ii) The strong states have effective administrative hierarchies, control the legitimate use of force, can mediate impending conflicts before they turn violent and are more capable of managing environmental degradation and change.[54] (iii) In strong liberal democratic states, both the structural conditions and livelihood factors that increase the risk of violent conflict are reduced.

Democracy gives people power to act to affect change - it creates opportunities that reduce the need for violent action to cause change and it tends to ensure a minimal level of welfare such that people are less likely to die from, for example famine.[52] Many violent conflicts in ecologically or economically marginal regions are evidence that relative poverty and poverty of opportunities due to inadequate access to the state may be a key cause of violence.[53] Another factor, large migrations have at times led to conflict and large migrations are likely as a consequence of climate change.[55] Goldstone (2001) said, 'migration is a factor in violent conflict'. So, understanding the way climate change may induce more migration. People rarely migrate for environmental reasons alone. So, understanding the way climate change may induce more migration also requires understanding the way it will interact with other factors.[56] In terms of migration, the influx of migrants into new areas have been a significant factor in many 'environmental conflicts'.[57]

Contraction in livelihoods, poverty, weak states and immigration are all risk factors in violent conflict and the research suggests that climate change may have direct and indirect effects on these risk factors. So, climate change may increase human insecurity and the risk of violent conflict. Given this uncertainty, there are dangers in speaking prematurely and vociferously about climate change in the language of security.[42]

Climate change may increase insecurity. A key aim of the research is to enhance understanding of climate insecurity to vulnerability of people's livelihoods due to climate change. In the developing world, there are many low income and resource dependent communities. The Assessment of Impacts and Adaptations to Climate Change (AIACC) project is an initial attempt at the kind of comparative vulnerability assessment process that is required to enhance understanding and improve policies to address climate insecurity. The AIACC project is revealing that the most potentially devastating impacts of climate change arise from a combination of multiple stresses acting in concert - of which climate stresses are but one and which also include ecosystem degradation, failed governance systems and economic decline.[58]

Present research on vulnerability to climate change investigates the diverse array of social and environmental factors operating over time and across an array of spatial scales, that structure vulnerability.[59] These factors include the sensitivity of resources such as freshwater, soils, reefs and fisheries to sudden and incremental changes in climate. Understanding the determinants of vulnerability in this way requires insights from a range of disciplines across the natural and social science. It also requires analysis of how policy agendas are shaped through science, policy interactions that can, at times reproduce Malthusian associations of resource scarcity with the risk of the violent conflict.[60]

Figure 2

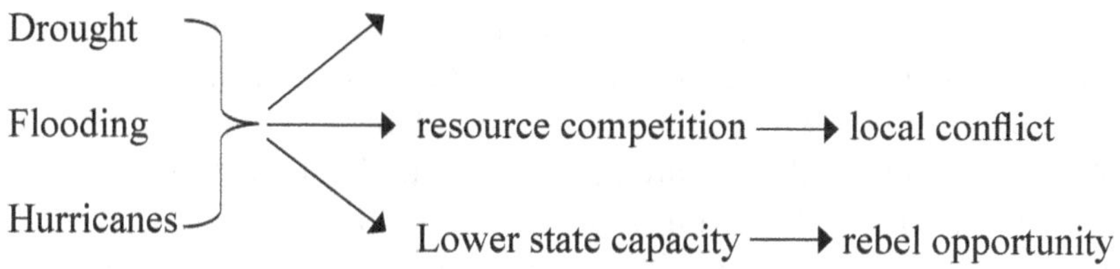

Drought

Flooding → resource competition ⟶ local conflict

Hurricanes → Lower state capacity ⟶ rebel opportunity

If the economically and politically powerful developed countries emit large amounts of greenhouse gases primarily understand vulnerability to climate change in developing countries as a risk to their national security through migration or violent conflict, then their response may be more weighted towards increased border protection and defence spending, rather than towards the reduction of emissions and efforts to foster adaptation.[42]

A second key aim of the resource to enhance understanding of climate insecurity should be to examine the consequences of livelihood insecurity and this outcome of climate change seems to be a factor that increases the risk of violent conflict. A question at the heart of most conflict research, while is, why do individuals choose violence? The answer is through psychological rather more than geographic research. There are many countries where climate change seems most likely to induce wide spread human insecurity and conflict, they are, Democratic Republic of Congo, Haiti, Sierra Leone, Sudan, Yemen. Within these countries, local level studies of low income resource dependent communities and low income urban communities, people could constrict their livelihoods, they have made when faced with change in the past and people are likely to make in the face of livelihood decline. They could also examine the local peace movements and their successes and constraints. There are multiple and over lapping institutions operating at various scales that are and will be directly involved in exacerbating or alleviating the adverse effect of climate change. These include local, national, regional and global governance institutions that make decisions and implement policies that directly such as the United Nations Framework Convention on Climate Change (UNFCCC) and indirectly such as development agencies affect capacity to adapt to climate change. 'There is no single agreed theory of institutional adaptation, nor are there clear criteria for what constitutes an 'effective' or a 'successful' institution.[61] A number of factors matter including an institution's legitimacy (moral and legal), its responsiveness to its constituents, its core values and its commitment to them, its ability to learn and experiment, the amount of resources available to it, its independence from short-- term political pressures, the quality of its management and the transparency of its decision making.[62] The role of the state is often critical in reducing both vulnerability to climate change and the risk of the violent conflict.

A third key aim of the research to enhance understanding of climate insecurity should be to examine challenges, climate change poses to states, including the capacity of the states to protect livelihoods and sustain peace, recognising of course that these may not be goals of some states. Various state institutions are involved in managing the security risks of climate change. These include climate-specific groups such as National Climate Change teams and environment and resource management agencies. They are engaged in management of problems that may arise from climate change e.g. increased rural-urban migration (land tenure institutions and urban planning agencies); increased morbidity (health service providers); increased climatic hazards (disaster management arrangements); increased demand for development assistance (diplomatic and development agencies) and increased violent disputes and crime (the judiciary and the police) are also important.

The Pacific Institute for Studies in Development Environment and Security describes that the security of individuals, communities, nations and the entire global community is threatened by non-military environmental threats. These threats are self-generated by fouling air and water and over harvesting land. These threats are not felt equally around the world. Southern countries face severe problems from desertification, while northern industrial countries deal with acid rain and Polar Regions see large depositions of persistent organic chemical pollutants. Climate change will cause uneven effects over the entire globe for the next 50 to 100 years, with some countries benefitting and others suffering.

Environmental issues are not high on the national security agenda. Those who study environmental problems e.g. de-forestation, loss of bio-diversity and climate change do not see the connection through to its higher order effects and those who study security problems e.g. non-proliferation, terrorism and civil conflict do not recognise the environmental roots and effects of these problems.

Environmental security (ecological security) reflects the ability of a nation or a society to withstand environmental asset scarcity, environmental risk or adverse changes or environment-related tensions or conflicts. The chart below illustrates the potential for economic activity to cause environmental changes that lead to conflict.

Figure 3: Environmental routes to conflict

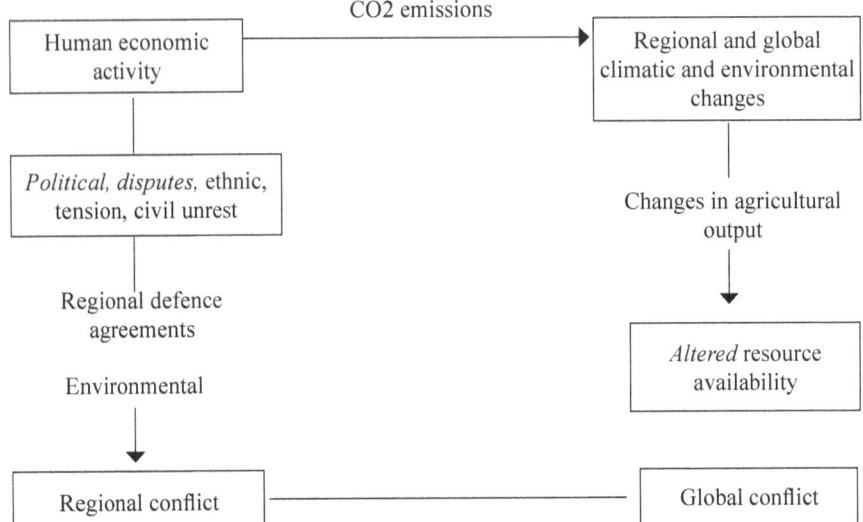

Environmental threats are real and ongoing. However, not every environmental issue will result in a security problem and most security problems are generated from complex situations involving environmental, political, social, and economic issues. Daniel Deudney first explained many such contradictions at the end of cold war (1991) at least in reference to traditional security definitions. It is arguable that potential obstacles have expanded since that time. In brief, these potential problem areas are:

- Complexity of variables and non-linear nature of relationships.
- Irreversible nature of environmental systems.
- Vision of environmental issues as external to political, economic, and social systems.
- Focus on state level analyses, imposes false divisions between relevant factors.
- Visions of environmental systems as natural and root causes of issues.

These issues have been presented in many of the attempts to address climate change and security in recent years, with abrupt climate change, first appearing as a security issue in 2003 Global Business Network (GBN) report commission by the US Department of Defence (Schwartz & Randall, 2003). Since that time, similar reports have been released by the Centre for Naval Analysis (CNA, 2007), Directorate for National Intelligence (NIC, 2008), Global Business Network (GBN, 2008), US Climate Change Science Program (CCSP, 2008), German Federal Ministry for Economic Cooperation & Development (2002), German Advisory Council on Global Change (WGBU, 2008), Swedish Development Agency (SIDA, 2008), European Commission (2008) and others. These reports have been valuable in raising awareness of climate risks. On IPCC data and projections, continued use of violent conflict as the measure of insecurity and use of the state as the unit and level of analysis, narrow our understanding of the risks of abrupt climate change in several crucial ways.

In climate terms, the most pressing questions concern what amount of Greenhouse Gases (GHGs) can the ecosystem absorb before a large scale shift in climate stability occurs? Atmospheric temperatures may rise gradually over the years, as per the IPCC projections. The National Research Council's 2002 report on abrupt climate change described several qualities of the system that creates 'abruptness'; first, the system is non-linear and shifts from one condition to another rapidly, perhaps within a number of years. Second, this change is irreversible as measured by human time scales, a condition made by the long term forcing of GHG emissions and atmospheric CO_{33} lifetimes. Third, the changes may occur due to second order effects of the global system, as positive feedback loops originally unrelated to the more obvious forcing (e.g. Arctic methane gas release caused by melting permafrost). Finally, abrupt changes are often categorised as such due to the inability of related systems to adapt, leading to the possible description 'dangerous' climate change. Climate change is very unlikely to lead directly to conflict, but may adversely affect social, political and economic systems. Most issues concerning resources and conflict, not from changes in the environment as a root cause, but rather the failure of political and economic systems to provide adequate resources or adaptation measures. At times such failure can be deliberate, as with resilience and livelihood targeting during violent conflicts.[63]

References
1. *Barnett 2003, Barnett and Adger 2003, Brown 1989, Edwards 1999, Swart 1996, Barnett, J. (2003), "Security and climate change", Global Environmental Change, 13, pp. 7-17.*
    Barnett, J & Adger, W. N., "Climate dangers and atoll countries", Climatic Change, 61, 2003, pp. 321-337.
    Brown, N., "Climate, ecology and international security", Survival, 31, 1989, pp.

519- 532.

Edwards, M., "Security implications of a worst-case scenario of climate change in the south- west pacific", Australian Geographer, 30, 1999, pp. 311-330.

Swart, R., Security risks of global environmental changes, Global Environmental Change. 6, 1996, pp. 187-192.

2. *Brauch 2002, Gleick 1992, Homer Dixon 1991, Van Ireland, Klaassen, Nierop and Vander wusten, 1996.*

3. *Homer-Dixon, T., "On the threshold: environmental changes as causes of acute conflict", International security, 16 (2), 1991, pp. 76-116.*

   Kaplan, R., "The coming anarchy", Atlantic Monthly, 273 (2), 1994, pp. 44-76.

   Myers, N., "Population, environment and conflict", Environmental Conservation, 14 (1), 1997, pp. 15-22.

4. *Barnett, J., and Dovers, S., "Environmental security, sustainability and policy", Pacifica review, 13 (3), 2001, pp. 157-169.*

   Shaw, B., When are environmental issues security issues? Environmental Change and Security

   Project Report 2, 1996, pp. 39-44.

5. *Rahman, A., Climate Change and Violent Conflicts, In Sulimans, M., (ed.), Ecology, Politics and Violent Conflict, Zed Books, London and New York, 1999, pp. 181-210.*

   Watson, R.2000. Presentation of the Chair of the Intergovernmental Panel on Climate Change to the Sixth Conference of Parties of the United Nations Framework Convention on Climate Change. November 13, Geneva. http://www.ipcc.ch

6. *Barnett, J., Global Warming and the Security of Atoll-Countries, Macmillan Brown Centre for Pacific Studies Working Paper, Christchurch, 2002.*

   Stripple, J., Climate change as a security issue, in Page, E., Redclift, M., (ed.), Human Security and the Environment: International Comparisons, Edward Elgar, Cheltnham, 2002.

7. *Brown, N., "Climate, ecology and international security", Survival, 31 (6), 1989, pp. 519-532. Edwards, M., "Climate change, worst-case analysis and eco-colonialism in the "Southwest Pacific", Pacifica Review, 8 (1), 1996, pp. 63-80.*

   Edwards, M., "Security implications of a worst-case scenario of climate change in the South- West Pacific", Australian Geographer, 30 (3), 1999, pp. 311-330.

   Ehrlich, P. & Ehrlich, A., "Population growth and environmental security", Georgia Review, 45 (2), 1991, pp. 223-232.

   Gleik, P., Effects of climate change on shared fresh water resources, in Mintzer, I., (ed.), Confronting Climate Change, Cambridge University Press, Cambridge, 1992, pp. 127-140; Homer Dixon -1991, Mathews 1990.

8. *Dabelko and Dabelko Environmental Security: Issues of Conflict and Redefinition', Environmental Change and Security Project Report, The Woodrow Wilson Center, Washington, DC, Vol. Spring, 1995, pp. 3-13.*

9. *Edward page, Department of politics and international studies University of Warwick, SGIR 7th pan-European, international relation conference Stockholm, September 9-11, 2010.*

10. *WCED Our Common Future, Oxford University Press, Oxford, 1987.*

    UNDP (United Nations Development Programme) 1994, Human Development Report (Geneva: United Nations):

    IPCC (2001), Climate Change 2001: Impacts, Adaptation, and Vulnerability,

Cambridge University Press, Cambridge.

Parry, M.L., et al., Technical Summary'. Climate Change 2007: Impacts, Adaptation and Vulnerability, Cambridge University Press, Cambridge, 2007, pp. 23-78.

UNDP (United Nations Development Programme) 2007. Human Development Report 2007/2008: Fighting Climate Change: Human Solidarity in a Divided World (Basingstoke: Palgrave).

11. IPCC (Intergovernmental Panel on Climate Change).2001a. Technical Summary: Climate Change 2001: Impacts, Adaptation, and Vulnerability, In Climate Change 2001: Impacts, Adaptation, and Vulnerability, Contribution of Working Group II to the Third Assessment Report of the Intergovernmental Panel On Climate Change, Cambridge University Press, Cambridge.

12. Hoegh-Guldberg et al., Pacific in Peril: Biological, Economic and Social Impacts of Climate Change on Pacific Coral Reefs, Greenpeace, Amsterdam, 2000, p. 4.

13. Berstein, P., et al., Effects of restrictions on international permit trading: the MS-MRT model, in weyant, J., (ed.), the costs of the Kyoto protocol: a multi-model evaluation, The Energy Journal, 1999, pp. 221-256.

14. Homer-Dixon, T., Environmental scarcity and violence, Princeton University Press, Princeton, 1999, And Baechler, 1999 a.

15. Wallensteen, P. and Sollenberg, M., "Armed Conflicts, Conflicts Termination and Peace Agreements 1989-96", Journal of Peace Research, 34 (3), 1997, pp. 339-358.

16. Gleditsch, N., Armed conflict and the environment, in Diehl, P., Gleditsch, N., (eds.), Environmental Conflict, Westview Press, Boulder, 2001, pp. 251-272.

Raknereud, A. And Hegre, H., "The hazard of war: reassessing the evidence for the democratic peace", Journal of Peace Research, 34 (4), 1997, pp. 385-404.

Vasquez, J., The war puzzle, Cambridge University Press, Cambridge, 1993.

17. Gough, M., Human security: the individual in the security question-the case of Bosnia. Contemporary security policy, 23, 2002, pp. 145-191.

Mochizuki. K., Conflict and people's insecurity: An insight from the experiences of Nigeria, 2000.

Ohlsson, L., Livelihood conflicts: linking poverty and environment as causes of conflict, 2000.

18. Homer-Dixon, 1999: 6-8.

19. Homer-Dixon, 1999: 166-8.

20. Homer-Dixon, 1999: 8-17, 1999: 177- 82; 2000.

21. European Commission 2008 - the report goes on to identify the Arctic, Latin America, Africa, Central Asia as most at risk from climate induced violent conflict .

22. Baechler G., Environmental degradation and violent conflict: hypotheses, research agendas and theory building, in Suliman, M. (ed.), Ecology, Politics And Violent Conflict, Zed Books, London and New York, 1999b, pp. 76-112.

23. Baechler, Environmental degradation in the south as a cause of armed conflict, in Carius, A., and Lietzmann, K. (ed.), Environmental Change and Security: A European Perspective, Springer-Verlag, Berlin, 1999a, pp. 107-130.

Homer- Dixon, t. and Percival, v., Environmental Scarcity and Violent Conflict: Briefing Book, American Association for the Advancement of Science, Toronto, 1996.

Wolf, A., 'Water wars' and water reality: conflict and co-operation along international waterways, in lonergan, S., (ed.), Environmental Change, Adaptation, *and Security, Kluwer Academic Publishers, Dordrecht, 1999, pp. 251-265.*

24. *Barnett, J., The Meaning of Environmental Security: Ecological Politics and In the New Security Era, Zed Books, London and New York, 2001b.*

25. *Eckstein, H., and Gurr, T., Patterns of Authority: A Structural Basis for Political Inquiry, Wiley, New York, 1975.*

26. *Boyce, J., et al., Power distribution, the environment and public health: a state-level analysis, Ecological Economics 29, 1999, pp. 127-140; Gledisch 2001.*

27. *IPCC (Intergovernmental Panel on Climate Change) 2001b. Summary for Policymakers: Climate Change 2001: Impacts, Adaptation, and Vulnerability, In Climate Change 2001: Impacts, Adaptation, and Vulnerability. Contribution of Working Group II to the Third Assessment Report of the Intergovernmental Panel on Climate Change, Cambridge University Press, Cambridge,*

28. *IPCC 2001 a.*

29. *Russell 'S., International Migration: Implications for The World Bank, Human Resources*
Development and Operations Policy Working Papers Number 54, World Bank, Washington, 1995.

30. *Van Ireland, E., Klassen, M., Nierop, T., Van Der Wusten, H., 1996. Climate Change: Socio-Economic Impacts and Violent Conflict. Dutch National Research Programme on Global Air Pollution and Climate Change, Report No. 410 200 006, Wageningen, Rahman, 1990.*
Nicholls, R, Hoozemans, F., Marchand, M., Increasing Flood Risk and Wetland Losses Due to Global Sea-Level Rise: Regional and Global Analyses, Global Environmental Change 9 (Supplementary Issue), 1999, pp. 69-87.

32. *Swain, A., The Environmental Trap: The Ganges River Diversion, Bangladeshi Migration and Conflicts in India, Department Of Peace and Conflict Research Uppasala University Report, Sweden, 1996.*

33. *Stern, N. 77. Economics of climate change: The stern review, Cambridge University Press, Cambridge, 2007.*

34. *www.migration drc. Org & C.R Parsons, R. Skeldon, T.L Walmsley and L.A Winters, International migration, Economic development and policy, Washington, The World Bank 2007, pp. 17- 58.*

35. *Ratha. D and Shaw. W: 2 - South-South migration and remittances, Washington DC, World Bank working paper no. 102, 2007.*

36. *Parsons et al., 2007.*

37. IPCC, 2007; Stern, 2007.

38. *Mc Carthy, J., Canziani, O., Leary, N., Dokkend. & White, K. (Eds.). Climate Change 2001: Impacts, Adaptation & Vulnerability, Cambridge University Press, Cambridge, 2001.*

39. *Oppenheimer & Alley 2004, Vellinga & Wood 2007, Schneider, Semenov & Patwardhan, 2007.*

40. *UNDP (United Nations Development Program), Ukun rasik a' an: East Timor human development report 2002. Dili: UNDP, 2002.*

41. *Adger, W. & Kelly. Social Vulnerability to Climate Change and the Architecture of Entitlements' Mitigation and Adaptation Strategies for Global Change 4, 1999, pp. 253-266.*

42. *Barnett, 2003.*

43. *Allen. M., & Lord. R, The blame game, Nature, 432, 2004, pp. 551-552.*

44. *Barnett, J. Climate Change, Insecurity and Justice, In W.N. Adger, J. Paavola, M.J. Mace, & S. Huq (Eds), Fainess In Adaptation to Climate Change, MIT Press, Cambridge, MA, 2006.*

45. *Myres 1993: 31.*

46. *Lonergan, Environmental degradation and population displacement, A VISO Bulletin, Issue No. 2. Global Environmental Change and Human Security Project, Vancouver, 1993, p. 29.*

47. *Ogata and Sen, 2003: 8; Gasper, 2005: 223-8.*

48. *UNDP, pp. 22-23.*

49. *Barnett, 2001 b.*

50. *UNDP, 1998.*

51. *Reno, W. 'Shadow States and the Political Economy of Civil Wars', In Berdal, M. and Malone, D. (Eds.) Greed And Grievance: Economic Agendas in Civil Wars, Lynne Rienner, Boulder, 2000, pp. 43-68.*

52. *Sen, A. Development as Freedom, Anchor Books, New York, 1999.*

53. *Goodhand, J. 'Enduring Disorder and Persistent Poverty: A Review of Linkages between War and Chronic Poverty', World Development, 31 (3), 2003, pp. 629-646.*
    Gough, M. 'Human Security: The Individual in the Security Question-The Case of Bosnia', Contemporary Security Policy, 23 (3), 2002, pp. 145-191.
    Keen, D. 'Incentives and Disincentives for Violence', In Berdal, M. and Malone, D. (Eds.) Greed and Grievance: Economic Agendas in Civil Wars, Lynne Rienner, Boulder, 2000, pp. 19-42.

54. *Eckstein, H., & Gurr, T. Patterns of authority: A structural basis for political inquiry, Wiley, New York, 1975.*
    Esty et al., State failure task force: phase II findings. Environmental Change and Security Project Report, 5, 1999, pp. 49-72.
    Hauge, W., and Ellingsen, T. Causal pathways to conflict, In P. Diehl, & N. Gleditsch (eds.), Environmental Conflict, West View Press, Boulder, 2001, pp. 36-570.

55. *Van Ire Land et al., Climate change: socio-economic impacts and violent conflict, Dutch National Research Programme on Global Air Pollution and Climate Change, Report No. 410200 006, Wageningen, 1996.*

56. *Meze - Hausken, E. "Migration Caused by Climate Change: How Vulnerable are People in Dryland Areas?" Mitigation and Adaptation Strategies for Global Change, 5 (4), 2000, pp. 379-406.*

57. *Baechler, G. Environmental Degradation in the South as a Cause of Armed Conflict, 1999, In A. Carius, & K. Lietzmann (Eds.), Environmental Change and Security: A European Perspective, Springer-Verlag, Berlin, pp. 107-130.*
    Klotzli, S., The Water and Soil Crisis in Central Asia - A Source For Future Conflicts? ENCOP Occasional Paper No. 11, Cetre For Security Policy and Conflict Research. Zurich, 1994.
    Peluso, N., & Harwell, N. Territory, Custom, and the Cultural Politics of Ethnic War in West Kalimantan, Indonesia, 2001, In N. Peluso, & M. Watts (eds.), Violent Environments, Lynne Rienner, Boulder, pp. 43-68..
    Swain, A., "Conflicts over Water: The Ganges Water Dispute", Security Dialogue, 24 (4), 1993, pp. 429-439.

58. *Leary et al., For whom the bell tolls: Vulnerability in a Changing Climate, A*

synthesis from the AIACC project, AIACC Working Paper No. 21, International START Secretariat, Florida, 2006.

59. *O' Brien et al., Mapping vulnerability to multiple stressors: climate change and globalisation in India, Global environmental change, 14, 2004, pp. 303-313.*

60. *Sarewitz, D, Pielke, R, and Keykhah, R, "Vulnerability and risk: some thoughts from a political and policy perspective", Risk Analysis, 23, 2003, pp. 805-810.*

61. *Adger, W. Institutional Adaptation to Environmental Risk under the Transition in Vietnam', Annals of the Association of American Geographers, 90(4), 2000, pp. 738-758.*

62. *Dovers, S. Institutions for sustainability, Tela. 7, the Australian Conservation Foundation, Melbourne, 2001.*
   Goodin, R. The theory of institutional design, Cambridge University Press, Cambridge, 1996.

63. *Briggs et al., 2009, Brown, 2004.*

2. CLIMATE CHANGE AND WATER SECURITY

Introduction

Climate change will have an impact on water and food security in significant and uncertain ways. The anticipated impacts of climate change are on agriculture and agricultural water management. It is clear that water availability will be a critical factor. The long term climate risk to agricultural assets and agricultural production that can be linked to water cannot be known with any certainty. Approximately, 70 per cent of the fresh water available on the planet is frozen in the icecaps and Antarctica and Greenland, leaving the remaining 30 per cent available for consumption.[1]

There are four main factors aggravating water security according to the IPCC: population growth, increased urbanisation with spiralling demands for water among more concentrated population, high level of consumption, and climate change will shrink the resources of fresh water. The impact of climate change on water and agriculture requires the distribution and extent of change in key variables that govern crop growth and water availability. Water management for agriculture encompasses all technologies and practices for plant growth; these range enhancing the capture and retention of rainfall to full scale irrigation of crops where there may be no rainfall at all. It also includes the provision of drainage and the avoidance and mitigation of flooding. Irrigated agriculture is the largest user of raw water. Agriculture develops more intensive use of land and water resources, its impact on natural ecosystems and damaging these ecosystems undermine the food producing system. The effective adaptations to the impacts of climate change on water and agriculture will require a sound understanding and integration of agronomic science with water management.

The food price crisis has led to the re-emergence of debates about global food security and its impacts on prospects for achieving the first millennium development goal: to end poverty and hunger.[2] The United Nations Development Programme (UNDP) warns that the progress in human development achieved over the last decade may be slowed down even reversed by climate change, as new threats emerge to water and food security, agricultural production and access, and nutrition and public health. The impact of climate change, sea level rise, droughts, heat waves, floods and rainfall variation by 2080, push another 600 million people into malnutrition and increase the number of people facing water security by 1.8 billion.[3]

Climate change, however, is considered as posing the greatest threat to agriculture and food security in the 21st century, particularly in many of the poor, agricultural based countries of sub Saharan Africa with their low capacity to effectively cope.[4] African agriculture is already under stress as a result of population increase, industrialisation and urbanisation, competition over resource use, degradation of resources and insufficient public spending for rural infrastructure and services.

The outlook for the coming decades is that agriculture productivity needs to continue to increase and will require more water to meet the demands of growing population. Ensuring equitable access to water and its benefits now and for future generations is a major challenge as scarcity and competition increase. The amount of water allocated to agriculture and water management choices will determine to a large extent, whether societies achieve economic and social development and environmental sustainability.[5]

While food and water security are largely determined by actions taken at the local and national levels, global factors, such as world food trade, global climate and climate change and competition for water also affect food and water security locally. Moreover, human alteration of land use patterns, urbanisation, elimination of wetlands, nutrient overloading in water systems and other biophysical changes could dramatically affect the ability of the global water cycle to support needed food production. The development of policies that mitigate adverse impacts, enhance positive impacts and support adaptation to climate and global change, together with enhancing local food and water security.

Thus, analysis of strategies for increased food and water security must take into account relevant hydrologic, agronomic, economic, social and environmental processes at global, regional, national and local levels. A project supposed by Germany's Federal Ministry for Economic Cooperation and Development entitled, "Food and Water Security under Global change, Developing Adaptive Capacity with a Focus on Rural Africa" has conducted research on adaption to climate change at various scales. This project is associated with the challenges the programme on water and food, under the Consultative Group on International Agriculture Research (CGIAR) deals with.

Policy makers are generally more interested in the development of adaption measures following political rather than hydrologic boundary. Vulnerability and adaption measures were also developed at the province and state levels for these two countries - Ethiopia and South Africa to discuss measures of vulnerability, adaption options and the role of information and various actors that is state, private sector and civil society in shaping adaption to climate change.

The impacts of climate change on crop production in the survey were simulated based on crop yield and production function models to assess the implications of climate change for local food security. The impact of climate change on water availability, water demands and irrigation was simulated to identity basin level adaptation strategies. To capture the interactions of climate change and adaptations at the national and regional (Sub-Sahara Africa) levels, a water and food productions model was updated to take into account the impacts of climate change of global change. The impact of global change on poverty and water and food security was assessed for case study countries and sub-Sahara Africa. Alternative adaptation strategies developed in using modelling framework, taking into account the local-level constraints and basin level challenges identified.

Water and climate change
Many of the anticipated impacts of climate change will operate through water changing rainfall and river flow patterns will affect all water users, threatened crop water, poor rained farmers and more vulnerable people by the floods, droughts, typhoons and monsoons causing water-borne disease incidence, glacier melt floods and sea level rise. These impacts of climate change are likely to fall predominantly on poor communities who are least able to cope- present as well as in future. Better management of water resource will also help to manage current climate variability and shocks, which are fundamental development issues in the world's poorest countries today.

Water is the early climate change impacts
The IPCC report makes it clear that once climate begins to change, water resources will be amongst the sectors most affected. It was first highlighted in the final statement of the scientific sessions of the second world climate conference, held in 1990, which recognised that: Among

the most important impacts of climate change will be its effects on the hydrologic cycle and on water management systems and through socio-economic systems.[6] Some of climate change impacts will reflect role of water in all life. So, rain-fed agriculture will have to adapt to new patterns of rainfall and health care systems will have to cope with diseases such as cholera and malaria due to changes in ecologies.

Changes in climate will be amplified in the water environment. Small changes in temperature are translated into large changes in river flows could have a major impact on the water supplies to growing urban communities with infrastructure built to meet their needs for shelter and transport. The temperature and rainfall patterns transmitted through media images of death and destruction like the devastation of New Orleans by Hurricane Katrina in 2005 and the floods in Bangladesh in 2007 caused by cyclone. Increasing biofuel production, which compounded water stress and hunger in many regions where commercial irrigation has efficiently harnessed water to generate jobs, threaten widespread unemployment and aggravate existing conflicts in the poor rural communities. Therefore, water will be at the heart of both the risks and responses to climate change adaptation. Water is an important part of the problem and an important part of the solution.

Water security
Water security is an early adaptation strategy for immediate benefits to vulnerable and underserved population through millennium development goals with strengthening systems and capacity for longer term climate risk management. Water security is the reliable availability of an acceptable quantity and quality of water for health, livelihoods and productions, coupled with and acceptable levels of water related risks.[7] Many societies will want to move beyond water security to take advantages of the economic, social and environmental benefits that can be derived from wise water use.

There are many societies where water security has not yet been achieved. The failure of the rains, of the water resources often exacerbates poverty and conflict in poor communities. In Asia, Central America and the Caribbean, it is the floods more often than droughts that have the most devastating impact on poor communities. From Vietnam and the Philippines to Honduras, Nicaragua and Cuba, the damage and flooding from hurricanes and typhoons, which bring too much rain, keeps people in poverty by wiping out assets and increasing economic vulnerability. Floods and droughts become more extreme and difficult to predict, climate change will make the achievement of water security more difficult.

In this context climate change represents serious threats. Water supplies in many countries are based on the assumption that the dams from which their water is taken will provide a certain yield. If the average rainfall declines or droughts expected those assumptions will no longer be valid and domestic supplies may be at risk. Farmers and other large water users face similar threats in countries from India and Nepal to Kenya and Uganda to Chile and Brazil, one of the more serious impacts of droughts is on the availability of electricity since less rain means less water though the turbines that generate it.

The achievement of water security is already a fundamental development challenge. There are many competing demands on water which is affected by the natural variability of the weather that drives the water cycle. These include growth in water demand due to increased requirements for industry, improved living standards and changes in diets and in patterns of production. In many countries, pollution from human and industrial wastes is also reducing the amount of usable water.

Investments in water security

Actions to implement water management are by nature adaption actions. Better water resources management means greater resilience today and more effective adaption in the future. Actions will be needed to be guided by sound information, science and best practices from both water and climate fields. Water policies and practices must aim to build institutions, information and capacity to predict plan for and cope with seasonal and inter annual climate variability as a strategy to adapt to long term climate change. To achieve the goals of water security and resources managers in an interactive way that enhances their ability to understand, cope with it and respond to new challenges as they emerge. It is important to establish effective, focussed institutions to manage water, with governance structures that support the engagement of the different stakeholders in decision making processes.

A water secure world will need investment across the following three: Better and more accessible Information; stronger and more adaptable Institutions; and natural and manmade Infrastructure to store, transport and treat water. These need at all levels in projects, communities, nations, river basins and globally information, consultation and adaptive management will be essential. The three E's (Equity, Environment and Economic) are essential for sustainable development broadly, and water management more. Specifically, the art of adaptation in water management will be to find the right mix of the three I's to achieve the desired balance between the three E's.

Agriculture, water and climate change

Small hold farmers, in majority of the world are poor. The International Fund for Agricultural Development (IFDA) estimates that there are 1.2 billion people who cannot meet their basic needs for sufficient food every day. The largest 800 million poor women, men and children often belonging to indigenous population, who live in rural environments. They often occupy marginal lands and depend heavily on rain-fed production systems. Hence strategies to reduce rural poverty will depend largely on improved water management in agriculture. The short term variability of rainfall and heavy rainfall is a major risk factor. Soil moisture deficits, crop damage and crop disease are all driven by rainfall and associated humidity. Increased evaporation and evapotranspiration with associated soil moisture deficits will impact rain-fed agriculture.[8] Increased evaporation of open water storage can be expected to reduce water availability for irrigation and hydro power generation. The Fourth Assessment Report of the Intergovernmental Panel on Climate Change (IPCC) predicts decreasing rainfall in northern and southern Africa, increasing rainfall over the Ethiopian/East African Highlands and a considerable increase in frequency of floods and drought.

A number of countries in Sub Sahara Africa (SSA) experienced water stress as a result of insufficient and unreliable rainfall, changing rainfall patterns or flooding. For Africa, it is estimated that 25 per cent of the population currently experience water stress with more countries expected to face high risks in future. Food security and rural livelihoods are linked to water availability and use. Food security is determined by the options people have to secure access to agricultural production and exchange opportunities. These opportunities are influenced by access to water. These water livelihoods linkages is important to understand the vulnerability of households to climate related hazards such as droughts and multi-faceted impacts that water security has on food and livelihood security. Climate change adaptation at local level means the populations at risk of water and food insecurity.

The water use and livelihood strategies are fundamental to the assessment of water stress and drought impacts will be key in assessment of climate change impacts. The concept of water security is increasingly used to describe the outcome of the relationship between the vulnerability of water, its accessibility and use. Water security is defined as 'availability of' and 'access to' water in sufficient quantity and quality to meet livelihood needs of all households throughout the year without prejudicing the needs of the users.[9] Calow et al., distinguish three links between water, health, production and household income. First lack of access to adequate water supply, both in equality and quantity for domestic uses can be a major cause of declining nutritional status and of disease and morbidity. Second, domestic water is often a production input. Such production is essential for direct household consumption or income generation Third, the amount of time used to collect water and related health hazards especially for women and girls has been well documented.[10]

For costal populations, water quality is likely to be affected by salinisation, or increased quantities of salt in water supplies. This will result from a rise in sea levels which will increase salt concentrations in ground water. Sea level rise will not only extend areas of salinity, but will also decrease freshwater availability in coastal areas. Saline intrusion is also a result of increased demand due in part to growing coastal population that leave ground water reserves increasingly vulnerable to contamination and diminishing water reserves.[11]

One of the most significant sources of water degradation results from an increase in water temperature. The increase in water temperatures can lead to a bloom in microbial populations, which can have a negative impact on human health. The rise in water temperature can adversely affect different inhabitants of the eco-system due to species sensitivity to temperature. The health of a body of water, such as a river is dependent upon its ability to effectively self-purify through bio-degradation which is in trouble when there is a reduced amount of dissolved oxygen. This occurs when water warms and its ability to hold oxygen decreases. Consequently, when precipitation events do occur, the contaminants are flushed into water ways and drinking reservoirs, leading to significant health implications.[12]

The agricultural implications of the IPCC working group I report
AR4 (IPCC, 2007) estimates a temperature rise in the range of 2 to 6 degree Celsius by 2100. This compares with temperature increase in the millennium assessment scenarios that fall in a lower range from 1.5 to 2.0 degree Celsius above pre-industrial in 2050 to between 2.0 and 3.5 degree Celsius in 2100.[13] The temperature estimated by AR4 the positive feedbacks that increase carbon dioxide concentrations in the atmosphere due to absorptive capacity of the seas and terrestrial vegetation and soils

and other feedbacks that the melting of polar and mountain ice caps at +4-5 degree Celsius, thawing of permafrost with release of large volume of methane and higher atmospheric retention of CO_2 in future at higher temperatures. These temperature and CO_2 concentration changes will have direct impacts on plant growth.

The balance of impacts from increasing temperature on aquatic eco-systems (rivers and lakes) is not clear. There are competing effects between reduced oxygen concentration in water at higher temperature and higher aquatic productivity with a likely net increase in oxygenation. There are attendant implications for aquaculture and flood production systems (deep water rice) as well as drains and drainage management, but these have not been sufficiently explored. Overall impacts on crops are a combination of direct temperature effects on respiration and photosynthesis increased water demand and the availability of soil moisture as determined by rainfall, runoff and applied water. The crops of different climates are classified into Agro-Ecological Zones (AEZ)[14] Irrigation is commonly found in major deltas in South, East and Southeast Asia, and their vulnerability in terms of displaced people as a result of current trends to 2050.

Rising temperature, rising potential evapo-transpiration rates and declining rainfalls conspire to increase the severity, frequency and duration of droughts. Large-scale land-use change is expected in all continents. AR4 estimates that some 75 million ha of land that is currently suitable for rain-fed agriculture, with a growing window of less than 120 days, will be lost by 2080 in Sub-Saharan Africa.[15] Worldwide cereal yields are expected to decline by 5 per cent for a 2°C rise in temperature and by 10 per cent for a rise of 4°C. Grain yields should decline above certain temperature thresholds, with grain number in wheat falling in temperatures above 30°C and flowering declining in groundnut when they are above 35°C.[16]

There is a rising consensus of opinion that Africa and South Asia are the most susceptible and vulnerable to climate change; both have large population of poor with meagre access to basic resources of water and productive land.

AR4 reported the following broad regional impacts:

- In Africa, by 2020, between 75 and 250 million people will be exposed to increased water stress and in some countries, yields from rain-fed agriculture could be reduced by 50 per cent.
- In Asia, by the 2050s, fresh water availability in Central, South, East and Southeast Asia, particularly in large river basins, will decrease.

The heavily populated mega deltas in the South, East and South east will be at risk due to increased flooding from the sea and rivers.
• In Latin America, productivity of some important crops will decrease and livestock productivity will decline with adverse consequences for food security.

The mid latitudes will suffer from declining yields because temperature change as a result of reduced water availability- for irrigation and rain-fed farming in the Mediterranean, South Europe, mid-west United States and the semi-arid to arid sub tropics.

Water, food security and climate change
Food security and climate change assess the likely impacts of climate change on four key dimensions of food security - availability, stability, access and utilisation.[17] FAO defines, "Food security is a situation that exists when all people, at all times have physical, social and economic access to sufficient, safe and nutritious food that meets their dietary needs and food preferences for an active and healthy life". Crop production is dedicated to food security. Industrial crops and beverage crops make no direct contribution to kilocalorie consumption by human beings although some industrial crop residues are used as livestock fodder. Water management achieves stability of crop production by maintaining soil conditions. Irrigation allows the cultivation of crops when rainfall is insufficient, ensures high value, high risk horticulture from failure and has played a major role, in achieving national and regional food security in Asia as well as improving individual livelihoods.[18]

In the future, food security strategies will be more complex. Higher temperatures will increase water demand and where rainfall declines, many will seek more irrigation to ensure food security and maintain livelihoods. At the same time water supplies available for irrigation will become more variable and will decline in many parts of the world. New agriculture demands the need to achieve better equity in access to reliable food supplies than in the past. As irrigation has been practiced on only 20 per cent at the world's cultivated land, there have been many poor, who have missed out on its benefits. The need to maintain viable aquatic eco-systems will place further stress on water resources, especially where the poor are dependent on them for their livelihoods. Water allocations to agriculture may fall in many parts of the world to the combined impacts of climate change, environmental needs and competition from higher value economic sectors. There will be strong pressure to produce more with less water and to spread the benefits of all water use more widely and wisely, because higher temperatures will reduce potential land and water productivity. Climatic variability in South-eastern Australia has had more profound impacts on water allocations and associated livelihoods in agriculture. If this magnitude of change occurs in developing countries, the impacts on poverty are expected to be much more profound.[19]

The rural population, mostly at risk, from anticipated climate change impacts in semi-arid and arid-zones that have few options for adapting to more water scarcity other than migration. Seasonal out migration is already a consistent feature of many rural communities of Sub Saharan Africa and South Asia where food security is no longer dependent upon locally grown produce. The mitigation of climate change can help reduce negative climate impacts on food security but due to lags in the climate change system, the effects of such actions will be felt only after 2050.[20] It should be stressed that amplified climatic variability and further ENSO fluctuations will generate extreme events that impact food production well before 2050. This means adaptation strategy, regardless of future emissions pathways, will be needed in coming decades to reduce the anticipated impacts of climate change. Food security will depend not only on climate and socio-economic impacts on food production, but also on economic growth, changes to trade flows, stocks and food aid policy.

Climate change affects all four dimensions of food security. They are:

i. Food production and availability

Climate change affects food production directly through changes in agro- ecological conditions and indirectly by affecting growth and distribution of incomes and thus demand for agricultural produce, changes in land suitability, potential yields such as CO_2 fertilisation and production of current cultivations are likely shifts in land suitability which are likely to lead to increase in suitable crop land in higher latitudes and declines of potential cropland in lower latitudes.

ii. Stability of food supplies

Weather conditions are expected to become more variable than at present, with increasing frequency and severity of extreme events. Greater fluctuation in crop yields and local food supplies can adversely affect the stability of food supplies and food security. Climatic fluctuations will be most pronounced in semi-arid and sub-humid regions and are likely to reduce crop yields and livestock numbers and productivity. As these areas are mostly in Sub-Saharan Africa and South Asia, the poorest regions with the highest levels of chronic undernourishment will be exposed to the highest degree of instability.

iii. Access to food

Access to food refers to the ability of individuals, communities and countries to purchase food in sufficient quantities and quality. Falling real prices for food and rising real incomes over the last 30 years have led to substantial improvements in access to food in many developing countries. Possible food price increases and declining rates of income growth resulting from climate change may reverse this trend.

iv. Food Utilisations

Climate change may initiate a vicious circle where infectious diseases, including water borne diseases, cause hunger, which makes the affected population more susceptible to those diseases. Results may include declines in labour productivity and an increase in poverty, morbidity and mortality.

A number of countries in sub-Saharan Africa already experience considerable water stress as a result of insufficient and unreliable rainfall, changing rainfall patterns or flooding, including increases stress in water availability, accessibility, supply and demand. It is estimated that the net balance of changes in the cereal production potential of Sub-Saharan Africa (SSA) resulting from climate change will be negative with net losses of up to 12 per cent. SSA countries will be at risk of significant declines in crop and pasture production due to climate change.[21]

Food security and rural livelihoods are linked to water availability and use. Food security is determined by the people who have a secure access to own agricultural production and exchange opportunities. These opportunities are influenced by access to water. Making these water livelihoods linkages is important for vulnerability of households to climate related hazards such as droughts, impacts of water security on food and livelihood security.

There are three links between water, health production and household income. First lack of access to adequate water supply, both in quality and quantity for domestic uses can be a major cause of declining nutritional status and of disease and morbidity. Second, domestic water is often a product input. Such production is essential for direct household consumption or income generation. Third, the amount of time used to collect water and related health hazards can be especially for women and girls.[22]

Vulnerability is dependent on nature of hazard. Vulnerability is not the poverty and poverty is not the vulnerability. Similarly, risks overlap with poverty, but they are not synonymous. All people face risks- how people especially the poor are able to deal with them.[23]

Impacts of climate change and water management

Climate change is severely impacting the hydrological cycle and consequently, water management. This will in turn have significant effects on human development and security.[24] Water is the primary medium through which climate change influences earth's eco-system, livelihood and well-being of societies. Global warming is likely to intensify, enhance or accelerate the global hydrological cycle.[25]

The water is affected by the climate change due to several reasons: *Changes in precipitation, which higher average temperatures and temperature extremes are projected to cause, will affect water resources availability through changes in form, frequency, intensity and distribution of precipitation, soil moisture, glacier and ice melt, river and ground water flows and lead to further deterioration of water quality. The global picture is complicated and uneven with different regions, river basins and localities being affected in different degrees and in a variety of ways.*

- Climate change affects the water cycle directly and through it, the quantity and quality of water resources available to meet the needs of societies and eco-systems. Climate change can result in an increased intensity in precipitation causing greater less ground water recharge. Receding glaciers, melting permafrost and changes in precipitation from snow to rain are likely to affect seasonal flows. Longer dry periods are likely to reduce ground water recharge, lower minimum flows in rivers and affect water availability, agriculture, drinking water supplies, manufacturing and energy production, thermal plant cooling and navigation. Increased intensity in rainfall, melting glacial ice and large scale deforestation is already increasing soil erosion and depriving the top soil of nutrients. Changes to the proper functioning of ecosystems will increase the loss of biodiversity and damage eco-system services.
- Rising sea levels will have serious effects on coastal aquifers, which supply substantial water to many cities and other users.[26] This phenomenon will also have severe impacts on food production in

major delta regions, which are the food bowl of many countries. Coastal ecosystem would also be profoundly affected including loss in productivity, changes in barrier islands, loss of wetland and increased vulnerability in coastal erosion and flooding.

- Global warming will impact water temperatures, which are expected to have substantial effects on energy flow. The composition and quality of water in rivers and lakes is likely to be affected in changing precipitation and temperature resulting from climate change. Changes in precipitation intensity and frequency influence non-point source pollution, making the management of waste water and water pollution more demanding and urgent.

- Climate change will directly affect the demand for water; for instance, changes in demands will derive from industrial and household use or from irrigation. Water demand for irrigation may increase as transpiration increases to higher temperatures. Depending on future trends in water use efficiency and the development of new power plants, the demand for water in thermal energy generation could either increase or decrease.

- Extreme weather events have become more frequent and intense in many regions resulting in a substantial increase of water related hazards. At the same time, demographic changes are exposing more people to increased flooding, cyclones and droughts. In the previous year's major flooding, which have resulted in many deaths and cost billions of dollars in damages, is an indication of what could lie ahead from increased variability.[27]

- The more intense droughts experienced in the past decade, which have affected an increasing number of people, have been linked to higher temperatures and decreased precipitation. The Fourth Assessment report of IPCC has concluded that there is 90 per cent probabilities that the extent of drought affected areas will increase.[28]

The management of water resources impacts almost all aspects of society and the economy including food production and security, domestic water supply and sanitation, health, energy, tourism, industry and the functioning of eco-systems. Under present climate variability, water stress is already high, particularly in many developing countries.[29] Managing water has always implied that societies deal with natural variability in the water supply. Climate change threatens to increase this variability, shifting and intensifying extreme weather patterns and introducing greater uncertainty in the quantity and quality of water supply over the long term. Adaptation to the current climate variability, while having direct benefits can also help society to prepare for the increased variability expected in the future.

Climate change is one of the main driving forces of change for water resources management, together with demographic, economic, environmental, social and technological forces.[30] Decision makers and policy makers in other disciplines have the solution to many water management problems. They need to recognise that all major decisions should take into account the potential impact on water, recognising water as the life-blood. While tackling these issues, decision makers should think beyond their own sectors and their decisions on water availability and the forces affecting it and should adopt a balanced, integrated and coherent approach.[31]

In the context dominated by worsening food security and malnutrition, increased energy shortages, spread of disease, humanitarian emergencies, growing migration, increase risk of conflict over scare land and water; and escalating ecosystem degradation, improved and integrated management of water and land resources become critical to sustainable development. The water scarcity will place on the environment and the importance of water in development, mitigation of impacts of water management on the environment will become increasingly difficult. Therefore, the establishment of new institutions, networks and better coordination and exchange of information will be necessary.

Agricultural systems dependent on water management

Rain-fed agriculture
In the arid and semi-arid regions, climate change impacts more than 80 per cent of global crop area on rain-fed lands and 60 per cent of global food output. The productivity of rain-fed agriculture in developing countries is lower than more secure irrigated conditions. Many sophisticated adaptations have been developed by farmers to allow cropping in arid and semi-arid conditions including mixed and companion cropping, flood water spreading and runoff harvesting. Rain-fed and pastoral agriculture dominate land use in many countries are major determinant of hydrology and runoff in a river basin.

African agriculture is predominantly rain-fed, with low rainfalls over a large portion of the continent and is dominated by small farms. There is growing realisation that the subsistence model of small-holder development cannot power rapid economic growth in Africa.[32] If African agriculture 'converges' with the rest of the world, there will be fewer farmers, larger urban and coastal populations as well as potential to increase labour productivity in agriculture and production. The transaction costs of farming are lower in commercial farming than in subsistence production, indicating opportunity for production growth in the continent.[33]

In the mid-latitudes, rain-fed agriculture generally has low productivity and is prone to drought and crop failures. Rain-fed staples such as wheat, maize, sorghum and millet are generally grown in conditions of deficient or sufficient rainfall. However, rice is naturally adapted to wet environments with excessive rainfall, requiring some measure of water control.

FAO (2002) concluded that better complementary benefits are derived from fertilizer and additional water both in rain-fed and irrigated conditions.

- The estimated average products of water and fertilizer increase with the volume of irrigation water applied and the proportional gains in yield can exceed the proportional increases in applied water and the amount of fertilizer;
- While irrigation and fertilizer inputs can generate substantial increase in crop yields above certain levels, the law of diminishing marginal returns applies to input use;
- Yield response to fertilizer varies with moisture conditions;
- The water use efficiency of supplemental irrigation increases with increasing nitrogen;
- Fertilizer use may tend to increase with the average number of irrigations;
- Both moisture and fertility management contribute to higher yields but neither moisture nor fertilizer alone will generate maximum yield.

Water availability can be improved through a range of well

established conservation techniques that enhance:
- Storage of rain water in the root zone both directly, and by collecting surface runoff from areas to crops.
- Soil-water conservation techniques such as zero tillage, have clear benefits for rain-fed crop production and are well adapted to mechanised farms, but further work is needed for smaller subsistence producers who rely on manual labour and animal power.[34]
- The intensification of water capture in rain-fed systems has

inevitable consequences for hydrology and runoff. Large scale water conservation activity in the upper catchments of Indian river basins has resulted in measurable depletion of runoff, sometimes to the detriment of established downstream users.[35]

Rain-fed cropping will remain inherently risky, and soil moisture conservation will become an important adaptation, if rains fail, crops will fail.

Irrigated agriculture

The pace of irrigation development accelerated rapidly after the Second World War (1945) and reached a peak in 1970s with heavy development financing. It has been credited with supporting the Green Revolution and economic development in Southern, Eastern and South eastern Asia, but there has been very little impact in Africa. Irrigated yields in developing countries are in general, two or three times higher than those of rain-fed yields.[36]

Irrigation is the prime means of intensification and will remain a keystone of food security policies in the face of climatic variability. Just over 300 million ha of the world's agriculture land is equipped for irrigation.[37] Just under 40 per cent of global equipped areas is now serviced by ground water sources illustrating the importance of local, privately accessed aquifers in buffering crop production.[38]

Inland fisheries and aquaculture

It is estimated that one billion people depends upon freshwater fish as the prime source of protein.[39] In many developing countries, most households in rural areas around lakes, rivers and wetlands are involved in fishing on at least a seasonal basis. Fishing has been neglected in terms of the management of water resources, especially for agriculture, with losses of habitat in streams, flood plains and deltas due to upstream development for irrigation.[40]

However, irrigation and associated waters storages may also provide new or alternative opportunities for both capturing fisheries and agriculture. Irrigation supplies have long been used to manage water quality for high value prawn production. The frequency of extreme droughts and floods will have a disproportionate effect on fish habitat and populations and the incidence of diseases is expected to rise. Climate included effects anticipated in the Mekong include a significant change in food chain because of declining water quality, changed flow in patterns of vegetation and saltwater intrusion in the lower deltas.

Forest land

Forests play key roles in both the global hydrologic cycle and in regulating runoff in basin scale. Trees can transpire water from considerable depths in the soil and thus play an important role in mediating surface runoff and ground water recharge. New forests generally generate less runoff than agricultural land use or mature forest because of high rates of growth and accompanying high rates of transpiration.[41] It is the principle adopted for forested catchments being preserved and managed for urban water supplies in many parts of the world from New York to Melbourne.

Deforestation and afforestation can have important impacts on catchment hydrology and South Africa has taken a unique step in considering (Commercial plantation) forestry as water user.[42] Forests contribute to local climate forcing through positive feedback on rainfall from high evapotranspiration. Forests will, therefore, play a key role in GHG mitigation strategies. Afforestation is an important means of earning carbon credits and has been piloted with some success through the clean development mechanism.

Irrigated agriculture has been able to capture large volumes of freshwater. The development irrigation to satisfy food needs has intensified the consumptive use of water to the detriment of the goods and services provided by natural eco- systems. Water borrowed from natural terrestrial and aquatic eco-systems can eventually undermine the modified systems that have taken their place.[43] In many different parts of the world, expanding cities have easily claimed water from agriculture by fair or foul means.[44] Cities will generate increasingly large amounts of effluent that will be recycled for agriculture, subject to water quality and health and safety considerations. The use of untreated waste water is already widespread with a variety of accompanying hazards to producers and consumers.[45]

In most parts of the world, water accounting and allocation systems are rudimentary and ill equipped to cope with the additional stress of climate change. Inter-sectoral competition and environmental consciousness will be exacerbated by climate change and will require much more sophisticated, detailed approaches to the specification and policing of water rights and allocations. As population increase, urbanisation progresses and climate change starts to bite the volume of water diverted directly for agriculture will decline and supply security for farming will become more erratic and viable as drought frequency and duration increase, with higher priority demands having to be met first.

Water demand in agriculture projections to 2030

Agricultural planning associated water resources assessment in most developing countries has turned a blind eye to climate change. The activities and investment associated with the 'no-change' prospective provides a convenient baseline for the analysis of the further impacts of climate change, using the projections of the IPCC Fourth Assessment Report (AR4) and subsequent analysis.

Current and projected trends in demand for food and agricultural production are given for 93 developing countries in FAO's world agriculture prospective study towards 2015/2030.[46] They are based on the United Nations Statistics Division Medium Population Projection, the World Bank income growth projections and FAO's own estimates of future agricultural productivity. It is estimated that total water use in crop production amounted to 7130 KM³ in 2000 and is likely to rise between 12,000 and 13,500 KM³ by 2050.47 An overall expansion in cropped area of 29 per cent is forecast to 2050 with rain-fed areas increasing from 549.812 million in 1998 to 698.743 million ha (27 per cent). Irrigated area is forecast to grow by 33 per cent from 2, 42,182 million ha to 318 million ha over the same period.[48]

Demand for water reflects irrigation needs, whereas water use in rain-fed agriculture is considered only in terms of evapotranspiration of water from the available rainfall. Locally, climate variability has significant impacts on crop area and crop production, especially in periods of drought or floods. Since the 1970s the extensive development of irrigation supplies and flood control has levelled out the impacts of climate variability and with the benefits of Green Revolution farming techniques, increased productivity to the point that commodity prices fell year on year in real terms until the early 2000s. During this period many countries maintained high carryover stocks of grains, but these increased in market demand and resulting in price increases, beginning in 2002 with Chinese buying.[49]

Raising water productivity

Water productivity is now considered in a wider landscape context, where both land and water resources are limiting.[50] Economic water productivity assesses the value of output per unit of water used and reflects crop choice, market conditions and the effectiveness of water management. There is considerable potential to raise the land and water productivity of rain-fed crops by improving soil-moisture status through better soil-water conservation and harvesting technologies or by providing supplemental irrigation where possible. A major reasoning behind widespread efforts to re-invigorate rain-fed farming in this way is that it will allow for mitigation of damage to natural ecosystems from: (i) abstraction of water for formal irrigation and (ii) expansion or rain-fed areas at the cost of forest and other natural ecosystems.[51] Water harvesting technologies can in dry periods, but provide less insurance against drought. Both the conditions for rain-fed cropping and the hydrology of rainwater harvesting and conservation will deteriorate in the tropics and semi-arid tropics under climate change.

There are many factors of yield that contribute to water-use efficiency, where the management of inputs is poor or unbalanced, water productivity may be low due to depressed yield for reasons other than water supply. Improving yield and water use efficiency in irrigated and rain-fed agriculture follow different pathways. In rain-fed cultivation, the focus is on increasing yield through good husbandry, best use of rainwater through soil moisture management and by the provision of more nutrients. In irrigated agriculture, the availability of water will dictate whether to focus more on yields or on water productivity resulting in different combinations of area and water use for an optimum level of production.

A number of reports have summarised the main trends and drivers in agricultural water management that mainly respond to economic factors relating to rising water scarcity.[52] Many recent reports on adaptation to climate change[53] anticipate a substantial increase in irrigated area in response to global temperature rise, higher rates of crop water use and declining and more variable rainfalls.

Five main climate change related drivers will affect the agriculture sector, they are: temperature rise, precipitation patterns including rainfall and snow; the incidence of extreme events i.e. floods and droughts; sea level rise; and the atmospheric carbon dioxide content. The impacts of climate change on water balance and implications for irrigation include the following:

- Reduction in crop yield and agricultural productivity where temperature constraints crop development;
- Reduced availability of water in regions affected by falling annual or seasonal precipitation;
- Exacerbation of climate variability in places where it is already highest;[54]
- Reduced storage of precipitation as snow and earlier melting of winter snow, leading to shifts in peak runoff away from the summer season where demand is high;[55]

- Inundation and increased damage in low-lying coastal areas affected by sea level rise, with storm surges and increased saline intrusion into vulnerable freshwater aquifers;
- Generally increased evaporative demand from crops as a result of higher temperature.

Climate change is adversely affecting water availability for agriculture and it is expected the competition with water management challenge. Allocations to cities, industry, rural water supply and sanitation are materially affected by climate change, but collectively they will reduce the quantity of water that can be allocated to agricultural use and hydro-environmental services.

Climate change challenges for water management

Physical impacts
There is a general consensus that climate change will have negative impacts on the global freshwater cycle. The IPCC tells that globally, the negative impacts of future climate change on freshwater systems are expected to outweigh the benefits. By the 2050s, the area of land subject to increasing water stress due to climate change is projected to be more than double the increasing water stress. The IPCC Technical Team on water and climate finds some key technical issues on climate change and water.
i. Changing rainfall patterns
    It is difficult to make good predictions about the future of rainfall and storms and more difficult to predict the impact of changing temperature and rainfall on water availability from rivers, lakes and underground sources. The relationship between the amount of rainfall and amount of water available in rivers, lakes or underground is a complex one.
ii. Runoff and stream flow
    IPCC summarised that by the middle of the 21st century, annual average river runoff and water availability are projected to increase as a result of climate change at high latitudes and in some wet tropical areas and decrease over some dry regions at mid-latitudes and in the dry tropics. Many semi-arid and arid areas are particularly exposed to the impacts of climate change and are projected to suffer a decrease of water resources due to climate change.

iii. Temperature, evaporation and aridity

Temperature increase means increase evaporation rates and the balance between evaporation and rainfall determines a climate. Its humidity or arid or aridity will tend to increase where rising temperatures are not matched by rising rainfalls. Changes in aridity will have a substantial impact on both surface water runoff and ground water recharge.

iv. Changing groundwater recharge and storage

The most difficult water resource management challenges monitoring and managing underground water which many communities depend on for their water supply. If the runoff from rainfall that flows into rivers and streams is affected by changes in temperature and land use in the infiltration of water into underground formations.

v. Water quality

The capacity of surface water resources to receive, dilute and remove human wastes is dependent on the volumes of water flowing. Changing runoff patterns and temperatures may result in water quality effects that either render water unusable or impose additional treatment costs on users. The intrusion of sea water into coastal fresh water systems is another possible consequence of climate change. This will not just occur in areas affected by sea level rise but also where reduced river flows are insufficient to prevent sea water from flowing upstream.

vi. Floods, droughts and storms

The impact of changing rainfall patterns will be the changing incidence of floods, droughts and storms. Higher temperature will lead to greater evaporation from the ocean and other sources, which will create more and more rainfall. Many impacts of climate change, climate related disasters are disproportionately affecting the poor. The 2008 UN Human Development Report stated that from 2000 to 2004, some 262 million people were affected by climate related disasters annually. Over 98 per cent of them lived in the developing world.

vi. Glacier and snow melt and loss of storage

The shrinking of glaciers and reduction in the volume of water 'stored' in snow fields is one of the earliest impacts on water resources expected from climate change. The areas currently act as huge natural reservoirs, collecting and storing water as snow in winter and releasing it gradually as water melts in summer. The water supply of one-sixth of the world's populations is dependent upon glacier and snow melt.[56] More people are dependent upon melt-fed rivers for their water supplies, agricultural water, navigation and hydro power. The loss of glaciers is particularly important in the Andean region of South America and the Western Himalayan region of South Asia. The International Centre for Integrated Mountain Development (ICIMOD) has identified over 200 glacial lakes in the Hindu Kush-Himalaya mountain range that are at risk of outburst.[57]

Water resource managers need the ability to track changes and to advise and support the implementation of appropriate responses. This requires extensive data and the ability to analyse and interpret it in order to guide planning and inform the broader community of its implications. The importance of hydrological monitoring has been highlighted at all United Nations conferences on water and sustainable development since 1977 Mar Del Plata conference, there has been worldwide decline in the availability of water resources over the past few decades. In many countries, the quality of hydrological data has deteriorated sharply since the Rio Summit in 1992. In 2008, international support was terminated for the Global Environmental Monitoring System (GEMS) program, which was the world-wide repository of water quality data.

Social and economic dynamics

Changes in the availability, timing and reliability of rainfall and the water resources that flow from it will impact on all water using sectors. These impacts will affect the broader dynamics of national economics as well as environmental and social needs, particularly in poorer societies. Water management is important for the achievement of many of the millennium development goals.

The changing in distribution and timing of rainfall will change patterns of access to water, creating to new surpluses in some areas and increased completion in others. Managing these evolving hydrologists will impose significant demands on water management. The increased variability of rainfall will impact growth potential and the costs of achieving water security. There is little correlation between water scarcity and economic development, a clear correlation exists with variability.[58] Ecosystem water use will be put under extreme pressure as the costs of water rise. Few countries have effective mechanisms to assure adequate water for ecosystems. Changing water security conditions will drive changes in economic activities. On balance, economic activity will be driven towards water secure areas and away from insecure areas. Climate change will impact on different dimensions of social and economic life as:

i.   Changing dynamics in urban areas

In Africa and Asia, the urban population is expected to double from 2000 to 2030. In 2030 over 80 per cent of urban dwellers will live in the cities of the developing world.[59] The impacts of climate change will not be limited to the capacity to supply potable water but will also apply to the capacity to dispose of wastes from large urban communities and the costs of building and maintaining other types of infrastructure.

- Water supply is a costly service to provide large settlements and if, availability of water is reduced by climate change, larger contributions will have to change their consumption patterns.
- Any increase in the intensity of rainfall and flooding as a consequence of climate change will increase the cost of roads and water drainage as well as flood protection works.
- If stream flows are reduced, the treatment must be intensified to maintain the water quality of receiving waters or make water available for reuse. Municipal waste water collection and treatment is already the most costly element of infrastructure required to meet MDGs for health, water and environmental protection.[60]
- Flood risks affect the area of land available for settlement as well as the cost of protecting vulnerable land from flooding.
- Bringing water from further afield not only increases the cost of water, but also expands the area affected by competition with cities for water. This will have economic impacts, whether through higher prices for rural products or the aggravation of rural unemployment, potentially leading to urban migration.

ii. Changing role of large industry, a major water user and polluter

Water intensive industries will also feel the impact of climate change. Water security is seen as a 'supply chain' issue in the private sector. Rising water demand and uncertainty regarding both the quantity and quality of water available to industrial users can threaten production. For industrial agriculture and agro processing, the agricultural water patterns and crop water requirements are essential. Many industries are now focussing on securing their water rights and minimising their water footprints and food. Traditionally, many large industries - from textile and leather processing to pulp, paper and steel production - not only used large volumes of water in their production processes, but also dispose large volumes of effluent. One response for industries is to seek to adapt their production methods to reduced water use and discharges.

iii. Changing dynamics in agriculture

Climate change will impact agriculture in many different ways. Dry land farmers are affected by both the short and long term variability in rainfall. Irrigation farmers have some mechanisms to protect them from short-term variation, but they may be more vulnerable when extreme events occur and their supplies fail. And all farmers are vulnerable to damage caused by extreme wind, storms and flooding.

In this context, agriculture is all about risk management. Every season, dry land, farmers have to face uncertainty in their decisions. If soil moisture is inadequate or dry early in season, the plant and seeding may fail to develop and they may become more vulnerable to pests and disease.

Large scale farmers address these uncertainties on the seasonal basis by using insurance and other financial instruments as well as technologies to protect themselves. In some areas, long-term weather trends can be predicted with some degree of accuracy, helping dry land farmers to make seasonal decisions. The value of long-term predictions tends to be of limited reliability. Longer term predictions are often so general as to be virtually useless for the planning of rain-fed production.

iv. Changing economics of irrigation

The threats of greater variability in rainfall and stream flow, the logical response is to invest in the capacity to manage water, essentially paying for greater security and reliability means storage in large dams. Storing water under ground is another local option to reduce runoff from rainfall, which may adversely impact downstream water users. While irrigation, whether supplied from 'ran of river' flows or from water stored naturally underground or artificially in manmade reservoirs appears to be more reliable than rainfall. Communities rich in underground water are fortunate in that they neither have to store water nor transport it to great distance. For this reason, communities have developed local ground water by better pumping technology.

The investments in water management made by farmers have enabled them to sustain increases in production, where farmers have made their own investments, they are often very conscious of the cost of their water and are more likely to practice more efficient water use. Irrigated agriculture is crucial to global food production and the FAO predicts its importance will increase. Developing countries are expected to expand irrigated areas by some 20 per cent before 2030.[61] This pattern of investment leading to greater productivity and increased ability to manage climate variability and risk is potentially an important driver of poverty reduction as well as contributing substantially to the world's ability to feed a growing population.

Compounding factors

The impacts of climate change on different sectors of society pose weighty challenges for water resource management. Other drivers of change in the short to medium term form an important part of the context within which water managers must respond to climate. These drivers include: - population growth, economic development and related changes in consumption patterns, technological developments, climate mitigation strategies and urbanisation and land use change.

Population growth

Most measures of 'water stresses' or 'water scarcity' are based on the amount of water available per person. Absolute amount of water available is often more closely associated with development performance, the implications of water scarcity are still important. Particularly in Africa, but also in Asia and Latin America, growing water scarcity is primarily a result of growing population competing for the same amount of water, rather than any change in the availability of the resource.

Economic development and related changes in consumption patterns

Where economic development results in rising personal incomes, people's consumption patterns change. Dietary changes in particular have implications for water demand. Most changes in consumption as a result of rising standards of living will have the effect of expanding individuals 'water footprints'.

Technological development

There are many areas in which technological change affects water and its management. For example, the developments of water saving technologies that reduce pressure on resource. However, there are many instances in which technological development has negative impacts on water resources. This is particularly the case with respect to water quality; many new chemical products and pharmaceuticals introduced into society are disposed of and disseminated through the water cycle with unpredictable consequences for human health.[62]

Urbanisation and land use change
Flooding is a major water resource management challenge even in the absence of climate change. The mounting costs of floods are due not only to changes in intensity and frequency, but to population, urbanisation and land use changes that push population and assets into vulnerable areas. So, urbanisation and land use planning can also be drivers that influence the way water impacts on society and society impacts on water.

Climate mitigation strategies
The idea that climate change mitigation strategies could themselves worsen the impact of climate change in the water sector is paradoxical but true. Pressures to expand hydroelectricity or to develop new sources of bio fuels will impact on water resources. Solving the energy challenge could, thus, aggravate pressures on water resources as well as putting pressure on food supplies. This is a good example of the need to consider impacts on water resources as part of the review of climate mitigation strategies.

Climate change adaptation to food and water security
Adaptation to climate change impacts should not be approached as a separate activity, isolated from other environmental and socio-economic concerns that also impact on the development opportunities of poor people.[63] In countries where the majority of people depends on agricultural income, proposed climate change adaptation strategies increasing agricultural productivity and making agriculture including livestock, fishery and forestry less vulnerable to climate stress and shocks.

Water management for agricultural production is a critical component that needs to adapt in the face of both climate and socio-economic pressures in the coming decades. Changes in water use will be driven by the combined effects of (i) changes in water availability; (ii) changes in water demand for agriculture as well as from competing sectors including urban development and industrialisation; and (iii) changes in water management.

With regard to agricultural production and water, climate change adaptation may include:-[64]
- Adoption of varieties and species of crops with increased resistance to heat stress, shock and drought;
- Modification of irrigation techniques, including amount, timing or technology;
- Adoption of water-efficient technologies to harvest water, conserve soil moisture and reduce sanitation and salt water intrusion;

- Improved water management to prevent water logging, erosion and nutrient leaching;
- Modification of crop calendars, i.e. timing or location of cropping activities;
- Integration of the crop, livestock, forestry and fishery sectors at farm and catchment levels;
- Implementation of seasonal climate forecasting;
- Additional adaptation strategies may involve land use changes that take advantage of modified agro climatic conditions.

Water related adaptation strategies will also affect the livestock sub-sector including improved rotation of pastures, modification of times of grazing, changing animal species and breeds integration of the crop and livestock systems adequate water supplies and crops.

Land users and rural communities adapt their land management practices to a number of political, economic, social, environmental and climatic changes. Sometimes, adaptation may be mal-adaptation such as: clearing forest land to gain additional arable land; increasing the cultivation of marginal land such as steep slopes. Leading to increased soil erosion; adoption of unsustainable cultivation practices as a result of dropping yields; introduction of new plants and animal species; or more intensive use of chemical inputs leading to population. All of this may increase land degradation and endanger biodiversity, possibly reducing the ability to respond to increasing climate risk in the future.

National Adaptation Plans (NAPAs) emphasise that irrigation will be a major adaptation approach in the agricultural sector. The problem with this strategy, however, is that adaptation practices that involve increased irrigation water use may place additional stress on water and environmental resources on the one hand, and will be influenced by changes in water availability resulting from climate change on the other.

The IPCC concludes that, if widely adopted, adaptation strategies in agricultural production systems have a substantial potential to negate climate change impacts and can even take advantage of positive ones. Policy attention by national governments and trans-national bodies will increasingly have to focus on the coordination of water uses across trans-boundary river basins and across different sectors and arbitration in increasing conflicts over water. Climate change and increased water demand for agriculture in future decades is anticipated to be an added challenge to transboundary frame work agreements, increasing the potential for conflict.

The measures for adapting to climate change related water shortages by, for example, increasing storage capacity upstream, increasing investment in irrigation infrastructure and efficient water-use technologies or revising land tenure and land use arrangements, can lead to increased competition for water resources.

Rain-fed systems will continue to offer the greatest scope of adaptation in terms of area, number of farmers and overall contribution to global food production, particularly since average cereal yields are low in developing countries. The combined effects of climate change and low adaptation capacity will increase vulnerability and local food insecurity in poor developing countries that depend on rain-fed production. In these regions, adaptation techniques and capacity building are most needed.

The irrigated systems associated with the major river basins that depend on some form of water control can be adjusted immediately to increase yields' performance. Performance gains in these systems will have the most impact on global food supply. The projected changes in global runoff in irrigation areas will indicate where water scarcity can be expected and where increased precipitation will present expanded opportunities about flood damage and water logging. Real-time monitoring of weather, water resources and farming systems responses' need to be undertaken to understand the effects of future climate dynamics.

Many potential adaptation options available for marginal change of existing agricultural systems are variations of existing climate risk management.[65] For some cropping systems, implementation of these options is likely to have substantial benefits under moderate climate change. Effective adaptation will need a comprehensive and dynamic policy approach covering a range of levels and issues, from farmer understanding of change in risk profiles to establishment of efficient markets that positively facilitate response strategies.

Forests can play a major role in adaptation strategies by regulating water flows, maximising water yield and ensuring water quality in watersheds. On the basis of IPCC projections, many drought prone and marginal areas are expected to become drier, requiring additional irrigation while the water supply itself will become less reliable. From a global overview of climate change projections and current irrigated areas, it can be concluded that in most developing country situations, except the humid tropics, both rain-fed and irrigated areas that receive sufficient water will be reduced unless there are opportunities for additional adaptation.

NAPAs (National Adaptation Plans)

The UN Framework Convention on Climate Change (UNFCCC) NAPA process is identified arid, semi-arid and dry sub-humid areas of the country as being the most vulnerable to drought. Agriculture was identified as the most vulnerable sector and in terms of Livelihoods, small scale rain-fed subsistence farmers and pastoralist were identified as the most affected. The NAPA process has identified and prioritised 11 project areas that address the immediate climate change adaptation needs in the country focussing on; human and institutional capacity building; improving natural resource management; enhancing irrigation agriculture and water harvesting; strengthening early warning systems and awareness raising of which the following are directly relevant to the water and agricultural sectors:

- Promoting crop insurance;
- Strengthening and enhancing drought and flood early warning systems;
- Development of small scale irrigation and water harvesting schemes in arid, semi-arid and dry sub-humid areas;
- Improving and enhancing rangeland resource management practices in pastoral areas;
- Community-based sustainable utilisation and management of wetlands;
- Community-based carbon-sequestration project;
- Realising food security through at least one multi-purpose large-scale water development project;
- Promotion of on-farm and homestead forestry and agro forestry practices in arid, semi-arid and dry sub-humid areas.

Global Water Partnership (GWP) at UNFCCC Bonn, 2010, Water Adaptation Development

GWP was an active and promoting its key themes at the latest meeting on the climate negotiations, held in Bonn. These ongoing discussions are incorporating elements of the Copenhagen Accord in draft UN text which was discussed at Cancun Mexico. GWP has held dialogues on water, climate change adaptation and development in Africa and China, to chart the way 'Beyond Cop 15'. The central message is, "building water security to support climate change adaptation for development needs to be accelerated as negotiations for a global climate deal progress."

Investment in water infrastructure is declining just at the time when the challenges to deliver are greater with climate change and increasing climate variability. Provision of additional finance for developing countries to tackle adaptation to climate change has been long agreed in principle and some funds have already been committed. There are major contested areas still to resolve on governance and sources of long term funding but potentially, Water should benefit as a central sector for agriculture, health and livelihoods as funds flow initially through fast track finance.

Institutionally, at international level and in many countries, key actors in the water and climate worlds have been working which was recognised in the GWP Bonn panel. It is now vital to construct ways in which interactions can take place leading to actions. GWP has rightly recognised, the main learning points from the dialogues are that water adaptation messages must be integrated within national development planning, there is need for increased organisational learning and investment must be accelerated.

There are many ways in which those tackling climate change adaptation, particularly for the poor and vulnerable countries which have special status in the climate negotiations, can learn from decades of experience in water management. Although progress on the MDGs has been mixed, considerable solutions on progress and decentralisation and participation have been a central thrust of water policy, but the climate change world has spent time re-learning the lessons through various pilot projects including on Community Based Adaptation.

A heavy contested issue in the climate negotiations is the relative roles of public and private finance. Developing countries want the funding to come from public finance in developed countries. Hence, developed country governments' are keen to engage the private sector, particularly on the low carbon side and private sector engagement is fundamental to deliver water infrastructure and services.

It is vital there is a combined effort between the water and climate change communities to devise strategies for growing more food, with less water and energy and what are the interactions between energy and water in low carbon pathways.

UNFCCC, COP 18 on water adaptation
In December 2012, during Cop 18, the parties adopted an ambitious three

year work programme of the Adaptation Committee to provide advice and coherence on adaptation issues under the UNFCCC. Adaptation efforts are currently scattered among various negotiation streams, such as the Nairobi Work Programme, the work programme on Loss and Damage, the Green Climate Fund, the Adaptation Fund, the Standing Committee on Climate Finance, the Technology Executive Committee and others.

Ensuring coherence and linking the adaptation divides is urgently required. Water security is central to successful adaptation as climate impacts are largely felt through the medium of water. Water is to adaptation efforts; energy for mitigation efforts- the connecting link.

Until the world puts water security at the centre of adaptation as part of national development, gains made in economic growth and development will continue to be undermined by water-related disasters such as droughts and floods, threatening food security, energy security and livelihood security worldwide.

This side event will explore the work of the Adaptation Committee and help define, how adaptation under the convention through a water security can support the work of the Adaptation Committee to connect efforts and enhance coherence on adaptation under the UNFCCC. This side event will also showcase how innovative approaches in water management by GWP and its partners in Africa, Asia, Latin America, the Caribbean and Central and Eastern Europe contribute to enhance climate resilience and offer opportunities for accelerating implementation of the Adaptation Committee's Work Programme.

Climate change mitigation to food and water security
The IPCC technical paper on water (Chapter 6) evaluates mitigation prospects of the different sectors to which water - related mitigation strategies in agriculture are related. Some of these strategies may be synergistic with adaptation. For instance, water management of paddy rice can have positive implications for both adaptation and mitigation, reducing emissions of methane as well as improving water-use efficiency.

At the same time, several water-related adaptation strategies, in particular those related to increased irrigation needs, may be counter to mitigation needs, in particular increased pumping of ground water using fossil fuel systems. Use of gravity or renewable energy sources for irrigation management may improve the scope for mitigation. Mitigation may also run counter to adaptation.

With regard to forests, their role in mitigation could be important. Forests respond slowly to changing climate and can sustain land

and water quality and regulate flows. However, there has been some controversy over the hydrological consequences in terms of afforestation. It has been projected that more than 50 per cent of the suitable area for CDM plantations at global scale would reduce runoff by about 60 per cent, so there are significant implications of CDM related plantations and a strong need to factor this into land-use change and catchment management in developing countries.

The IPCC considers how agriculture can contribute to mitigation of global emissions of greenhouse gases, to help stabilise future carbon dioxide levels and restrain global warming. Agriculture contributes greenhouse gases and also, cycles and stores carbon through the photosynthesis and biological accumulation of carbon in plant matter and soil organic matter.

The IPCC (2007) estimates that there is good potential in agriculture, particularly in the tropics to mitigate greenhouse gas emissions. Chapter 6 of this document is devoted to mitigation of greenhouse gas accumulation through agriculture.

FAO could assist member countries in understanding the implications of climate change on water resources and agriculture and in developing better regional and local projections of impacts in order to develop planned adaptive strategies, improve water governance and build specific capacity in water management. Given the instrumental value of water to all economic sectors, agriculture cannot act alone. National water management actions will need to be focussed at national level but supported by regional and international initiatives.

FAO report - climate change, water and food security
FAO focusses comprehensive report on water management as climate change conditions evolve. The focus is on irrigated systems that currently produce 40 per cent of global food output from 20 per cent of the global stock of cultivated land and withdraw more than 70 per cent of the volume of water used for human benefits. The FAO reports are:-
- Reductions in river runoff and aquifer recharge are expected in the Mediterranean basin and in the semi-arid areas of the Americas, Australia and Southern Africa, affecting water availability in regions that are already water-stressed.
- In Asia, the large contiguous areas of irrigated land that rely on snow melt and high mountain glaciers for water will be affected by changes in runoff patterns, while highly populated deltas are at risk from a combination of reduced inflows, increased salinity and rising sea levels.
- Irrigation provides approximately 40 per cent of the world's food,

including most of its horticultural output, from an estimated 20 per cent of agricultural land.

- Aquifers are depleted in many parts of world where they are most important. China, India and the United states- sometimes facilitated by perverse incentives of subsidised energy and support prices for irrigated products.
- Future global food demand is expected to increase by some 70 per cent by 2050, but will approximately double for developing countries.
- The world has a large stock of underperforming canal irrigation infrastructure and a vibrant ground water sector that is competitively depleting its own lifeblood.
- Temperatures are expected to rise at a faster rate in the upper latitudes, with slower rates in equatorial regions.
- Water-scarce areas of the world will generally become drier and hotter.
- Increased atmospheric temperature will extend the length of the growing season in the northern temperate zones, but will reduce it almost everywhere else.
- Worldwide cereal yields are expected to decline by 5 per cent for 2°C rise in temperature and by 10 per cent for a rise of 4°C.
- Cereal production is expected to fall by between 9 and 11 per cent in the developing country regions and Australia/New Zealand, but it is expected to increase by as much as 11 per cent in the developed countries including Russia, thereby reinforcing existing disparities in food production.
- However, the temperature increases that open up new growing seasons for cereals in the higher latitudes and the associated increases in evapotranspiration rates will increase the demand for irrigation.
- Where irrigation is already common place and rainfall declines such as in Southern Europe, crop water productivity will have to increase or crop areas will contact.
- Rain-fed farming will become more precarious in the mid and low latitudes while productivity may rise for a time in the higher latitudes.
- Large scale land use changes are expected on all continents.
- In rain-fed systems, if potential evaporation rates increase, available root zone moisture content will be more rapidly depleted, requiring either shorter season crop varieties or acceptance of lower yields and more frequent crop failure.

- Agriculture accounts for 69 per cent of all water withdrawals in the world, but offers the lowest economic return per unit of water.

Human rights and climate change

Impacts of climate change on implementation of human rights to food and water
Climate change has visible and predominantly negative impacts on global crop yields and the availability of potable water. Climate change poses the challenges for the implementation of the human rights to food and water. The increase in temperature will reduce the food production and the climate change will shift rainfall patterns, leading to many areas with low rainfall. As a result, the productivity of rain-fed agriculture will significantly decrease in some regions of the world. Changing rainfall patterns will affect the availability of drinking water. The natural seasonal melting of snow in summer, the driest season of the year, supplies water to many rivers which are central to fresh water supply. However, the melting of glaciers caused by global warming will lead to the disappearance of these natural water reservoirs and water shortages. The sea level rise increases the salinity of ground water and soils in coastal regions and small islands.

The impact of climate change will be much more severe in the tropical and sub-tropical climate zones. Many people, who live in those regions, lack the physical, social, economic and technological resources necessary to adapt to the changes that will be brought by climate change. Vulnerable people are socially disadvantaged and already suffer from hunger and poverty due to their gender, age, disease or belonging to the minority group.

Small family farmers, landless peasants and day workers of rural communities are most vulnerable along with women and girls who are particularly most exposed to the adverse impacts of climate change.

Principles of international environmental and human rights law
The foundations of International climate change law were laid at the United Nations Conference on Environment and Development (UNCED), held in Rio de Janeiro in 1992 With UNFCCC which can be regarded as the constitution of international climate policy. The 'Earth Summit' was the Rio Declaration on Environment and Development, Agenda 21 and the Convention on Biological Diversity.

The international environmental law is based on the following principles:
- The polluter pays principle, stating that the polluter should, in

principle, bear the cost of pollution (Rio Declaration on Environment and Development, Principle 16).

- The principle of limited territorial sovereignty, according to which state parties have the responsibility to ensure that activities within their jurisdiction or control do not cause damage to the environment of other states or of areas beyond the limits of national jurisdiction (Preamble of the UN framework convention on climate change; Rio Declaration on Environment and Development Principle 2).
- The principle of common but differentiated responsibilities, stating that industrialised countries should assume the leadership of the fight against climate change and its negative effects (UN Framework Convention on Climate Change Art 3(1)).

The main pillar of the human rights system is the International Bill of Human Rights of the United Nations, which consists of the Universal Declaration of Human Rights adopted by the General Assembly in 1948, the International Covenant on Civil and Political Rights (ICCPR) and the International Covenant of Economic, Social and Cultural Rights (ICESCR), both adopted in 1966.

The obligations of states are stipulated in the treaties and the principles of the organs of the United Nations human rights system. The main and key obligations of states are:

- The obligations to respect means that states must refrain from undertaking measures, as for example, displacements, which curtail the enjoyment of human rights.[66]
- The obligations to protect require states to take steps in order to prevent both corporations and individuals from abusing human rights.[66]
- The obligation to fulfil means that states must take positive action to achieve progressively the full realisation of human rights. For instance, they have the duty to facilitate people's access to water and other resources necessary to assure their livelihood. If an individual or a group of people are not able to realise their rights to adequate food and water with their obligations to take immediate action to correct such situations.[66]
- The obligations to take steps, individually and through international assistance and cooperation, especially economic and technical, to the maximum of its available resources, with a view to achieving progressively the full realisation of the rights recognised in the International Covenant on Economic, Social and Cultural Rights.[67]

- The obligations to act promptly and expeditiously, making an efficient use of available resources, in order to undertake specific and purposeful measures to ensure the enjoyment of human rights.[68]
- The obligations to guarantee, regardless of the extent of available resources, that the rights to food water can be exercised without discrimination (Prohibition of discrimination). In addition, special attention must be given to members of vulnerable and disadvantaged segments of population.[68]
- The obligation of states to guarantee the access to at least minimum levels of food and water to all people under their jurisdiction.[68]
- The obligation to facilitate the participation in the political decision making process of those people who are affected in their rights by the adopted policies.[66]

The obligation to cooperate internationally is usually given a subsidiary role. In the first instance, states are bound to implement international human rights on their own territories. The task of international community is to make the climate regime consistent with the existing human rights obligations of states.

Climate change, states and human rights
Climate change is caused by human activities and has extensive negative impacts on the fulfilment of human rights. International human rights, therefore, constitute an obligation of states to promptly undertake measures to reduce greenhouse gas emissions and global warming on the international protection of human rights. The human rights require state actions to focus on those who are most vulnerable to human rights abuses. Thus, international environmental law must mandate and realise emissions reductions' of a magnitude that is adequate to protect the rights of these people from climatic changes. A fundamental question arises about the need of large emission reductions that hundreds of millions of people in developing countries are still affected by hunger, poverty and lack of access to proper sanitation among others. In order to fulfil the human rights of people in poor countries, it is therefore, indispensable to build a system of green energy supply.

According to the UN Framework Convention on Climate Change, the obligation to cooperate internationally recognised in the ICESCR, it is the duty of industrialised countries to take the lead in the reduction of emissions as well as to assist developing countries to reduce their emissions, through financial aid and the transfer of low carbon technologies. Specific mechanisms and measures to reduce emissions must respect human rights, resulting from their obligation to respect and protect; states must ensure that such measures- for instance, the expansion of agro fuels, the construction of dams or the utilisation of forests as carbon stores- do not curtail the enjoyment of human rights of people both on their own national territories and beyond. Measures to protect the environment shall not lead to people losing their means of existence; they rather must recognise and respect the traditional rights of indigenous people and local communities to land and other resources. When affected by such measures, these groups shall be fully and effectively involved in all levels of decision making process.

To minimise the impacts of climate change; (i) States must protect people living on their territory from these impacts and ensure the realisations of minimum essential levels of the rights recognised in the ICESCR under changing climatic conditions. (ii) Each state must conduct studies on the regional impacts of climate change. (iii) States must take steps to improve their possibilities to adapt better to climate change, giving special attention to the vulnerable segments of population. (iv) States are obliged to ensure access to information on short-term and long-term climatic changes to all people (early warning system) as well as to facilitate their participation in the political process in which the adaptation measures are developed and decided. (v) States must also guarantee that all persons whose individual rights are being affected by the impacts of climate change are protected by the law.

Many developing countries are particularly hit by the negative impacts of climate change and that the costs of adaptation greatly exceed their financial means, they depend on international cooperation to fulfil their human rights obligations. States committed themselves to engage in such cooperation in article-2 of the ICESCR. The human rights obligation of industrialised countries to help developing countries through financial aid and technical cooperation to protect their citizens from the impacts of climate change is supported by the environmental 'law principle of common but differentiated responsibilities' in the UN Framework Convention on Climate Change. Measures undertaken by the international community to increase the adaptive capacity of people to climate change must be oriented towards human rights standards. The international community must find a solution to the issue of climate refugees. More and more people are forced to leave their homes as a result of extreme weather events and floods. But in the present international law, these people currently do not have a legal status and are not under the responsibility of the office of the United Nations High Commissioner for Refugees.[69] This gap in international law must be closed without delay.

A human rights - based approach

A Human Right-Based Approach (HRBA) aims towards ensuring all human rights for everyone. Human rights provide over aching frame works for national laws, regulations, government planning and policies. Development process and goal setting should be a process towards fulfilling human right norms and standards as expressed in human rights declaration and treaties, guided by principles like human dignity, equity and non-discrimination, participation and inclusion, accountability and the rule of law.[70] HRBA may be applied to all policy sectors and development planning including agriculture, forest and development policies.

The indigenous peoples have collective right to the territories and make their own development strategies; they depend on agriculture for their livelihoods within these areas. According to the United Nations Declaration on the Right of Indigenous Peoples (UNDRIP), which was adopted by the large majority of the world's countries at the UN General Assembly in September 2007, indigenous peoples have a right to self-determination and also a right to free, prior and informed consent if laws, policies or activities are planned that may affect them and their interests.

The international human rights have their origin in the UN that these international standards are discussed and state conduct evaluated. Human rights norms and standards, the situation of vulnerable groups and individuals, were marginal topics at the UN - based negotiations on climate change in Copenhagen 2009 and also at the summit on food security in Rome in November 2009. This issue was brought by non-governmental organisations and indigenous peoples. The World Food Summit in Rome in November 2009 and the climate negotiations of the UNFCCC (Cop 15) in Copenhagen in December 2009 seemed to take place in relative isolation from each other.

A HRBA commands a holistic and integrated strategy[71] and calls for the adoption of appropriate policies at both national and international levels, oriented to the eradication of poverty and the fulfilment of all human rights for all.[72] It provides a framework for analysis and action through which adverse effects on marginalised populations can be identified and counteracted. In this approach, in the case of traditional people depending on land, territories and natural resources for their livelihoods, the right to food[73] should be understood as overlapping with the right to land. The right to food would also need to be understood, in context with the special rights of indigenous peoples' self-determination and to free, prior and informed consent with regard to laws, policies and measures that are likely to affect them.

References

1. *IPCC 2007, Climate change 2007, the Fourth IPCC Assessment Report. Cambridge University Press, at http://www.ipcc.ch/.*

2. *Wiggins S., 'Rising Food Prices - A Global Crisis', Briefing Paper No. 37, ODI, London, 2008.*

3. *UNDP, Fighting Climate Change - Human Solidarity in a Divided World, UNDP, New York, 2008.*

4. *Shah, M., Fischer, G., and Van Velthuizen, H., Food Security and Sustainable Agriculture, The Challenges of Climate Change in Sub-Saharan Africa, Laxenburg: International Institute for Applied Systems Analysis, 2008, Nellemann, C., MacDevette, M., Manders, T., Eickhout, B., Svihus, B., Prins, A., and Kaltenborn, B. (eds.), the Environmental Food Crisis, The Environment's Role in Averting Future Good Crises, A UNEP Rapid Response Assessment, Arendal, UNDP, 2009.*

5. *Molden, D. (ed.), Water for Food, Water for Life, Water Management Institute, London: Earthscan and Colombo, 2007.*

6. *In UN general assembly, 45 session, the report of Secretary General, November 8, 1990, p.q.*

7. *David grey and Claudia W. Sadoff, water security for growth and development water policy, 2007.*

8. *Bates, B.C, Kundzewicz, Z. W., Wu, S. and Palutikof, J. P. (Eds.), 'Climate Change and Water', Technical Paper of the Intergovernmental Panel on Climate Change, IPCC Secretariat, Geneva, 2008.*

9. *Callow, R., Mac Donald, A., Nicol, A. And Robins, N. (na), 'Ground water security and drought in Africa - linking water availability access and demand' (unpublished manuscript).*

10. *Magrath, P. and Tesfu, M., Meeting the needs for water and sanitation of people living with HIV/AIDS in Addis Ababa, Ethiopia, (Addis Ababa: Water Aid Ethiopia, 2006).*

11. *IPCC 2007, Climate Change 2007: Impacts, Adaptation and Vulnerability, Contribution Of Working Group II to The Fourth Assessment Report Of The Inter-governmental Panel on Climate Change, M.L., Parry, O.F. Canziani, J.P Palutikof, P.J Van der Linden and C.E Hanson, Eds., Cambridge University Press, Cambridge, UK, 2007.*

12. *IPCC, Climate Change 2007: Mitigation of Climate Change. Contribution of Working Group III to the Fourth Assessment Report of the Intergovernmental Panel on Climate Change, Cambridge University Press, Cambridge, UK and New York, USA, 2007.*

13. *Alcamo, J., D. Van Vuuren, C. Ringler, W. Cramer, T. Masui, J. Alder and K. Schulze, "Changes in nature's balance sheet: model - based estimates of future worldwide*

ecosystem services", Ecology and Society, 10 (2), 2005, p. 19.

14. *FAO, Global Agro - Ecological Zones, Version 1.0. FAO Land and Water Digital Media Series 11, CDROM, Rome, 2000.*

15. *IPCC 2007.*

16. *Smith, P., D. Martino, Z. Cai, D. Gwary, H. Janzen, H. Kumar, B. Mc Carl, S. Ogle, F. O' Mara, C. Rice, R. Schools, O. Sirotenko, M. Howden, T. Mc Allister, G. Pan, V. Romanenkov, U. Schneider, S. Towprayoon, W. Wattenbach and J. Smith, Greenhouse Gas Mitigation in Agriculture, Phil. Trans. R. Soc. B (2008) 363, pp. 789-813.*

17. *Schmidhuber, J. and F. Tubiello, Global Food Security under Climate Change, PNAS, 104 (50), 2007, 19703 - 19708.*

18. *Hussain, I., Pro-Poor Intervention Strategies in Irrigated Agriculture in Asia-Poverty in Irrigated Agriculture: Issues, Lessons, Options and Guidelines, Final Synthesis Report Submitted to the Asian Development Bank, International Water Management Institute (IWMI), Colombo: Sri Lanka 2005.*

19. *Sperling, F. (ed.), Poverty and climate change: reducing the vulnerability of the poor through adaptation. Washington, DC: AFDB, ASDB, DFID, Netherlands, EC, Germany, OECD, UNDP, UNEP and the World Bank (VARG), 2003.*

20. *Tubiello, F. And G. Fischer., Reducing climate change impacts on agriculture: Global and regional effects of mitigation, 2000-2080 Technological Forecasting and Social Change 741030-1056, 2007.*

21. *Fischer G., M. Shah, F. Tubiello and H. Van Velthuizen, Socio-Economic and Climate Impacts on Agriculture: an Integrated Assessment, 1990-2080, Phil. Trans. R. Soc. B (360) 2067-283, 2005, Shah et al., 2008.*

22. *Magrath and Tesfu, 2006.*

23. *Ludi and bird, Offer great understanding in distinguishing between poverty and vulnerability, Vulnerability is dependent on the nature of the hazard, 2007, at http:// climateemergencyinstitute. com/food_sec_subsaharan_mburia.html.*

24. *Intergovernmental Panel on Climate Change Secretariat (IPCC) 2007: Impacts, Adaptation and Vulnerability, Contribution of Working Group II to the Fourth Assessment Report of the IPCC.*

25. *IPCC 2008 Technical Paper VI In: Climate Change and Water, Geneva IPCC Secretariat, pp. 210.*

26. *Intergovernmental Panel on Climate Change Secretariat (IPCC), 2008: Technical Paper VI, In: Climate Change and Water, Geneva, IPCC Secretariat, p. 210.*

27. *United Nations Development Programme (UNDP), Human Development Report, 2007/2008- Fighting Climate Change: Human Solitarity in a Divided World, UNDP, New York, 2007.*

28. *IPCC, Secretariat, Impacts, Adaptation and Vulnerability, Contribution of Working Group II to the Fourth Assessment Report of IPCC, 2007.*

29. *IPCC, Secretariat, 2007: Impacts, Adaptation and Vulnerability. Contribution of Working Group II to the Fourth Assessment Report, IPCC.*

30. *World Water Assessment Programme, 2009: The United Nations World Water Development Report 3: Water in a Changing World. Paris UNESCO and London, Earthscan.*

31. *World Water Assessment Programme: The United Nations World Water Development Report 3: Water in a Changing World. Paris UNESCO, and London Earthscan.*

http://www.unesco.org/water/wwap/wwdr/wwdr3/

32. Schmidhuber, J., J. Bruinsma and G. Boedeker. *Capital Requirements for Agriculture in Developing Countries to 2050. Expert meeting on how to feed the World in 2050 Food and Agriculture Organisation of the United Nations. Economic and Social Development Department. Rome, 2009.*

33. Collier, P. and S. Dercon. *African Agriculture in 50 Years: Small holders in a rapidly changing world? Expert meeting on how to feed the world in 2050 Food and Agriculture Organisation of the United Nations. Economic and Social Development. Rome, 2009.*

34. Laxmi, V., O. Erenstein and R.K. Gupta, *Impact of zero tillage in India's rice - wheat systems, CIMMYT, Mexico, D.F, 2007.*

35. Batchelor C., A. Singh, M.S. Rama Rao and J. Butterworth. 2005. *Mitigating the potential unintended impacts of water harvesting. http://www.nri.org/projects/ WSSIWRM/ eports/ water% 20 harvesting % 20 impacts.pdf.*

36. Ruane, J., A. Sonnino, P. Steduto and C. Deane. 2008. *Coping with water scarcity: What role for biotechnologies? Land and water discussion paper NO. 7. FAO. Rome.*

37. FAO. 2010. AQUASTAT database *http://www.fao.org/nr/water/aquastat/main/ index.stm.*

38. Siebert, S., P. Doll, S. Feick, K. Frenken and J. Hoogeveen. *Global map of irrigation areas version 4.0.1. (University of Frankfurt (Main), Germany, and FAO, Rome, Italy, 2007).*

39. IRD. 2010. *Tropical marine ecosystem programme. http://www.mpl.ird.fr/suds-en-ligne/ecosys/ ang-ecosys/intro 1.htm.*

40. CA. *Water for Food, Water for Life. The Comprehensive Assessment of Water Management in Agriculture D. Molden (Ed) London, Earthscan and Colombo, International Water Management Institute, 2007.*

41. Hofer, T. and B. Messerli. 2006. *Floods in Bangladesh. History, Dynamics and Re-thinking the role of the Himalayas. UNU Tokyo. p. 468; Calder;*
   Brown et al., *"A review of paired catchment studies for determining changes in water yield resulting from alterations in vegetation", Journal of Hydrology, 310 (1-4), 2005, pp. 28-61.*

42. Dye, P. and D. Versfeld. *"Managing the hydrological impacts of South African plantation forests: An overview", Forest Ecology and Management, 251 (1-2), pp. 121-128.*

43. MA (Millennium Ecosystem Assessment) 2005. *Ecosystem services and human well-being: wetlands and water synthesis, World Resources Institute, Washington, DC, 2007, p. 68, CA.*

44. Molle, F. and J. Berkoff, *Cities versus Agriculture: Revisiting Intersectoral Water Transfers, Potential Gains and Conflicts. Comprehensive Assessment of Water Management in Agriculture Research Report 10, International Water Management Institute, Colombo, 2006.*

45. Scott, C.A., N.I. Faruqui and L. Raschid-Sally. 2004. *Waste Water Use in Irrigated Agriculture Confronting the Livelihood and Environmental Realities. Oxford University Press. SBN13: 9780851998237.*

46. FAO (2003) *World Agriculture: Towards 2015/2030', Rome: FAO, 2002.*

47. de Fraiture, C. *Assessment of Potential of Food Supply and Demand Using the Waterism Model, International Water Management Institute, Columbo, 2005.*

48. *Bruinsma J. (ed.) .2009. The Resource Outlook To 2050: By how much do land, water and crop yields need to increase by 2050? Expert meeting on how to feed the world in 2050 Food And Agriculture Organisation of the United Nations. Economic and Social Development Department. Rome.*

49. *FAO (2006) SOCO; The demand for the products of irrigated agriculture in Sub-Saharan Africa. FAO Water Report 31. Rome.*

50. *Molden, D., T. Oweis, P. Steduto, P. Bindraban, M. Hanjra and J. Kijne, 2010. Improving agricultural water productivity: Between optimism and caution. Agricultural Water Management 97: 528-535.*

51. *Rockstrom, J., J. Barron and F. Fox. 2001. Water Productivity in Rain-fed Agriculture: challenges and opportunities for small holder farmers in drought-prone tropical Agro- Ecosystems.*

52. *FAO, 2003; CA, 2007; FAO, 2007a: Agricultural trends to 2050. FAO Rome.*

53. *Fischer, G., F.N. Tubiello, H. Van Velthuizen and D.A. Wiberg. 2007. Climate change impacts on irrigation water requirements: Effects of mitigation, 1990-2080. Technological forecasting and social change 74(2007) 1083-1107; Nelson, G.,M. Rosegrant, J. Koo, R. Robertson, T. Sulser, T. Zhu, C. Ringler, S. Msangi, A. Palazzo, M. Batka, M. Magalhaes, R. Valmonte-Santos, M. Ewing and Le. D. 2009. Climate Change impact on agriculture and costs of adaptation. (Washington, DC: IFDRI).*
Padgham, J. Ed. 2009. Agriculture Development under a changing climate. Opportunities and challenges for adaptation. Joint department discussion paper-issue1. World Bank. Washington;
Climate Adaptation Working Group, 2009. Shaping Climate Resilient Development: A Framework for Decision Making.

54. *Peel, M-C., T.A. McMahon and B.L. Finlayson. 2004. Continental differences in the variability of annual runoff-update and reassessment. Journal of Hydrology; Peel, M-C., G-C. Pegram and T.A. McMahon. 2004a. Run length analysis of annual precipitation and runoff. International Journal of Climatology.*

55. *Barnett T.P., J.C. Adam, D.P. Lettenmaier. "Potential impacts of a warning climate on water availability in snow dominated regions", Reviews. Nature, vol. 438:17, pp. 303-308.*

56. *Ibid.*

57. *ICIMOD, Inventory of Glaciers, Glacial Lakes and Identification of Potential Glacial Lake Outburst Flood (GIFs) Affected by Global Warming in the Mountains of Himalayan Region (Kathmandu: ICIMOD, 2007). DVD/CD-ROM.*

58. *Casey Brown and Upmanu Lall, "Economic Development: The Role of Variability and a Framework for Resilience", Natural Resources Forum, 2006.*

59. *UNFPA, State of World Population 2007: Unleashing the Potential of Urban Growth. (New York: United Nations Population Fund, 2007).*

60. *J. Winpenny, 2003, Financing water for all.*

61. *FAO (2002) World Agriculture: Towards 2015/2030, an FAO Perspective, FAO/Earthscan Publishers, Rome/London, 2002.*

62. *"It's Time to Learn from Frogs" by Nicholas D. Kristof, New York Times, June 28, 2009.*

63. *OECD (2003) Poverty and Climate Change. Reducing Vulnerability of the Poor through Adaptation, OECD, Paris, 2003,*

64. *Bates et al., 2008.*

65. *Howden, et al., 2007: Adapting agriculture to climate change, available on: http://pubs.giss.nasa.gov/abs/ho03300x.html*

66. *CESCR (Committee on Economic, Social and Cultural Rights (1999): General Comment No- 12. The Right to adequate food (Art-11), Geneva.*
   CESCR (Committee on Economic, Social and Cultural Rights (2002): General Comment No- 15. The Right to Water (Art 11 and 12 of the International Covenant on Economic, Social and Cultural Rights) Geneva.

67. *ICESCR (Art 2, Para-1): International Covenant on Economic, Social and Cultural Rights. 1996. New York.*

68. *CESCR (Committee on Economic, Social and Cultural Rights) (1990): General Comment No 3: The Nature of States Parties Obligations (Art 2, Para-1), Geneva.*

69. *UNHRC (United Nations Human Rights Council) 2009: Report of the Office of the United Nations High Commissioner for Human Rights on the Relationship between Climate Change and Human Rights.*

70. *UN, 2003.*

71. *According to the Vienna declaration and programme of action (world conference on human rights) 1993, article 5: all human rights are universal, indivisible and interdependent and interrelated.*

72. *UN (2003), The Human Rights Based Approach to Development Cooperation, towards a common understanding among UN agencies.*

73. *From the International Covenant of Economic, Social and Cultural Rights (1966).*

3. CLIMATE CHANGE AND FOOD SECURITY

Introduction

Agriculture is a vital and fundamental human activity, providing human societies with food, clothing, medicine and other useful products with vital ecosystem services including biodiversity, soil formation, water regulation and carbon sequestration. The world population will be reached at 9.1 billion by 2050 and agricultural production needs to grow accordingly to meet this growing demand. The international Food Policy Research Institute (IFPRI) estimates that climate change could reduce irrigated wheat and rice yields by 30 and 15 per cent respectively.[1] 75 per cent people living in rural areas depend upon agriculture, forestry and fisheries. Thus the climate threats are very significant for the livelihoods of large populations of the worlds. Climate change impacts on agriculture and other land based sectors. About one-third of the global emissions can be attributed to the Agriculture, Forestry and Other Land Use (AFOLU) sectors. All GHG emissions about 13.5 per cent for Agriculture and 17.4 per cent for land use and forestry.[2] The AFOLU sectors can make an important contribution to reach the necessary targets for reducing the threat of climate change.

At the beginning of 2010, a new project, mitigation of climate change in Agriculture (MICCA) was established at the Food and Agriculture Organisation (FAO) of the United Nations to support efforts to mitigate climate change through agriculture in developing countries and to remove GHG emissions. If changes are implemented in production systems, emissions can be reduced and sinks created in biomass and soils while resilience and productivity of the agricultural systems are increased.

According to the 4th Assessment Report from Intergovernmental Panel on Climate Change (IPCC), agriculture, forestry and other land uses account for approximately 30 per cent of the total anthropogenic GHG emissions. These agriculture accounts for about 60 per cent of

N_2O and 50 per cent of CH_4 emissions, whereas deforestation and land use change are mainly causing CO_2 emissions.[3]

The Agricultural Forestry and Other Land Use Mitigation Project (AFOLU MP) is an overview of all agriculture and forestry based mitigation projects. Its objectives pointed out that many agricultural projects or even rural development projects are developed that do not have agricultural mitigation as their main goal; however, their activities indirectly also provide climate change mitigation benefits through the implemented agricultural activities.

The agriculture, forestry and fisheries sectors can offer significant opportunities to address the fight against climate change within sustainable development paths, especially in Least Developed Countries (LDCs) by reducing negative impacts on land and water resources, enhance ecosystem management and services, improved food security and generate income opportunities, leading to production systems and rural livelihoods for better resource use efficiency.

The food and agriculture organisation (FAO) defines food security as a "situation that exists when all people, at all times have physical, social and economic access to sufficient, safe and nutritious food that meets their dietary needs and healthy life".[4] This definition indicates four key dimensions of food supplies: availability, stability, access and utilisation. The first dimension relates to the availability of sufficient food that is an overall ability of the agricultural system to meet food demand. Its sub dimensions include the agro- climatic fundamentals of crop and pasture production.[5] And the entire range of socio-economic and cultural factors that determine where and how farmers perform in response to markets. The second dimension, stability, relates to individuals who are at high risk of temporarily or permanently losing their access to consume adequate food. An important effect of unstable climate variability on landless agricultural labourers who depend on agricultural wages in a region damaged due to heavy rain is that they lose their food and if the labourers who fall sick and cannot earn their daily wages, lack of stable food and they cannot take proper insurance against illness. The third dimension access covers access by individuals to adequate resources to acquire appropriate foods for a nutritious diet. The key element is the purchasing power of consumers and the evolution of real income and food prices. However, these resources need not be exclusively monetary but may also include traditional rights e.g. to a share of common resources. Finally, food safety and quality aspects of nutrition are related to health at global, regional, national and local levels.

Climate change will mean higher average temperatures, changing rainfall patterns and rising sea-levels. There will be more and more intense, extreme events such as droughts, floods and hurricanes and they pose a major threat to agricultural systems. Developing countries are particularly vulnerable because their economies are closely linked to agriculture and a large proportion of their populations depend directly on agriculture and natural ecosystems for their livelihoods. Many agricultural systems in developing countries are at crisis point. Rising global population is taking a heavy toll on farmlands, range lands, fisheries and forests. Agriculture is not only a source of the commodity, food, but equally important as the source of income.

The Intergovernmental Panel on Climate change (IPCC) estimated that 31 per cent of total emissions in 2004 came from agriculture and forestry.

The climate is changing and agricultural systems must also change to avoid catastrophe. Farming, fishing and forest communities will need to adapt their livelihood systems, while mitigation efforts must address both the contribution of agriculture to the climate change problem and management practices in reducing greenhouse gases in the atmosphere. The 15 research centres are supported by the CGIAR (Consultative Group on International Agricultural Research).

A project supported by Germany's Federal Ministry for Economic Cooperation and Development entitled, "Food and Water security under Global change; Developing Adaptive capacity with a focus on Rural Africa" has conducted research on adaptation to climate change at various scales. This project is associated with the programme on water and food under the Consultative Group on International Agricultural Research (CGIAR).

Policy-makers are generally more interested in the development of adaptation measures following political rather than hydrologic boundary. Vulnerability and adaptation measures were also developed at the province and state levels for these two countries- Ethiopia and South Africa to discuss measures of vulnerability, adaptation options and the role of information and various actors- that is the state, private sector and civil society in shaping adaptation to climate change.

The impact of climate change on crop production in the survey was simulated based on crop yield and production function models to assess the implications of climate change for local food security. The impact of water availability, water demands and irrigation was simulated to identify basin-level adaptation strategies. To capture the interactions of climate change and adaptation at the national and regional (Sub-Sahara Africa) levels, a water and food projections model was updated to take into account the impacts of climate change. The impact is that their many partners have been working to help poor farming, fishing and forest communities achieve sustainable livelihoods in the face of variable and uncertain weather. Adopting many of the existing technologies in adaptation and mitigation climate change in agriculture includes such as improved crop, soil and water management practices. The challenge programme on Climate Change, Agriculture and Food Security (CCAFS) launched in nearly 2010, unites the world's best researchers in agricultural science, climate science and earth system science to address the climate change - food security problem.

Agriculture and food supply chains are essential to meet current and future threats to food security and environmental resilience. The global food system does not yet provide adequate calories or nutrition to everyone on the planet, yet it enables some part of population to over consume. In the coming decades, global agriculture must produce more food to feed a growing population while adapting to climate change which remains an increasing threat to agricultural yields.[6] Extreme weather events such as droughts and floods are predicted to become more frequent, adding to the global burden of hunger caused by poverty, weak governance, conflict and poor market access.[7] Agricultural practices suitable in different regions can reduce net GHG emissions while maintaining or improving yields and adapting to more extreme weather.[8]

United Nations Framework Convention (UNFCCC), the Group of 20 nations (G20) and the United Nations Convention on sustainable development: the organising body of the Rio + 20 Earth summit held in June 2012 adopted appropriate policy and financial actions to support implementation of these solutions on a global level. The Commission on Sustainable Agriculture and Climate Change was convened in February 2011 to identify practical evidence- based policy action to achieve food security in the context of climate change.

Food insecurity and climate change
Food insecurity afflicts communities throughout the world wherever poverty prevents assured access to food supplies, causing widespread human suffering, food insecurity contributes to degradation and depletion of natural resources, migration to urban areas and across borders and political and economic instability. The food system faces global population grows to around 9 billion by 2050.[9]

Africa's population is projected to double from just over 1 billion in 2010 to about 2 billion by 2050.[10] For low-income population, food insecurity negatively affects future livelihoods through the forced sale of assets that are difficult to rebuild and through reduced expenditure on education.[11] Food insecurity can lead to long term health issues. In Sub-Saharan Africa, poor health reduces agricultural productivity and some agricultural practices contribute to health problems such as malaria, pesticide poisoning and disease transmission from animals to human.[12] In crowded cities, food security is weakened by the lack of suitable, nutrient-rich soil as well as growing space available for local families.[13]

Progress has been made towards the Millennium Development Goal of reducing the global poverty rate below 23 per cent, yet there is great disparity in income growth both between and within countries.[14] Poverty is influenced by household-level conditions such as health, education, harvests, assets and expenses as well as by regional-level conditions such as infrastructure, markets, economic growth, enabling institutions and conflict or disasters.[15] The low and middle income countries are facing a double burden of malnutrition among children, overweight and obese people and diet-related chronic diseases.[16] For households that are chronically under-nourished or at risk of food insecurity, food choices are extremely limited and are largely driven by prices relative to incomes.

Food insecurity is a growing concern throughout, particularly for poor women and children. In 2010, approximately 925 million individuals were undernourished.[17] A recent study on the future of food and farming identified six key drivers of change affecting the global food system: - a growing global population; changing diets; food system governance including globalisation of markets, subsidies and trade restrictions; completion for resources, particularly land, water and energy; consumer values and ethics; and the impacts of climate change.[18]

Vulnerability to climate change and food insecurity is a function to climate hazards, sensitivity to climate-related shocks and stresses and capacity to adapt. The climate hazard is often determined by geography. For example, communities in coastal areas experience much higher exposure to sea level rise and related effects of ground water. Sensitivity is often determined by livelihood strategies, with households dependent on rain-fed agriculture, pastoralism, fisheries or other natural resource based livelihood strategies' representing particularly vulnerable groups.

People's ability to maintain food security in the face of climate change will depend on their adaptive capacity. Adaptive capacity is significantly influenced by access to and control over critical resources, such as information and knowledge on climate change, natural resources such as land water for agriculture and opportunities for earning sustainable income.[19] Structural and relational factors such as inequitable policies, power relationships and cultural norms also play an important role in determining adaptive capacity. Socially excluded groups including female headed households, orphans, persons suffering from HIV/AIDS and land-less people are highly vulnerable to the impacts of climate change and food insecurity.

In many contexts, women may face higher risks of food insecurity due to gender inequality. They play a critical role in agriculture and in managing household food supplies, but may lack access to services and control over important resources and decisions affecting food security. As a result, they become trapped in vicious cycle, with food insecurity.

A reduction in the production potential of tropical developing countries, many of which, are already due to poor land and scarcity of water resources, exacerbates serious food insecurity. Food insecurity will be further affected by loss of cultivated land and nursery areas for fisheries.[20] Food and stability will be changed due to extreme weather event such as flooding and droughts. Food access might be impacted through economic problems and direct health - effects of climate change. Health problems can be released to reduced access to clean water, flooding, drought, sea level rise or increased precipitation. Increasing exposure to vector borne - malaria and water borne - cholera diseases.[21]

Food security and climate change
Food security exists when all people at all times have physical or economic access to sufficient, safe and nutritious food to meet their dietary needs and food preferences for an active and healthy life.[22] Food security is influenced by four key dimensions: availability of sufficient food; economic, physical and social access to the resources needed to acquire food; stability of this availability and access; and utilisation including nutrition, food safety and quality.[23]

Achieving food security for all requires a coordinated effort that incorporates preventive, promotional, protective and transformative measures. Preventive measures aim to help people avoid food insecurity and include social insurance systems. Promotional measures aim to reduce vulnerability to food insecurity by enhancing incomes and capacities. Protective actions are relief measures, required when preventive and promotional measures fail. These three types of action are transformative measures that seek to address issue of social inequity and exclusion.[24]

It is estimated that food production will need to increase by 50 per cent by 2030 just to keep up with the demands of a growing global population.[25] At the same time, climate change is projected to cause decrease in global cereal production of 1-7 per cent by 2060. These reductions will be greatest in developing countries and particularly in South Asia and Sub-Saharan Africa. Reduced production leads to higher food prices and increasing food insecurity, particularly for rural families in developing countries who are net buyers of food.[26] WFP estimate that globally, 10-20 per cent more people will be at risk of hunger by 2050 without climate change. In Sub-Saharan Africa, it is estimated that 10 million more children will be malnourished as a result of climate change.[26]

The risk of hunger resulting from climate change is the result of both direct impacts on food systems and indirect impacts that affect the different dimensions of food security. The direct effects listed in Table-1 have indirect consequences for all four dimensions of food security: availability, access, utilisation and stability as shown in Table-2 given below:

Table 1: Direct consequences of climate change for food systems

Climate change impact	Direct consequences of food systems
Increased frequency and severity of extreme weather events.	• Crop failure or reduced yields. • Loss of livestock. • Damage of fisheries and forests. • Destruction of agricultural inputs, such as seeds and tools. • Excess or shortage of water. • Increased land degradation and desertification. • Disruption of food supply-chains. • Increased costs for marketing and distributing food.
Rising temperatures. Shifting agricultural seasons and erratic rainfall.	• Increased evapotranspiration, resulting in reduced soil moisture. • Great destruction of crops and trees by pests. • Greater threats to human health such as disease and heat stress that reduce the productivity and availability of agricultural labour. • Greater threats to livestock health. • Reduced quantity and reliability of agricultural yields. • Greater need for cooling/refrigeration to maintain food quality and safety

Sea level rise.	<ul><li>Greater threat of wildfires.</li><li>Reduced quantity and quality of agricultural yields and forest products.</li><li>Either an excess or shortage of water,</li><li>Greater need for irrigation.</li><li>Damage to coastal fisheries.</li><li>Direct loss of cultivable land due to inundation and salinisation of soil.</li><li>Salinisation of water sources.</li></ul>

Table 2: Indirect consequences of climate change for food system[27]

Dimensions of food security	Indirect consequences of climate change Impacts
Availability of sufficient food.	<ul><li>Reduced agricultural production locally and globally.</li><li>Decreased availability of fishery and forest products.</li><li>Increased pressure on food reserves.</li><li>Decreased exports and increased imports.</li></ul>
Access to the resources needed to acquire food.	<ul><li>Increased food prices.</li><li>Loss of incomes due to damage to agricultural production or interruption in livelihoods activities.</li><li>Need to adjust agricultural practices and other livelihood strategies to manage uncertainty of changing hazards and conditions.</li><li>Increasing migration to urban and semi- urban areas.</li></ul>
Utilisation, including nutrition, food safety and quality.	<ul><li>Health impacts including food - borne diseases and malnutrition.</li><li>Dietary and nutritional changes based on changing availability of or access to preferred foods.</li><li>Abilities to utilise food may be affected by disease.</li><li>Persons living with HIV & AIDS may experience difficulties in maintaining anti - retro viral therapies and may be more susceptible to infections.</li><li>Impact on food safety resulting from water pollution, increased temperatures and damage to stored food.</li></ul>

Stability of availability and access.	• Instability of food supplies affecting both availability and pricing of food. • Insecurity of incomes from agriculture and fisheries. • Population displacement and migration. • Potential for increased conflict over resources.

Impacts on food production

Climate change affects agriculture and food production in complex ways. It affects food production directly through changes in agro-ecological conditions and indirectly by affecting growth and distribution of incomes and thus demand for agricultural produce. Changes in temperature and precipitation associated with continued emissions of greenhouse gases will bring changes in land suitability and crop yields. The IPCC considers four families of socio-economic development and associated emissions scenarios (SRES) A2, B2, A1 and B1.[28] Depending on the SRES emission scenario and climate models, global mean surface temperature is projected to rise in a range from 1.80C to 4.00C by 2100.[29]

In temperature latitudes, higher temperatures are expected to bring predominantly benefits to agriculture: the areas potentially suitable for cropping will expand, the length of growing period will increase and crop yield may rise. Heat waves and droughts in the Mediterranean region increased heavy precipitation events and flooding in temperature regions including the possibility of coastal storms.[30] In drier areas, climate models predict increased evapotranspiration and lower soil moisture levels.[31] Temperature rise will also expand the range of many agricultural pests and increase the ability of pest populations to survive the winter and attack spring crops.

Another important change for agriculture is the increase in atmospheric Carbon dioxide (CO_2) concentrations. Higher CO_2 concentrations will have a positive effect on many crops, enhancing biomass accumulation and final yield.

However, the magnitude of this effect is less clear with important differences depending on management type and crop type.[31] A number of recent studies have estimated the changes in land suitable, potential yields and agricultural production on the crops and cultivation. These estimates include adaptation using available management techniques and crops. These studies are in essence based on the FAO/International Institute for Applied Systems Analysis (IIASA), Agro-Ecological zone (AEZ) methodology.[32]

Impacts on stability of food supply

Global and regional weather conditions are also expected to become more variable than at present, with increases frequency and severity of extreme events such as cyclones, floods, hailstorms and droughts.[33] By bringing greater fluctuations in crop yields and local food supplies and higher risks of landslides and erosion damage, they can adversely affect the stability of food supplies and food security. If climate fluctuations become more pronounced and wider-spread droughts and floods, the dominant causes of short term fluctuations in food production in semi-arid areas, will become more severe and more frequent. In semi-arid areas, droughts can dramatically reduce crop yields and livestock numbers and productivity.[34] Most land of Sub-Saharan Africa and parts of South Asia, the poorest regions with the highest level of chronic under-nourishment will also be exposed to the highest degree of instability in food production.[35] These impacts will crucially depend on whether such fluctuations can be countered by investments in irrigation, better storage facility or higher food imports.

Impacts on food utilisation

Climate change will also affect the ability of individuals to use food effectively by altering the conditions for food safety and changing the disease from vector, water and food borne diseases. The main concern about climate change and food security is that changing climatic conditions can initiate a vicious circle where infectious diseases cause or compounds hunger which makes the affected populations more susceptible to infectious diseases. Due to climate change, the drought, higher temperatures or heavy rainfalls have an impact on the disease and these changes affect food safety and food security.[36]

Extreme rainfall events can increase the risk of outbreaks of water-borne diseases particularly where traditional water management systems are insufficient to handle the new extremes.[36] Likewise, the impacts of flooding will be felt most strongly in environmentally degraded areas and where basic public infrastructure including sanitation and hygiene is lacking. This will raise the number of people exposed to water-borne diseases such as cholera and thus lower their capacity to effectively use food.

Impacts on access to food

Access to food refers to the ability of individuals, communities and countries to purchase sufficient quantities and qualities of food. Over the last 30 years, real prices for food and rising real incomes have led to substantial improvements in access to food in many developing countries. Increased purchasing power has allowed a growing number of people to purchase not only more food but also more nutritious food with more protein, micronutrients and vitamins.[37] In East Asia, it was income growth that provided the basis for the boost in demand for food, which was largely produced in the region; in the near East North American region, demand was spurred by exogenous revenues from oil and gas exports and additional food supply came largely from imports. But in both regions, improvements in access to food have been crucial in reducing hunger and malnutrition.

The agro-ecologic and economic models have stressed the impact of climate change on agricultural Gross Domestic Product (GDP) and prices. At the global level, the impacts of climate change are likely to be very small; under a range of SRES and associated climate change scenario, the estimates range from a decline of -1.5 per cent to an increase of +2.6 per cent by 2080. At the regional levels, the importance of agriculture as a source of income can be much more important and in these regions, the economic output from agriculture will be important contributor to food security. The strongest impact of climate change on the economic output of agriculture is expected for Sub-Saharan Africa which means that the poorest and already most food insecure region is also expected to suffer the most affecting the largest contribution of agricultural incomes.

Impacts on food prices
The SRES development paths describe a world of robust economic growth and rapidly shrinking importance of agriculture in the long run underway for decades in many developing regions. SRES scenarios describe a world where income growth will allow the largest part of the world's population to address possible local production short falls through import, safety and stability issues of food supplies.[38] Real incomes rise more rapidly than the real food prices, where income levels are low and shares of food expenditures are high, higher prices for food may still create or exacerbate a possible food problem. There are number of studies to measure impacts of climate change on food prices and the messages from these studies are: first, on average, prices for food are expected to rise moderately in line with moderate increase of temperature until 2050. Second, after 2050 and with further increase in temperature, prices are expected to increase more substantially. Third, price changes expected from the effects of global warming are on average much lesser than price changes from socio-economic development path.

Global challenge on food security

To achieve food security in low-income and middle-income countries, a challenge with the complications of climate change, will require early investment to support small holder farming systems and the associated food systems that supply poor consumers. The Research programme on Climate Change Agricultural and Food Security (CCAFS) of International Agricultural of the Consortium Research Centres (CGIAR) is working across research disciplines, organisational mandates and temporal levels to assist immediate and longer-term policy actions.

Science and policy are united in recognition of the serious global challenges of making enough food available for growing population and changing dietary patterns under climate change.[39] Food systems vary enormously around the world and different consumers' access food differently. Most of the world's poor rural population continues to rely for their sustenance and livelihoods primarily on local food and local economies that are properly integrated into global markets.

The World Bank presents cross-country econometric evidence to show that investment in agriculture, in which small hold farmers participate as managers and labourers, has doubled the impact on poverty reduction as in any other sector.[40] Future impacts of climate change on the incomes and food security of poor households will very much depend on losses in agricultural yields that are local or widespread.[41] Climate change is not the only determinant of food security: rapid environmental, economic and political changes may be connected globally but have disparate impacts in different locales.[42] Agriculture is also a major contributor to greenhouse gas emissions both directly[43] and a proximate driver of land use change.[44] The challenge is to mitigate these emissions without compromising food and livelihood security, particularly of the poor rural majority.

The research in climate agriculture and food systems is to address highly local contexts with requisite attention to wider scale institutional mechanisms for spreading solutions, developing shared visions of the future and negotiating differential roles and responsibilities. Global society needs both local and global action to accelerate sharing of institutions, practices and technologies for adaptation and mitigation.

CCAFS works

- The objectives of CCAFS (Climate Change of Agriculture and Food Security) are: to identify and test pro-poor adaptation and mitigation practices, technologies and policies for food systems, adaptive capacity and rural livelihoods and to provide diagnosis and analysis that will ensure cost effective investments, the inclusion of agriculture in climate issues in agricultural policies, from the sub-national to the global level in ways that bring benefits to the rural poor.

- CCAFS will focus place-based work and policy engagement in a series of regions, each with a Regional Program Leader. Regions are selected according to multiple criteria, particularly vulnerability of the food system to climate change. Work is underway in 2011 in three regions - West Africa, East Africa and South Asia-which are home to 139 million people, whose food systems are highly vulnerable.
- Work is organised under four themes. These are:
  ○ Adaptation to progressive climate change.
  ○ Adaptation through managing climate risk.
  ○ Pro-poor climate change mitigation.
  ○ Integration for decision making ensures effective engagement with policy communities. The CCAFS provides wider context of biophysical and socio-economic change and demand-driven tools and database.
- CCAFS seeks to understand how gender relations and other social disparities influence responses to climate change and to formulate strategies to enable equitable access. Capacity enhancement is an integral part of research design. CCAFS is putting major effort into policy engagement and communications at all levels.[45]
- CCAFS was conceived and designed as a partnership between the global environmental change community and the CGIAR (Consultative Group on International Agricultural Research) which ensures the research on farming systems, ecosystems and food markets with the expertise of the global environmental change community, for example, scenarios for emissions and climate,[46] Global environmental governance,[47] and land cover change.[48]
- CCAFS is active in the Agricultural Model Inter-comparison and Improvement Project[49] and supporting the development of the IMPACT models that links climate change, crop and global trade models, giving the capacity to forecast impacts of different emissions scenarios on the prevalence of malnutrition.[50]
- CCAFS provides, future opportunities to incorporate models of livelihood systems, land use and adaptation investments.

- For agriculture and food systems, the divide between adaptation and mitigation is artificial. CCAFS deliberate integrative and cross disciplinary approach brings adaptation and mitigation options together and considers both technical and institutional solutions.[51]
- This approach will align with the needs of farmers and governments by providing evidence based guidance on the trade and food security and adaptation-mitigation and technical-institutional options.[52]
- The greenhouse gas emissions associated with agricultural practices, agriculture makes major contribution to greenhouse gas emissions as a driver of tropical deforestation, forest degradation and land use change from grass lands and wet lands.[53]
- CCAFS work drawing on the lessons of REDD+[54] will be crucial to successful land based mitigation that does not compromise[55] livelihoods and food security. Fisheries and livestock[56] are subject to trade and benefit from holistic analysis within CCAFS.
- CCAFS recognises the need to span boundaries across research and policy domains. To link knowledge and action entails involvement of policy makers in all stages of the research cycle and an understanding of policy as dynamic and policy centric across the public, private and civil society sectors.
- CCAFS is working directly with decision makers and agencies that are implementing the carbon payments, food crisis response and delivery of climate services.
- CCAFS has a strategic approach to work over multiple temporal and spatial levels in order to address uncertainty, to provide relevant science ahead of major long term change and to optimise the chances for finding solutions to complex environmental and social issues that require multi-level actions.
- CCAFS is co-developing a range of risk management strategies including financial instruments, climate forecasting services and institutional support to indigenous risk-management practices.
- CCAFS addresses connections between agriculture, food security and climate change policies globally, recognising that the distribution of costs, benefits and access to decision making varies with levels.
- To allocate resources effectively, the CCAFS strategic choices are:-
  - Stakeholder exercises to select subsets of priority thematic issues in each region.
  - Field work at limited number of benchmark sites shared across themes and where possible, shared with other research programmes.

o Working in a subset of countries in any region.
o Focus on select set of policy processes at national, regional and global levels each year.

Climate change impacts on agriculture

Agriculture is highly sensitive to climate, both in terms of longer-term trends in the average conditions of rainfall and temperature, which determine the global distribution of food crops, but also in terms of inter-annual variability and the occurrence of droughts, floods, heat waves, frosts and other extreme events.[57] A changing climate is associated with increased threats to food safety, post-harvest losses and pressure from species, pests and diseases.[58]

Extreme weather events and climate change will affect the food production systems and the natural resources- particularly in environment to degradation and desertification, in areas of widespread or water stress and wherever poverty undermines the capacity of rural people to take the needed preventive steps.[59] Farmers can longer rely on historical averages of temperature and rainfall, making it harder for them to plan and manage production when planting seasons and weather patterns are shifting. Climate change is likely to change rainfall patterns, resulting in shorter growing seasons in the future, particularly the farmers in Africa and parts of South Asia who rely on rain-fed agriculture.[60]

In-efficiency in food supply has negative impact on the environment, lower productivity and waste food. Current farming practices including and clearing and inefficient use of fertilizers and organic residues make agriculture a significant contributor to GHG emissions.[61] Every year an estimated 12 million hectares of agricultural land, which could potentially produce 20 million tonnes of grain, are lost to land degradation, adding to the billions of hectares that are already degraded.[62]

Climate change impacts on agriculture sectors such as crop production; animal production; fisheries; food handling; processing; and trading.

Crop production. Crop production is extremely susceptible to climate change. It has been estimated that climate changes are likely to reduce yields or damage crops in the 21st century[63] in the different parts of the world. Climate change affects the microbial population of macro-environment (soil, air and water and the population of pests or other vectors). It is the contributing factor of biotic diseases such as fungi, bacteria, viruses and insects. The biotic factors such as nutrient deficiencies, air pollutants and temperatures/moisture extremes also affect plant health and productivity. The impacts of biotic factors on crop production and food security are more obvious, it is important to have significant impact on the safety of food crops.

Animal production. Climate change, in particular rising temperatures, can have both direct and indirect effects on animal production. Heat stresses caused by the inability of animals can have a direct and detrimental effect on health, growth and reproduction. Changes in the nutritional environment can have an indirect effect. Climate change may affect diseases and infections which are naturally transmitted between animal and man in a number of ways. It may increase, the transmission cycle of many vectors; and the range and prevalence of vectors and animal reservoirs. In some regions, it may result in the establishment of new diseases, changes in feeding practices, and changes in the ecological situations in which animals increase these effects.

Fisheries. With higher temperature, global fisheries production may change and migration of fish from one region to another in search of suitable conditions. Other climatic changes impacting on fisheries include surface winds, high CO_2 levels and variability in precipitation. Climatic changes could affect productivity of aquaculture systems and increase the vulnerability of cultured fish to diseases and reduce returns of farmers. Extreme weather events could result in escape of farmed stock and contribute to reduction in genetic diversity of wild stock affecting biodiversity.

Food handling, processing, trading. Climate change impacts not only on primary production but also on food manufacturing and trade. Emerging hazards in primary production could influence the design of safety management systems required to effectively control those hazards and ensure the safety of the final product. Increasing average temperatures could increase hygiene risks associated with storage and distribution of food commodities. Reduced availability and quality of water in food handling and processing operations will also give rise to new challenges to hygiene management. These risk management measures and adaptation strategies will post greatest challenge for developing countries.

Agriculture is important for food security in two ways: it produces the food people eat and it provides the primary source of livelihood for 36 per cent of the world's total work force. In the heavily populated countries of Asia and the Pacific, this share ranges from 40 to 50 per cent and in sub-Saharan Africa, two-thirds of the working population still make their living from agriculture.[64] If agricultural production in the low-income developing countries of Asia and the Africa is adversely affected by climate change, livelihoods of large number of the rural poor will be put at risk and their vulnerability to food insecurity increased.

Agriculture, forestry and fisheries are all sensitive to climate. Their production processes are likely to be affected by climate change. The food security implication of changes in agricultural production patterns and performance are of two kinds:

- Impacts on the production of food will affect food supply at the global and local levels. Globally, higher yields in temperate regions could offset lower yields in tropical regions. However, in many low income countries with limited financial capacity to trade and high dependence on their own production to cover food requirements, it may not be possible to offset declines in local supply without increasing reliance on food aid.

- Impacts on all form of agricultural production will affect livelihoods and access to food such as the rural poor in developing countries. acquisition, preparation and consumption are as important for food security and agricultural production. Technological advances and the development of long distance marketing chains that produce more and packaged foods system performance far less dependent on climate, it was 200 years ago.

GHG in the atmosphere will continue to create threats of serious social, economic and ecological consequences. The planet is experiencing more extreme weather such as heavy precipitation events, coastal high water, geographic shifts in storm and drought patterns, and warmer temperatures.[65] Global temperatures are dramatic and urgent reductions in GHG emissions across a wide range of human activities including the burning of fossil fuels and land use. In the coming decades, global climate change will have an adverse overall effect on agricultural production and will bring critical threshold in many regions. To reduce the effect of climate change on food supplies, livelihoods and economies greatly increased adaptive capacity in agriculture- both to long-term climatic trends and to increasing variability in weather patterns- is an urgent priority.

Food chains. Agriculture is highly sensitive to climate, both in terms of long-term trends in the average conditions of rainfall and temperature, which determine the global distribution of food crops, but also in terms of inter-annual variability and the occurrence of droughts, floods, heat waves, forests and other extreme events.[66]

Inefficiencies in food supply chains have a negative impact on the environment, lower productivity and waste food. Current farming practices including land clearing and inefficient use of fertilizers and organic residues, make agriculture a significant contributor to GHG emissions.[67] Every year, an estimated 12 million hectares of agricultural land, which could potentially produce 20 million tonnes of grain, are lost to land degradation, adding to the billions of hectares that are already degraded.[68]

Globally, agriculture is both a part of the problem and a part of the solution to climate change. Agriculture continues to expand into forested and other lands in a number of regions. Land use change, deforestation is responsible for as much as 15 per cent of global GHG emissions and another 12-14 per cent is associated with direct agricultural GHG emissions including from fertilizers and livestock.[69]

GHGs are emitted across the food supply chain. The largest sources of emissions are related to agricultural production through new land for cultivation, use of nitrogen fertilizer and methane from livestock. Drives for many production systems occur throughout the supply chain are influenced through global and national policies.[70] There is large potential for reducing net food system emissions, per unit of food consumed as well as in absolute terms through efficiency measures in production and also through demand management.[71] The biophysical potential of agriculture mitigation has been estimated based on highly aggregated data and implementation has been limited due to financial and policy constraints.[72] Some types of food production system destabilise the natural resources base, drive the loss of biodiversity and contribute to GHG emissions with the potential to damage environment and to compromise the world's capacity to produce food in the future.

Agricultural production systems are associated with a series of interconnected natural resource management challenges. Agriculture consumes 70 per cent of total global 'blue water' withdrawals from available rivers and aquifers, and will increasingly compete for water with pressures from industry domestic use and the need to maintain environmental flows.[73] Some modern agricultural practices adversely affect soil quality through erosion, compaction, acidification and salinisation, and reduce biological activity as a result of pesticide, excessive fertilisation and loss of organic matter.[74] Food loss in low-income countries occurs in the production, storage and distribution stages of supply chains, whereas there is significant waste at the consumption stage in medium and high income countries. Extreme weather events caused by climate change will damage infrastructure, resulting in detrimental impacts on food storage and distribution, to which the poor will be most vulnerable. In developing countries, nearly 70 per cent of people live in rural areas, where agriculture is the largest supporter of livelihood and income.[75]

The Third Assessment Report (TAR) emphasised the ways how climate change impacts food, fibre and forest around the world. The elevated level of CO_2 will have effects on plant growth and yield. The impacts of climate change on food production, water, forestry and fishing resources are discussed below:

Effects on food production Fourth Assessment Report (FAR) estimates that temperature increase from 1 to 3^0C will have a net positive impact on the potential for food production, but this effect will become negative above this range of warming. Plants response to elevated CO_2 alone, without climate change is positive. These effects depend on a variety of other influencing factors: species, growth stage, management regime, and water and nitrogen application.[76]

Many recent studies confirm that temperature and precipitation changes in future decades will modify and limit the direct CO_2 effects on plants. The IPCC concludes with high confidence that the projected changes in the frequency of severity of extreme climate events will have more serious consequences for food and forestry production, and for food security will change in projected means of temperature and precipitation.[77] This relates to the increased frequency of crop loss due to extreme events such as droughts or heavy precipitation.

More frequent extreme events may lower long-term yields by directly damaging crops at specific developmental stages such as temperature thresholds during flowering. Crops do have threshold responses to their climatic environment, which affect their growth, development and yield.

Other important factors for food production are the impact of weed and insect pests' in causing diseases on plants and animal health. Land available to agriculture will be affected through climate change. The reduction of precipitation in certain areas will have the consequence that larger areas are no longer suitable for agriculture. Many crops will be affected and will no longer be able to be planted in certain regions.

Adaptation policies will also have an influential effect on the availability of water in certain regions in the future. If water harvesting and irrigation systems are developed and become more efficient, the total net effect will be very different from the 'no response' scenario.

Effects on forests: Forests are ecosystems with a dense tree canopy, covering 30 per cent of earth's surface worldwide. They cover 42 per cent of the surface in the tropics, 25 per cent in temperate areas and 33 per cent in the boreal zone.[78] Forests are needed both for climate change mitigation and for agricultural usages. The deforestation and degradation in tropical and sub- tropical regions is responsible for about 17 per cent of anthropogenic greenhouse gas emissions.[79] Forests are the biggest storage of terrestrial ecosystems carbon stocks. Forests provide a huge number of non-timber forest products, most of which are important for subsistence livelihoods.[80]

Modelling studies that examine the impact of climate change on forests estimate an increase in global timber production, forests will benefit from the fertilising effect of increased CO_2 concentrations in the atmosphere. In drought prone areas, the effects will be different from boreal areas where productivity might increase due to higher temperature and elevated CO_2 concentrations. In the forests, there is evidence of both regional increase and regional decrease of fire activities linked to climate change. Climate change might influence drought and other risks in the forest industry: pulp and paper production are affected. For many forests types, pests and diseases are a major source of concern.

The IPCC Fourth Assessment Report has collected information on the effects of extreme climate events on commercial forestry and reduced access to forest land, increased costs for road and facility maintenance and direct damage to trees by wind, snow, frost or ice. Indirect damages are wild fires, insect outbreaks and thaws on logging rates.[81] Deforestation rates are high in some tropical and sub-tropical regions, particularly in South East Asia and Amazon.

The socio-economic impacts of climate change on forests will be dominated by the relocation of forests' economic activities through a shift in production preferences such as bio-fuels. The most important socio-economic effects will be linked to income opportunities and forest based communities. Although these communities are likely to impact on global wood production, they may be especially vulnerable because of their limited ability as rural, resource dependent communities to respond to risk in a proactive manner.[82]

Non Timber Forest Products (NTFP) are an equally important source of income for the livelihood of rural communities.[83] NTFP cover fuel, forest food, medical plants and other collected goods. The FAO estimates that in many rural communities of sub-Saharan Africa, NTFP may supply over 50 per cent of farmers' cash incomes and provide the health needs for over 80 per cent of the population.[84]

Effects on fisheries: World fishing resources are already overused or depleted in many regions of the oceans and aquaculture will become increasingly important. Around 2.6 billion people receive an important part of their protein supply from fish.[85] The IPCC compares aquaculture with animal husbandry because they share similar vulnerabilities and adaptation needs to climate change. These similarities include ownership issues in access and use of land and water, the control of inputs, diseases and predators.[86] Aquaculture is also dependent on captured fish, because fish is often needed as a food source for the aquacultures.

The negative impacts of climate change on aquaculture and fresh water fishing include stress from increased temperature; oxygen demand and increasing acidity; uncertainty of future water supplies from river flows for fish farms; the increase in extreme weather events; the increased frequency of diseases; the expected sea-level rise; and the conflict of interest with coastal defence needs. The positive impacts are seen in increased growth rates, an increased length of growing seasons due to increasing ocean temperature, and range expansion for fish species and access to new areas of the oceans due to the decrease in recover.[87]

Ocean acidification can become a profound source of change in the long term productivity of oceans. Some studies estimate that fish growth can increase due to higher temperature and other studies show that temperature increases in the wrong season could negatively affect fish populations. The IPCC concludes that aggregate level effects of changes in primary production due to climate change related impact of food chain.

Effects on water: The availability of water will be an essential factor influencing food security effects of climate change. Water is not only for food production, but it is an important good, needed for survival and almost all human activities. At the end of 2002, the UN Committee for Economic, Social and Cultural Rights adopted the general comment to the International Covenant on Economic, Social and Cultural Human Rights (ICESCR) concerning the right to water. Improved water management is seen in many circumstances as an essential element for the long term availability of water to people.

The impact of climate change on fresh water will be huge. Climate change will change precipitation patterns and increase the variability of precipitation. The risks of flooding and droughts will increase in many areas.[88] The expected sea-level rise will extend the ground water, resulting in a decrease of freshwater available for humans, agriculture and ecosystems in coastal areas. Semi-arid and arid regions are particularly exposed to the impact of climate change on fresh water.[89]

Adverse effects of climate change on fresh water systems might aggravate the impact of other water stress factors such as population growth, changing economic activity (both the irrigation for agriculture and industrialisation), land use changes and urbanisation. Changes in the amount and variability of rain fall, as well as melting glaciers will have major impacts on water availability, agriculture and food security. Around 70 per cent of all water consumed worldwide is used in agriculture.

Small holder agriculture: The impact of climate change will be particularly substantial to small holder agriculture. Their livelihood systems, in low latitudes, will be greatly affected by climate change. The farming system will be affected by changes in temperature, elevation of CO_2 and precipitation of yields of both food and cash crops. The productivity of livestock and fisheries system will be affected as well as the potential income from collecting activities in forests.

The climate change impacts on small holder farmers are the decreasing water-supply through the reduced flow of rivers which feed irrigation systems, the effects of sea-level rise on coastal areas, the increase of tropical storms and changes in other environmental conditions such as forest fires.

Many small hold livelihoods are comprised of a variety of income sources from forests, remittances and other non-agricultural income strategies. These livelihoods can also involve systems requiring several crop and livestock species. Government support can also play a role but many small hold farmers are marginalised in national and international agricultural policies.

Climate change, vulnerability and food insecurity
The report is on the effects of climate change on food security and nutrition on the most affected and vulnerable regions and populations.[90]

- People who are poor are vulnerable to hunger because they lack the resources to meet their basic needs on a daily basis.
- Vulnerability of agricultural systems, communities, households and individuals to climate change and these vulnerabilities could lead to increased vulnerability to food insecurity. Vulnerability is influenced by the degree of exposure of systems, communities, households and individuals to climate change.

- Household level vulnerability is most often associated with threats to livelihoods. Livelihoods are seriously affected by low productivity due to too little land and lack of fertilizer and livelihoods are risky and susceptible to collapse due to droughts that cause harvest failure. The majority of food producing small holders in many countries are net buyers' of food which leaves them vulnerable and market related risks. Livelihood shocks can affect individuals - illness, accidents, retrenchment; or entire communities- floods, epidemics, livestock disease; or entire economies- financial crises, natural disasters, conflict, widespread food price hikes.[91]

- Potential impacts of climate change on food security include both direct nutritional effects and livelihood effects. Both biophysical and social vulnerability are critical as one considers the impact of climate change on food security. Social vulnerability examines the demographic, social, economic and other characteristics of the population that affect their ability and cope with negative shocks.

- Climate change affects vulnerability to food insecurity through its biophysical effects on crop, livestock and farming system productivity. Changes in temperature and precipitation means an increased variability in food production and effects on income for food producers in rural areas and for urban consumers and increased variability in agricultural production leading to more price and income fluctuations.

- Climate change has different dimensions as the focus turn from plants and animals through agricultural systems to individuals, households, communities and countries. Individual plants and animals have relatively well defined vulnerabilities to change in climate.

- The poor are likely to be particularly vulnerable to food insecurity brought about by climate change. But food insecurity occurs today even in the richest countries as a result in degradation of the environment and will make more people susceptible to food insecurity from climate change in future.[92]

- Young children and older populations can also be at greater risk due to their dependency, financial or physical on other household members. Communities are not considered poor in the financial sense; they may still be vulnerable to food insecurity due to their lower social status.

- Access to food is also conditioned by power imbalances in the social and political sphere. For example, support for community led initiatives such as food banks and state financed food distribution systems may be reduced during times of economic hardship induced by climate change.

• There is likely to be substantial overlap between the poor and vulnerable, those who are food insecure and those affected by climate change.

• Women and children are most vulnerable to climate change. Highly variable and unpredictable climatic conditions would lead to food shortages at the household level and thus, children and elderly are most likely to be affected by undernourishment.

Local environmental systems are where the immediate effects of climate change are felt and key actors in societal responses to climate change. But global, national and local, social and political institutions will all play important roles in managing the effects of climate change on food security.

Role of women in agriculture and food production
The direct climate change threats to agriculture, policies and programmes must target all those who are involved in agricultural production. A joint report by the World Bank, the Food and Agricultural Organisation of United Nations and the International Fund for Agricultural Development (2009) estimated that women account for 60 to 90 per cent of total food production.

In developing countries as a whole, women constitute approximately 43 per cent of the agricultural labour force, ranging from 20 per cent in Latin America to 50 per cent in South Eastern Asia and Sub-Saharan Africa.[93] Providing women, who manage agricultural operations in a very cost-effective way the information about improved farming practices generally, and climate change responses in particular, yet is rarely adopted in practice.[94]

Women are typically disadvantaged in other aspects of farming such as access to productive inputs and services and land ownership. Women are less likely to enjoy the same level of access to agricultural inputs as men.[95] Among all agricultural land holders in West Asia and North Africa, less than 5 per cent are women while this figure is approximately 15 per cent for Sub-Saharan Africa. At a regional level, Latin America has the highest average share of female agricultural holds. A recent study found that overall incidence of land ownership in the rural population in the state of Karnataka, India was 39 per cent for men and only 9 per cent for women.[96] Evidence suggests that on average, female headed households own smaller plots than male headed households These gendered constraints directly affect women's farm productivity.

In many contexts, women may face higher risks of food insecurity due to 'gender inequality'. They play a critical role in agriculture and in managing household food supplies, but may lack access to services and control over important resources and decisions affecting food security. CARE is a particularly focussed on promoting gender equality and empowering women to build the resilience of their families and communities to adapt to climate change and achieve food security.

CARE is committed to both food security and climate change adaptation as programming and policy advocacy priorities.[97] Food security to be a basic human right and a critical element of household livelihood security. Food security focusses on empowering poor women and girls to realise food and nutrition security. It addresses four dimensions of food security including (i) protecting and promoting resilient livelihoods to ensure adequate food availability and access; (ii) improving utilisation with a focus on nutritional status; (iii) enhancing stability through vulnerability; and (iv) risk reduction and management. Food security approach incorporates transformative activities that emphasise equity, women's empowerment rights and appropriate governance. Promoting environmental sustainability and enhancing adaptive capacity are key elements of the approach.[98]

CARE's approach is focussed on increasing the capacity of people, particularly the most vulnerable groups to adapt to climate change including support for climate resilient livelihoods; disaster risk reduction; and empowerment, advocacy and social mobilisation to address the underlying causes of vulnerability including poor governance, gender inequality and inequitable access to resources and services.[99] The objectives of climate change, adaptation and food security are:

- Increasing agricultural productivity, climate resilience and sustainability, particularly for small hold farmers, for example, by promoting conservation of agriculture practices, restoration of degraded soils and agricultural biodiversity.
- Promoting rights of vulnerable people, particularly women to critical livelihood resources such as land and water.
- Integrated water resource management.
- Sustainable land use management and ecosystem services.
- Technology transfer such as irrigation, conservation and sustainable agriculture, and biogas technology.

- Disaster risk reduction strategies.
- Enhancing government capacity to implement social protection schemes.
- Linking emergency food assistance to longer term food security responses.
- Promotion of savings and insurance schemes.
- Assessing vulnerability to and impact of climate change on the different dimensions of food security.
- Improvement of food security relates to gender equality, nutrition and climate variability and change.
- Partnership with other humanitarian development and environmental organisations, research institutions, government and private sector to identify practical and effective responses' to climate change and food insecurity.
- Knowledge management and sharing across sectors, communications and awareness rising.

Food security and adaptation to climate change
The following practices for adapting to climate change in the food and agriculture sector are described below:
- Protecting local food supplies, assets and livelihoods against the effects of increasing weather variability and increased frequency and intensity of extreme events through:
  o General risk management.
  o Management of risks specific to different ecosystems- marine, coastal, inland water and floodplain, forest, dry land, island, mountain, polar, cultivated.
- Avoiding disruptions or declines in global and local food supplies due to changes in temperature and precipitation regimes, through:
  o More efficient agricultural water management in general.
  o More efficient management of irrigation water on rice paddies.
  o Improved management of cultivated land.
  o Improved livestock management.
  o Use of new, more energy-efficient technologies by agro-industries.
- Protecting ecosystems, through provision of such environmental services as:
  o Use of degraded or marginal lands for productive planted forests or other cellulose biomass for alternative fuels.
  o Clean Development Mechanism (CDM) carbon sink tree plantings.
  o Watershed protection.

o Prevention of land degradation.
o Protection of coastal areas from cyclones and other coastal hazards.
o Preservation of mangroves and their contribution to coastal fisheries.
o Biodiversity conservation.

FAO has defined the following elements in a framework for climate change adaptation.[100]

- Legal and institutional elements;
- Policy and planning elements;
- Livelihood elements;
- Cropping, livestock, forestry, fisheries and integrated farming system elements;
- Ecosystem elements; and
- Linking climate change adaptation processes with technologies that promote carbon sequestration and substitutes for fossil fuels.

Climate change also requires adaptive management that focusses on modifying behaviours over the medium to long term to cope with gradual changes in precipitation and temperature regimes. These modifications are likely to concern consumption patterns, health care, food and agricultural production practices, sources and use of energy and livelihood strategies. Strengthening resilience for all vulnerable people involves adopting practices that enable them to:

- Protect existing livelihood systems;
- Diversify their sources of food and income;
- Change their livelihood strategies; and
- Migrate if there is no other option.

To strengthen resilience the agriculture based livelihood systems include:

- Research and dissemination of crop varieties and breeds adapted to changing climatic conditions;
- Effective use of genetic resources;
- Promotion of agroforestry, integrated farming systems and adapted forest management practices;
- Improved infrastructure for small-scale water capture, storage and use; and
- Improved soil management practices.

Food security through mitigation of climate change

The practices include wider adoption of best practices for mitigation in the food and agriculture sector providing new employment opportunities in the commercial agriculture sector as well as enhancing the sustainability of vulnerable livelihood systems. Such practices include:

- Reducing emissions of CO_2 such as through reduction in the rate of land conversion and deforestation, better control of wildfires, reduction of emissions from commercial fishing operations, and more efficient energy use by forest dwellers, commercial agriculture and agro-industries. UNFCCC and Kyoto Protocol recognise the potential role of forests in providing a variety of adaptive ecosystem services to mitigating climate change through carbon sequestration. These services include biodiversity preservation, watershed protection on mountain slopes, control of desertification, and maintenance of environmental integrity of fragile coastal zones. Cyclical loss and regrowth of trees and forests is a natural process. Forests are regularly ravaged by the spread of plant, pests and fire. The natural burning of trees and other organic matter releases CO_2 into the atmosphere. Changes in temperature ranges and precipitation, attributable to climate change can harm forests further. Droughts and forest fires are expected to increase with devastating effects on forests that are already stressed by human activity. Forests' capacity to play their natural role in maintaining climatic stability is closely linked to food systems response to the challenge of climate change. Action is needed on several fronts through an integrated approach that the global demand for additional land to produce food and fuel, the dependence on forests as a source of livelihood for many rural people in developing countries and the economic value of ecosystem services provided by forests. The actions required include creating economic alternatives of using forest resources unsustainably, promoting second generation bio-fuels to avoid land clearing for bio-fuel crops and enforcing strictly from wild-fires to clear land for commercial development.

- Reducing emissions of methane and nitrous oxide, such as through improved nutrition for ruminant livestock, more efficient management of livestock waste and of irrigation water of rice paddies, more efficient applications of nitrogen fertilizer on cultivated fields and reclamation of treated municipal waste water for aquifer recharge and irrigation are needed.

Reducing methane emissions from ruminant livestock: methane emissions per animal and per unit of livestock product are high when the animals' diet is poor. There are several technologies for reducing methane release from enteric fermentation. The basic principle is to increase the digestibility of feed stuffs, by either modifying feed or manipulating the digestive process. Most ruminant in developing countries, particularly in Africa and South Asia have a very fibrous diet. However, some techniques are often beyond the reach of small holder livestock producers, who lack the capital and sometimes the knowledge to implement changes. Another approach is to increase the level of starch or rapidly fermentable carbohydrates in the diet, thereby reducing excess hydrogen and the subsequent formation of methane.

The non-ruminant sources of animal protein in the diet can mitigate emissions from enteric fermentation and contribute to food security by improving the livelihoods of livestock-dependent households and adding diversity to the diet.

As the world population increases, reducing rice agriculture remains largely untenable as a strategy for reducing methane emissions from paddy rice fields. However, substantial reductions are possible through a more integrated approach to rice paddy irrigation and varietal selection. Many rice varieties can be grown under much drier conditions than those traditionally employed, with large reductions in methane emission without any loss in yield.

Reducing methane emissions from manure

Manure is the residue from animals' digestive processes; it contains important amounts of nitrogen, phosphates and potassium that provide valuable soil nutrients when applied to farmers' fields. Poor manure management can increase the loss of pollutants to the environment. Nitrogen in manures can be lost as nitrate, nitrous oxide or ammonia. If manure is managed as a liquid substance, it decays and forms methane. In the wild areas, manure is spread over a wide area and decomposes in the oxygen in the natural environment. The capture and burning of methane released from animal wastes is an increasingly applied form of energy generation and forms the basis for several carbon reduction and trading projects.

It is assumed that manure emissions in cool climates could be reduced by 50 per cent through adoption of an alternative management option to replace the storage of manure as liquid slurry in open pits. In warmer climates, where methane emissions from liquid slurry are estimated to be more than three times as high, a reduction potential of 75 per cent is considerably reasonable.[101]

Reducing nitrous oxide emissions from agricultural soils

A major direct source of nitrous oxide from agricultural soils is the wide spread increase in the use of synthetic nitrate-based fertilisers, dried by the need for greater crop yields and by more intensive farming practices. The widespread and often poorly controlled use of animal waste as fertiliser can also lead to substantial emissions of nitrous oxide from agricultural soils.

Net nitrogen use in farming affects climate change, because it is linked to nitrous oxide emissions, and water pollution, because nitrates pollute soil, fresh and marine waters. Net nitrogen use can be measured quite easily by recording the amounts of nitrogenous fertilisers and manures that are used on the farm. The best way to manage human interference in the nitrogen cycle is to maximise the efficiency of nitrogen uses.[102] Rapid incorporation and shallow injection of livestock wastes reduce nitrogen loss to the atmosphere by at least 50 per cent, and deep injection into the soil essentially eliminates the loss.[103] Options for reducing emissions from grazing systems are also important adding nitrification inhibitors to urea or ammonium fertiliser compounds before application can substantially reduce emissions of nitrous oxide.[104] Land drainage is another option for reducing nitrous oxide emissions before nitrogen enters.

• Sequestering carbon, such as through improved management of soil organic matter, with conservation agriculture involving permanent organic soil cover, minimum mechanical soil disturbance and crop protection; improved management of postures and grazing practices on natural grasslands, including by optimising stock numbers and rotational grazing; introduction of integrated agro-forestry systems that combine crops, grazing lands and trees in ecologically sustainable ways: use of degraded, marginal lands for productive planted forests or other cellulose biomass for alternative fuels; and carbon sink tree plantings.

Carbon sequestration involves increasing the carbon storage in terrestrial systems above or below the ground. The main efforts in agriculture is to manage greenhouse gases through planting trees. Carbon sequestrations have additional benefits, including increased root biomass, soil organic matter, water and nutrient retention capacity and land productivity. Investments in improved land management leading to increased soil fertility and carbon sequestration can often be justified by their contributions to agronomic productivity, national economic growth, food security and bio-diversity conservation.[105]

Recent studies have shown that well-managed grass lands and conservation agriculture can work as well or better as techniques for sequestering carbon.[106] Carbon sequestration explores four feasible options: reforestation and afforestation, rehabilitating degraded grasslands, rehabilitating cultivated soils and promoting conservation of agriculture.

Food and agriculture within UNFCCC

The Rio Declaration (RD) and UNFCCC guiding principles relate directly to the food and agriculture sectors, which include agriculture forestry and fisheries - in agreement with FAO definitions. The UNFCCC recognises among its primary concerns the need to ensure that ecosystems are not disrupted and food production is maintained (Article 2). Five specific references are made in relation to agriculture, forests and ecosystems:

- Promotion of GHG abatement technology development and transfer in all sectors including agriculture, forestry (4.1c);
- Promotion of sustainable management, conservation and enhancement of GHG sinks and reservoirs including biomass, forests and oceans as well as other terrestrial, coastal and marine ecosystems (4.1d);
- Cooperation in preparing for adaptation to the impacts of climate change;
- Developing and elaborate appropriate and integrated plans for coastal zone management, water resources and agriculture, and for the protection and rehabilitation of areas, particularly in Africa, affected by drought and desertification as well as floods (4.1e); and
- A commitment to support developing countries to address climate change impacts and respond with a focus on arid and semi- arid areas, forested areas and liable to forest decay and areas with fragile ecosystems.

Despite such important explicit references to food and agriculture activities, UNFCCC makes no reference to rural development and only one to Least Developed Countries (LDCs) (Article 4-9). Yet, rural development is fundamental to allow small holders and communities in LDCs achieve efficient use of land and water resources while implementing climate change responses.[107] The RD makes no reference to the terms agriculture, forest, fisheries, food, hunger, rural development - while the term ecosystem is mentioned only once (Principle- 7).

The Kyoto Protocol (KP) came into effect on December 11, 1997, formalises rules for operationalising key principles of UNFCCC in relation to emission reduction commitments of Annex 1 parties as well as establishing flexible financial mechanisms and international emission trading. The KP mentions the term agriculture and promotes sustainable forest management practices, afforestation and reforestation; sustainable forms of agriculture in the light of climate change considerations and increased use of renewable forms of energy. The agriculture and ecosystems at large, through promotion of efficient use of biomass resources for energy are to achieve low carbon growth in a resource efficient and socially inclusive manner.

The Kyoto Protocol addresses the food and agriculture sectors in Article 3.3 and 3.4 and Annex 16 to CP1. In particular, Articles 3.3 and 3.4 regulate the national reporting of GHG emissions related to land use, land use change and forestry (LULUCF), limiting mandatory reporting of land carbon sources and sinks to afforestation, reforestation and deforestation activities.[108]

The other reference to agriculture and forestry in the KP is, to seek to support regional programmes containing measures to mitigate climate change and measures to facilitate adequate adaption to climate change including in the agriculture, forestry and waste management sectors.

Adaptation is fundamental in limiting the adverse effects of climate change in coming decades, increasing the resilience of vulnerable systems to climate shocks. The implementation of adaptation actions are based on Article 4-8 and 4-9 of the UNFCCC and Article 10 of the KP and include Decision 5/CP.7, 2001 and Decision 1/CP.10, 2004. National adaptation programmes of Action (NAPAs) prioritise urgent and immediate adaptation needs for LCDs (Article 4.9).

Successful adaptation not only depends on governments but also on the active and sustained engagement of stakeholders (Nairobi Work Programme) including national, regional, multilateral and international organisations, the public and private sectors, civil society. The objective of Nairobi Work Programme is to help countries to improve their understanding and assessment of the impacts of climate change. The UNFCCC maintains a coping strategy database to facilitate the transfer of knowledge from communities already coping with specific hazards under current or evolving climate change.

Developing countries require international assistance to support adaptation (Article 4.4, 4.8, 4.9), which includes funding technology transfer and capacity building. Funding for adaptation is provided through the financial mechanism of the UNFCCC, currently operated by the Global Environmental Facility (GEF) and the Adaptation Fund Board (AFB). Funding opportunities include: The GEF Trust Fund including support for vulnerability and adaptation assessments as part of national communications; Least Developed Countries Fund (LDCF) and Special Climate Change Fund (SCCF) under UNFCCC; Adaptation Fund (AF) under the Kyoto Protocol managed by the Adaptation Fund Board (AFB). In operation terms, the UNFCCC Adaptation Fund Board began calls for project funding in 2010; only one such project, focussed on reducing vulnerability from coastal erosion in Senegal.

The post-2012 UNFCCC agreements were elaborated in the Copenhagen Accord and formalised via the Cancun Agreements at COP on December 16, 2010 - that developing countries, especially fast growing economies, must contribute to internationally monitored emission reductions via projects approved and funded via NAMAs (CA5). Importantly, these documents provide support for REDD+, reducing emission from deforestation and forest degradation and the need to enhance removals by forests (CA.6); and promote scaled up, new and additional, funding to enable and support enhanced action on mitigation, adaptation, technology development, technology transfer and capacity building. The level of funding for all of these activities is specified through the Copenhagen Green Climate Fund (GCF), approved in Cancun, at the level of US $30 billion annually for the period 2010-2012 and US $100 billion annually by 2020. This represents the cost of fighting climate change in developing countries through adaptation and mitigation actions expected by UNFCCC.[109]

At COP 16 in Cancun, these principles were re-affirmed and strengthened in terms of Cancun Adaptation Framework (CAF), by aiming at building resilience of Socio-economic and ecological systems including through economic diversification and sustainable management of natural resources (CP 16). As stated in the Cancun agreements, climate responses must be extended in scope, sectorally via programmes of activities in order to reach the scale of emission reductions necessary to stabilise global climate to below 2 degree Celsius. Developing countries must be brought into future agreements, through development of nationally appropriate mitigation actions (NAMAs), to be monitored nationally for the most part including large scale mitigation projects through GCF (Green Climate Fund). FAO could take the lead in helping the international community design and implement expanded options for agriculture that are relevant to future climate agreements.

Greening the economy with agriculture
The concept of Green Economy refers to increased attention towards green activities and jobs, with a main focus on renewable energy and low-carbon processes along the value chain from production to consumption. Green Economy must be aligned with Article 2 of UNFCCC. According to the recent UNEP GE book[110] a green economy in practice means: (i) low-carbon growth; (ii) resource efficient; and (iii) socially inclusive. Greening the Economy with Agriculture (GEA), FAO defines, "Greening the Economy with Agriculture refers to increasing food security (in terms of availability, access, stability and utilisation) while using less natural resources, that is increasing nutrient and energy efficiency throughout the food value chain".

The definition implies that a GEA framework for action should include climate change adaptation and mitigation responses in agriculture and forestry with a support to increasing food security, promoting sustainable use of natural resources, enhancing ecosystems' resilience and generating rural development opportunities. In the context of climate mitigation component into its strategic goals, GEA will need to promote sustainable activities in both intensive and extensive agricultural systems.

A specific focus on food security, poverty reduction and rural income creation requires policy activity and aimed at increasing formal references to agriculture, food security and rural development in future climate agreements, promoting eligibility of food and agriculture projects for adaptation and mitigation. FAO GEA program, focussing for instance on developing new CDM (Climate Development Mechanism) methodologies for the food and agriculture sectors, aimed at expanding the existing limited range-narrowly focussed on methane capture and new methodologies include methane emission reduction from rice fields through more efficient water use; reductions of NO_2 emissions through more use of fertilisers; promotion of ago-forestry systems for enhanced resilience and carbon sequestration.

FAO GEA initiative could provide significant support in devising, implementing and monitoring sustainability principles for adaptation and mitigation project activities in the food and agriculture sectors with reference to issues relevant to LDCs such as food security, gender, biodiversity, conservation, rural development. FAO could promote the creation of an Agriculture Panel or working group, serving the international climate policy and technical support in order to develop new agriculture methodologies with food security and sustainable rural development criteria.

GEA can help the international community to move beyond carbon as the main climate mitigation activities, for example by promoting payments or funding based on a range of ecosystem and social services[111] that are highly relevant to climate change responses such as improved water availability and quality, preservation of biodiversity, soil conservation, reduced use of chemical fertilisers, increased income opportunities.

The AWG-KP (Ad-hoc Working Group on Kyoto Protocol) work on LULUCF, has succeeded in extending REDD+ to REDD+ - reducing deforestation and forest degradation, conservation of forest carbon stocks, sustainable forest management, and enhancement of forest carbon stock- by considering forest management.[112] REDD+ should be further expanded to include sustainable agricultural practices and food production techniques at least within and near forested areas with a focus on conserving land, bio- diversity and ecosystems, while providing enhanced livelihood and economic opportunities to local communities in LDCs.

GEA effort could address for ecosystem and social services in agriculture, in connection with REDD+, agricultural activities away from forested areas and the linkages between climate responses on the one side, and food security and rural development on the other. The key work on REDD+, GEA could promote, design and develop agricultural project activities generating credits without permanence problems such as associated to non-CO_2 GHG emission reductions rather than to carbon sinks.

FAOs work on adaptation spans five overarching, interlinked themes:
- Data and knowledge for impact and vulnerability assessment and adaptation planning.
- Institutions, policies and financing to strengthen capacities for adaptation.
- Sustainable and climate-smart management of land, water and bio-diversity.
- Technologies, practices and processes for adaptation.
- Disaster Risk Management.

Agriculture and UNFCCC beyond COP 18

The Durban decision presents a key opportunity for all stakeholders to promote resilient, sustainable and humane agricultural landscapes that ensure food security for the 21st century.[113] Agriculture now appears decoupled from other sectors in negotiations,[114] parties choose, including Subsidiary Body for Scientific and Technological Advice (SBSTA) work program on agriculture, policy and finance in agriculture must support multiple social and environmental goals and incorporate, respect, prioritise and further the principles outlined here in, including:-

- **Inclusion and support of all stakeholders:** Policy and finance in agriculture must include marginalised stake holders, including women, small hold farmers, pastoralists, small-scale fishers, and indigenous people as well as civil society groups advocating on behalf of animals. These groups are all important stakeholders in the agriculture discussion. Their inclusion should be enhanced and valued in policy debated and decisions regarding food security, agriculture and agriculture related funding. As an important matter of process, the Durban call for submissions included civil society and other stake holders. All future UNFCCC decisions on agriculture should further this goal of equitable and inclusive decision making.

- **Addressing climate change adaptation and mitigation:** Climate change, agriculture, development and other policies now need to be developed.

- **Ensuring animal welfare, food security and other social and environmental outcomes:** As the world faces increasing challenges from climate change, it is imperative to seek and implement solutions that fulfil multiple social goals. Given the large number of animals raised for food globally, particularly in welfare-depriving systems, the welfare of these animals should be evaluated, enhanced and safeguarded in agricultural climate solutions.

- **Ensuring consistency across UNFCCC decisions and mechanisms:** Parties must ensure that any initiatives, arrangements, rules or mechanisms that might be established by the conference of parties to the UNFCCC or meeting of the parties to the Kyoto Protocol including NAMAs, the CDM, and a REDD+ mechanism, are elaborated and implemented.

SBSTA (Subsidiary Body for Scientific and Technological Advice):
Addressing Agriculture in the UNFCCC on June, 2013

The SBSTA invited parties and admitted observer organisations, on their views, how to enhance the adaptation of agriculture to climate change impacts while promoting rural development, sustainable development and productivity of agricultural systems and food security in all countries, particularly in developing countries.

The importance of agriculture: Global food production and food security are threatened by climate change. Every person in the world depends on agriculture; most of the rural poor depend on agriculture for their livelihoods while agriculture has tremendous importance as a means of driving sustainable development. Local and mostly small-scale food producers feed the vast majority of world population.[115] The sustainability of agriculture and enhancement of food security, now and into the future, is of absolutely vital importance. Agricultural activities contribute about 15 per cent of global greenhouse gas emissions and to achieve UNFCCC goal of limiting average global temperature increase to 2^0C.

Policy goals: The policies should be designed and implemented to meet four goals:

- In sustainable ways, maintain and increase the security of food supplies for food insecure people, particularly in developing countries;
- Enable small-scale food producers and other vulnerable populations to become more resilient to climate change;
- Sustainably reduce emissions from the agricultural sector; and
- Reduce emissions from the conversion of other land to agriculture.

Guiding principles: Countries agreed under the UNFCCC to prevent dangerous climate change: to allow ecosystems to adapt naturally to climate change, to ensure that food production is not threatened and to enable economic development to proceed in a sustainable manner; climate policies that encompass agriculture must include safeguards and approaches that:

- Protect and promote ecosystems and biodiversity.
- Protect and promote rural people's gender equitable access to natural resources.
- Protect and promote food security and the right to food.
- Protect and promote the rights of indigenous people and local population.
- Promote poverty reduction and climate adaptation.
- Protect and promote farm animals' health and ability to express natural behaviour.

- Protect and promote the rights of vulnerability groups by requiring sufficient transparency, consultation and active involvement of affected communities by supporting: adaptation policies and rights-based approach during design and implementation of adaptation policies, ensuring the active involvement of the affected communities.

Strengthening agriculture through UNFCCC

- Climate policies relating to agriculture should be in line with the guiding principles identified and reflect recommendations from relevant international institutions including the committee on World Food Security (WFS) and the International Assessment of Agricultural Knowledge, Science and Technology for Development (IAASTD).
- In the current UNFCCC agenda, discussions on climate and agriculture should be coordinated and consistent with discussions relating to adaptation, technology, mitigation, LULUCF, REDD+ and flexible mechanisms.
- Systems for bio-diverse, socially and gender equitable and resilient agriculture need to be developed, demonstrated, tested and implemented agricultural systems into ones which improve the health of ecosystems, communities and cultures even in the face of changing climate.
- Small scale food producers should be enabled to practice farming systems that are resilient in the face of climate change, bio-diverse and strengthen the ecosystems of which they are part.
- Agro-ecological small-scale food producers and other forms of sustainable, ecological and climate resilient food production should be promoted.

References

1. *Nelson, G.C. Are Biofuels the Best Use of Sunlight? In Handbook of Bioenergy Economics and Policy, ed. Madhu Khana and David Zilberman, 10, Springer, New York, 2009.*
2. *IPCC, climate change 2007: The Physical Science Basis, Contribution of Working Group I to the Fourth Assessment Report of the Intergovernmental Panel on Climate Change, Cambridge University Press, Cambridge, UK, 2007.*
3. *IPCC (Intergovernmental Panel on Climate Change), "Summary for Policymakers". In Climate Change 2007: Impacts, Adaptation and Vulnerability. Contribution of Working Group II to Fourth Assessment Report of the Intergovernmental Panel on Climate Change, M. L. Parry, O.F. Canziani, J.P. Palutikot, P.J. Van der Linden, and*

C.E. Hanson, eds., Cambridge University Press, Cambridge, UK, 2007.

4. Food and Agriculture organisation (2002). The State of Food Insecurity in the World 2001, Food and Agriculture Organisation, Rome.

5. Tubiello FN, Soussana J-F, Howden SM (2007) Proc Natl Acad Sci USA 104:19686-19690.

6. Foresight 2011; INRA/CIRAD, 2011; IAASTD 2009; Lobal et al., 2011; The Hague conference 2010.
   INRA/CIRAD. 2011. Agrimonde: Scenario and Challenges for Feeding the World in 2050. Versailles: Editions Quae.
   International Assessment of Agriculture Knowledge, Science and Technology for Development (IAASTD). McIntyre BD, Herrenhr, Wakhungu J, Watson Rt, eds. 2009. Agriculture at a Crossroads: A Synthesis of the Global and Sub-Global IAASTD Reports, Island Press Washington, DC.
   Lobell DB et al., 2011. Climate trends and global crop production since 1980, science 333:617- 620.

7. Beddington JR, Asaduzzaman M, Clark ME, Fernandez, Bremauntz A, Guillou MD, Howlett DJB, John MM, Lin E, Mamo T, Negra C, Nobre CA, Schools RJ, Van Bo N, Wakhungu J. 2012. What Next for Agriculture after Durban? Science 335:289-290; IPCC 2007.

8. Pretty et al., "Sustainable intensification in African agriculture", International Journal of Agricultural Sustainability, 9 (1), 2011, pp. 5-24.

9. United Nations Population Division. World population prospects: the 2010 revision, United Nations Department Of Economic And Social Affairs, United Nations Population Division, New York, 2010, http://esa.un.org/wpp/unpp/panel-population. htm,

10. UNDP 2006.

11. FAO, 2010. Global hunger declining but still unaccepting high, Retrieved from: www.fao.org/docrep/olz/al390e00.pdf

12. World Bank, (2008): Strategic Climate Fund, World Bank, Washington, DC, June 2008, Available at: http://siteresource.worldbank.org/INTCC/Resources/strategic_ climate_fund_final.pdf#stratgic_climate_fund (Accessed September 15, 2008).

13. World Watch institute, 2011. State of the world: innovations that nourish the planet, WW Norton & Company, New York, 2011.

14. United Nations, The Millennium Development Goals Report 2011, United Nations, New York, 2011.

15. IFAD, 2011. Rural poverty report: new realities, new opportunities for tomorrow's generation. International Fund for Agricultural development, Rome.

16. World Bank (2008); World Development Report 2008, WHO, Washington, DC, 2011.

17. FAO (2010), Global hunger declining, but still unacceptably high. Retrieved from:www.fao. org/docrep/012/aL39oe/aL390e00.pdf.

18. Foresight, (2011); the Future of Food and Farming: Executive Summary. The Government Office for Science, London. Retrieved from: www.bis.gov.uk/assets/ bispartners/foresight/docs/ food-and-farming/11-547-future-of-food-and-farming-summary.pdf.

19. FAO and WFP (2009): The state of food insecurity in the world: Economic crises -impacts and Lessons Learned.

20. *FAO, 2003 a. conceptual framework for national, agricultural, rural development, and food security strategies and policies, by K. Stamoulis and A. Zezza. Rome.*

21. *Easterling WE, Aggarwal PK, Batima P, Brander KM, Erda L, Howden SM, Kirilenko A, Morton J, Soussana JF, Schmidhuber J, Tubiello FM (2007) "Food, Fibre and Forest Products". In Climate Change 2007: Impacts, Adaptation and Vulnerability. Contribution of Working Group II to the Fourth Assessment Report of the Intergovernmental Panel on Climate Change, Palutikof, P.J. van der Linden and C.E. Hanson, Cambridge University Press, Cambridge, UK.*

22. *FAO (1996), Rome Declaration and World Food Summit Plan of Action. Retrieved from:www. fao.org/docrep/003/x8346E/X8346e02.htm#p1-10.*

23. *Foresight (2010). Synthesis report C11: ending hunger. Foresight project on global food and farming futures. The government office for science, London. Retrieved from: www.bis.gov.uk/ assets/bispartners/foresight/docs/food-and-farming/ synthesis/11-631-c11-ending-hunger.pdf.*

24. *Devereaux, S. and Sabates-Wheeler R. (2004); Transformative social protection. IDS working paper 232. Retrieved from: www.ntd.co.uk/idsbookshop/details. asp?id=844. The order has been changed to reflect the need to focus on preventive and promotional measures needed when these fail.*

25. *World Bank (2008). World Development Report, 2008, Washington, DC.*

26. *Parry et al., Climate change and hunger: Responding the challenge, world food programme (Rome, Italy, 2009).*

27. *Adapted from: FAO (2008). Climate change and food security: A framework documents.*

28. *IPCC, Special report on emissions scenarios, Cambridge University Press, Cambridge, UK, 2000.*

29. *IPCC, 2007.*

30. *Rosenzweig C., Tubiello FN, Goldberg RA, Mills E, Bloomfield J (2002) Global Environ Change, vol. 12, pp. 197-202.*

31. *IPCC online. 2001. Glossary of Terms used in the IPCC of Working Group II to Third Assessment Report. Available at: www.ipcc.ch/glossary/index.htm.*

32. *Fischer G, Shah M, Van Velthuizen H, Climate Change and Agricultural Vulnerability. A Special Report Prepared as a Contribution to the World Summit on Sustainable Development, (Laxenburg, Austria: International Institute for Applied Systems Analysis, 2002).*

33. *IPCC, 2007; IPCC, 2001.*

34. *IPCC, 2001.*

35 *Bruinsma J, (ed.) (2003), World agriculture: Towards 2015/2030, A Food and Agriculture Organisation Perspective, Earthscan, London.*

36. *IPCC, 2007*

37. *Schmidhuber J, Shetty P (2005) Acta Agri Scand, 2/3-4, pp. 150-166.*

38. *Fischer et al., 2002, climate change and agricultural vulnerability*

39. *Godfray HCJ Pretty J, Thomas SM, Warham EJ, Beddington JR: Linking policy on climate and Food Schience 2011, 331, pp. 1013-1014.* Foley J.A, Ramankutty, N.Brauman, K.A, Cassidy, E.S, Gerber, J.S, Johnston, M., Mueller, N.D, O'comell C., ,Ray, D.K West, P.C Balzer, C., Bennet, E.M, Carpenter, S.R., Hill, J., Monfreda, C., Polas Ky, S., Tilman, D and Zaks, D.P.M. (2011),Solutions for a cultivated planet. Nature 478 (7369) (October 2012):337-342. *http://dx.doi.org/10.1038/nature 10452.*

40. *World Bank: World Development Report 2008: Agriculture for Development.*
41. *Hertel TW, Rosch SD, 2010: Climate change, Agriculture and poverty. Appl Econ Perspect Policy 2010, 32:355-385 doi:10.1093/aepp/ppq016.*
42. *Ingram J. Ericksen P, Liverman D (Eds): Food Security and Global Environmental Change, Earthscan, London, UK, 2010.*
43. *Barker T, Bashmakov I, Bernstein L, Bogner JE, Bosch PR, Dave R, Davidson OR, Fisher BS, Gupta S, Halsnaes K et al., Technical Summary. In Climate Change, 2007: Mitigation Contribution of Working Group III to the Fourth Assessment Report of the Intergovernmental Panel on Climate Change. Edited by Metz B, Davidson OR, Bosch PR, Dave R, Meyer LA, Cambridge University Press, Cambridge, United Kingdom/New York, NY, USA, 2007.*
44. *Harvey M, Pilgrims, 2011; food policy: The new competition for land: food, energy and climate change.*
45. *CCAFS: Proposal for CGIAR Research program 7: Agriculture and food security, Copenhagen, 2011.*
46. *Moss RH, et al., 2011: The next generation scenarios for climate change research and assessment.*
47. *Lahsen M, et al., 2010: Impacts, adaptation and vulnerability to global environmental change.*
48. *Verbury PH et al., 2011: changes in using land use and land cover data for global change studies.*
49. *USDA. The Agricultural Model Inter comparison and improvement projects, 2011.*
50. *Nelson GC et al., 2009: climate change impact on agriculture and costs of adaptation. Food policy report.*
    Nelson GC et al., 2010: Food security, Farming and climate change to 2050.
51. *Jarvis A, et al., 2011: An integrated adaptation and mitigation framework for developing agriculture research: synergies and trade-offs.*
52. *Vermeulen SJ, et al., 2010: Agriculture, food security and climate change.*
53. *Defries R, Rosenweig C; 2010; Towards a whole-landscape approach for sustainable land use in the tropics.*
54. *Negra C, et al., 2011 from REDD+ for Agriculture (CCAFS) Copenhagen.*
    Kissinger et al., 2011, Copenhagen, Denmark: CCAFS from REDD+ Agriculture.
55. *Badjeck et al., 2010: Impacts of climate variability and change on fishery-based livelihoods.*
56. *Thomton Pk, Gerber PJ; 2010: Climate change and the growth of the livestock sector in developing countries.*
57. *IPCC 2012. Managing the risks of extreme events and disasters to advance climate change adaptation, Cambridge University Press, Cambridge, 2012.*
58. *Costello et al., 2009. IAASTD, 2009, Vermeulen et al., 2012. I*
59. *PCC, 2012.*
60. *World Bank, 2008.*
61. *IPCC 2007; The Hague conference 2010.*
62. *United Nation Convention to combat desertification, 2011.*
63. *IPCC, 2007.*
64. *ILO, 2007. Chapter 4. Employment by sector. In key indicators of the labour market (KILM), 5th edition. Available at: www.ilo.org/public/english/employment/strat/ kilm/download/kilm04. pdf.*

65. *IPCC, 2012.*
66. *IPCC, 2012.*
67. *IPCC 2007, The Hague conference, 2010.*
68. *United Nations convention to combat desertification, 2011;* Bai et al., 2008.
69. *Royal Society 2009, Foresight 2011.*
70. *Foresight 2011. Future of food and farming. London.*
71. *Foresight 2011, INRA/ CIRAD, 2011.*
72. *Vermeulen S.J, Aggarwal PK, Ainslie A, Angelone C, Campbell BM, Challinor AJ, Hansen JW, Igram JSI, Jarvis A, Kristjanson P, Lau C, Nelson GC, Thorton PK, Wollenberg E. 2012. Options for support to agriculture and food security under climate change. Environmental Science and Policy, 15, pp. 136-144.*
73. *Foresight 2011, The future of food and farming: Executive summary. The Government office for science. London. Retrieved from: www.bis.gov.uk/assets/bispartiners/11-547-future.docs/ food-and-forming-summary.pdf.*
74. *NAS 2010.*
75. *FAO, 2003b. World agriculture: toward 2015/2030, chapter 13, Earthscan, Rome.*
76. *Ainsworth and long-2005(2005) what have we learned from 15 years of tree-air CO2 enrichment (FACE)? A mela-Analysis of the responses of Photosynthesis, canopy properties and plant production to rising COS.In: New phytol, 1652005, pp. 35/-372.*
77. *Easterling WE, Aggarwal PK, Batima P, Brander KM, Erda L, Howden SM, Kirilenko A, Morton J, Soussana J-F, Schmidhuber J et al., Food, Fibre and Forest Products. In Climate Change, 2007: Impacts, Adaptation and Vulnerability. Contribution of Working Group II to the Fourth Assessment Report of the Intergovernmental Panel on Climate Change. Edited by Parry ML, Canziani of, Palutikof JP, Van der Linden PJ, and Hanson CE, Cambridge University Press, Cambridge, UK, 2007, pp. 273-313, IPCC.*
78. *Sabine, C.L. et al., 2004: The oceanic sink for anthropogenic CO2. In: science, 3052004, pp. 367-371.*
   Fischlin, A./Price, V./Leemans, R./Gopal, B./Turley, C./Rounsevell, M.D.A./ Dubo, O.P./ Tarazona, J./ Velichko, A.A. (2007): Ecosystems: their properties, goods and services. In: Parry, M.L./Canziani, O.F./Palutikof, J.P./Van der Linden, P.J./Hanson, C.E. (eds.): Climate Change 2007: Impacts, Adaptation and Vulnerability. Contribution of Working Group II to the Fourth Assessment Report of the Intergovernmental Panel on Climate Change, Cambridge University Press, Cambridge u.a, 2007, pp. 211-272.
79. *Barker T, Bashmakov I, Bernstein L, Bogner JE, Bosch PR, Dave R, Davidson OR, Fischer BS, Gupta S, Halsnaes K et al., Technical Summary: In Climate Change 2007: Mitigation, Contribution of Working Group III to the Fourth Assessment Report of the Intergovernmental Panel on Climate Change. Edited by Metz B, Davidson OR, Bosch PR, Dave R, Meyer LA, Cambridge University Press, Cambridge, UK/ NewYork, NY, USA, 2007.*
80. *Shvidenko, A. et al., Regional certificate full carbon account. Fusion of Remotely sensed Data, on-ground Information and Ecological Modelling. Paper presented at the EGUO5 General Assembly of the European Geosciences Union, Vienna, April 22-27, 2005.*
81. *Easterling et al., 2007.*

82. Davidson, D.J/Williamson T./Parkins, J.R. "Understanding Climate Change risk and vulnerability in northern forest-based communities", In: Canadian Journal of Forest Research, 33 (II), 2003, pp. 2252-2261.

83. Easterling et al., 2007.

84. FAO, 2004a. Carbon sequestration in dry land soils. World soil resources reports, No. 102. Rome. Available at: ftp://ftp.fao.org/agl/agll/docs/wsrr102.pdf.

85. FAO, (2008) GM food safety assessment: tools for trainers. In press. Expert Meeting on Global Perspectives on Fuel and Food Security: Technical Report. February 18-20, 2008, Rome. 86. Easterling et al., 2007, 5, 4.

87. Easterling et al., 2007, 5, 4, 6.

88. Kundzewicz ZW, Mata LJ, Arnell NW, Doll P, Jimenez B, Miller KA, Oki T, Sen Z, Shiklomanov I A (2007, 3.3.1). Freshwater Resources and their Management. Climate Change 2007: Impacts, Adaptation and Vulnerability. Contribution of Working Group II to the Fourth Assessment Report of the Intergovernmental Panel on Climate Change, M.L.

89. Kundzewicz et al., 2007, 3.4.2.

90. FAO, 2008 provides a useful discussion of the basic concepts of food security including Vulnerability.

91. de Janvry and Sadoulet, E. "Subsistence farming as a safety net for food-price shocks", Development in practice, 21 (4-5), 2011, pp. 449-456.

92. Pieter, G., Maluf, R.S. And Da Silva Rosa, T. (2011). Mudancas Climaticas, Vulnerabilidades e Adaptacao. Rio de Janerio: COEP. 2011.

93. FAO, 2011a: These numbers would imply that labour productivity for women is much higher than for men with little evidence to support that.

94. IAASTD, 2008. Agriculture at a crossroads: The synthesis Report Science and Technology. Washinton, DC, USA: International Assessment of Agricultural knowledge, Science and Technology for Development. http://www.agassessment. org.

95. Dey- Abbas 1997; Quisumbing, 1996; Thapa 2008; Agarwal, 2011

96. Swaminathan, H., Suchitra, J.Y. and Lahoti, R. KHAS: Measuring the Gender Asset Gap, Indian Institute of Management Bangalore, Bangalore, 2011.

97. CARE International (2010): Global food security strategy.

98. CARE International (2010): Global food security strategy, Draft June 2010.

99. CARE (2010), what is adaptation to climate change?

100. FAO, 2007a. Adaptation to climate change in agriculture, forestry and fisheries: perspective, framework and priorities. Report of the FAO Interdepartmental Working Group on Climate Change. Rome.

101. IPCC, 2007b. Climate change 2007 - mitigation of climate change. Contribution of working group III to the fourth assessment report of IPCC, Cambridge University Press, Cambridge, UK, 2007.

102. Smill, V. "Nitrogen in crop production: an account of global floods", Global Biogeochemical Cycles, 13 (2), pp. 647-662.

103. Rotz, C.A.2004. Management to reduce nitrogen losses in animal production. Journal of Animal science 82 (e.SUPPL) E 119-E137

104. Monteny, G.J., Bannink, A. and Chadwick, D. "Greenhouse gas abatement strategies for animal husbandry", Agriculture, Ecosystems and Environment, 112, 2006, pp. 163-170.

105. *FAO, 2004a.*
106. *Mannetje, L.t, 2006. The role of grasslands and forests as carbon stores, University of Wageningen, Wageningen, the Netherlands, 2006.* 107. *UNEP GE; GEA.*
108. *16. CMP 1: The implementation of land use, land use change and forestry activities contribute to the conservation of bio diversity and sustainable use of natural resources.*
109. *UNEP further estimates that over USD 200 billion annually are needed from now till 2050, in order to promote sustainable growth in the agriculture sectors of LCDs.*
110. *UNER, 2011, Towards a green economy: Pathways to sustainable Development and poverty Eradication.*
111. *Ottaviani and Scialabba, 2011.*
112. *Consideration of further commitments for Annex I parties under Kyoto Protocol, Cancun, CMP 6.*
113. *Food security includes four pillars- availability, access, nutritional quality and stability.*
114. *Draft decision (COP 18/ CP-17) outcome of the work of the Ad Hoc working Group on long- term cooperative Action under the convention.*
115. *ETC Group, 2009.*

4. CLIMATE CHANGE AND HEALTH SECURITY

Introduction

Climate change affects public health through environmental consequences such as sea-level rise, changes in precipitation resulting in flood and drought, heat waves, changes in hurricanes and storms and degraded air quality. The main four diseases (e.g. cardiovascular disease, malnutrition, diarrhoea and malaria) as well as floods, the World Health Organisation (WHO) estimated 1, 66,000 deaths and about 5.5 million disabilities were attributable to climate change in 2000.[1]

It is a study about climate change effects from a largely ecological and metrological base to one that focusses on the human health consequences of climate change mitigation and adaption. Climate change is a significant and emerging threat to public health. WHO has an active and long standing programme on protecting health from climate change. WHO provides evidence and supports capacity building and implementations project to strengthen the health system to climate change. Climate change is threatened to forest, rising sea-levels, changing rainfall patterns which will exist economic, political and humanitarian stresses and affect human development in all parts of the world. The UNDP with national, regional and local planning bodies to help them responds effectively to climate change and promote low-emission, climate- resilient development.

Gender differences occur in health risks that are directly associated with meteorological hazards. Women and men differ in their roles, behaviour and attitudes regarding actions that could help to mitigate climate change. Surveys show that men are more energetic than women; particularly in private sector women are often responsible for most of the household consumer decisions including food, water and household energy. There are gender differences in relation to the health and safety risks of new technologies to reduce greenhouse gas emissions. The women fulfil their roles and responsibilities as careers of their families during extreme climate events.

Human-induced climate change poses great and growing risk to human, social and biological well-being. Human-induced climate change will affect future levels of population health, patterns of disease and death, attempts to reduce the rich-poor health gap, social stability and geo-political security. Climate change will cause novel health disorders and mostly prevailing risk to health. For example, the ongoing rapid urbanisation in most low and middle income countries is expanding the numbers exposed to the extra-local heats caused by the "urban heat island" effect and global warming will amplify that heat exposure.[2]

Since 1990, WHO has published a series of reports on climate change and has participated in review processes such as the intergovernmental panel on climate change. These activities have outlined four key characteristics of the health risks generated by warming and a more variable climate. First, the hazards increased risks of extreme weather such as fatal heat waves, floods and storms, having potentially more serious effects on infectious disease. Dynamics shift to long- term drought conditions in many regions, melting of glaciers that supply fresh water to large population centres and sea level increases leading to salinisation of sources of agriculture and drinking water. Second, the health impacts of climate change are potentially huge. Malaria, diarrhoea and protein-energy malnutrition together cause more than 3 million deaths each year.[3] Third, the greenhouse gases that cause climate change originate mainly from developed countries, but the health risks are concentrated in the poorest nations, which have contributed least to the problem.[4] Finally, many of the projected impacts on health are avoidable through a combination of public health interventions in the short-term, support for adaptation measures in health-related sectors such as agriculture and water management, and a long-term strategy to reduce human impact on climate.

Climate change effects and impacts on population health
Global climate changes have occurred naturally due to various astronomical cycles, variations in solar energy output and volcanic activity. The human actions are changing atmospheric composition, causing global climate change over the past few decades.[5] Human induced climate change increase in the atmospheric concentration of GHGs are amplifying the greenhouse effects. The excessive increase in fossil fuels burning agricultural activities and several other economic activities have critically augmented greenhouse gas emissions.

Today, many infectious diseases such as HIV/AIDS, Hantavirus, Hepatitis, SARs etc. are circulating. This reflects the combined impacts of rapid demographic, environmental, social, technological and other changes in our ways-of-living. Climate change will also affect infectious disease-occurrence worldwide.[6] World Health Organisation estimated, in the world health report 2002, that climate change was estimated to be responsible in 2002 for approximately 2.4 per cent of world-wide diarrhoea, and 6 per cent of malaria in some middle income countries.[7] Effects of climate change on health will have impact on majority population in the coming decades and put the lives and well-being of billions of people at increased risk.[8] IPCC states that, "climate change is projected to increase threats to human health". Climate change directly effects hazards such as heat waves, floods and storms and more complex pathways of altered infectious disease pattern, disruptions of agricultural and other supportive eco-systems and potentially population displacement and conflict over resources such as water, fertile land and fisheries.[9]

Climate change affects the social and environmental determinants of health-clean air, safe drinking water, sufficient food and secure shelter etc. are necessary to live a healthy life for mankind. Global warming that has occurred since 1970s caused over 1, 40,000 excess deaths annually. Diarrhoea diseases, malnutrition, malaria and dengue are highly climate sensitive and lead to a prodigious climate change. The quantities of Carbon dioxide and other greenhouse gases increase heat in the atmosphere and affect the global climate. In the last 100 years, the world has been warmed by approximately 0.75°C and over the last 25 years, the rate of global warming has accelerated, at over 0-18°C per decade.[10]

Now, climate change is happening faster than expected risk of tipping points and upper end of IPCC scenarios. Climate change impacts health in many ways, directly or indirectly with heat, extreme weather, floods, drought, wildfires, water and food-causing diseases and vector-borne diseases. Climate change affects life systems on air, water, food, shelter and security. The Copenhagen diagnosis on climate change, that:-

- Surging greenhouse gas emission.
- Recent global temperatures demonstrate human-based warming.
- Acceleration of melting of ice-sheets, glaciers and ice-caps.
- Rapid Arctic sea-ice decline.
- Current sea-level rise underestimated.
- Sea-level prediction revised.
- Delay in action risks irreversible damage.
- The turning point must come soon.

Climate change affects social determinants of health- clean air, safe drinking water, sufficient food and secure shelter. Extreme high air temperatures directly impact on cardiovascular and respiratory diseases. In Europe 2003 heat air of summer caused 70,000 deaths as recorded.[11]

Rising sea levels and increased extreme weather events will destroy homes, medical facilities and other essential services. Heavy rainfall affects the supply of fresh water and increases the disease of diarrhoea which kills 2.2 million people every year.[12] Floods contaminate fresh water supplies, increase water-borne diseases and create insects such as mosquitoes. It also damages home, disrupt the supply of medical and health services and decrease the production of floods in many of the poorest regions- by up to 50 per cent by 2020 in some African countries.[13] Climate change is projected in the area of china where the snail-borne disease Schistosomiasis occurs.[14] The vulnerable people are mostly affected by climate change in Small Island developing states, coastal regions, megacities, mountains and Polar Regions. Population growth is playing a vital role in global change to which the contribution of global emissions is obvious.[15] United Nations projections on today's population is that 7 billion will increase to 9.3 billion by 2050.[16] Adverse global influences on health such as rising food prices and some infectious diseases have also impeded attainment of the United Nations Millennium Development Goals.[17] Future global health goals must be better integrated with the fundamental influences of poverty, inequity, illiteracy, climate change, land use patterns and food insecurity on health.

Climatic health impacts on women
- **Migration and displacement:** Climate change can affect migration in three ways. First, the effects of warming and drying in some regions will reduce agricultural potential and undermine 'eco-system services' such as clean water and fertile soil. Second, the increase in extreme weather events in particular, heavy precipitation and river floods in tropical regions- will affect even more people and may generate mass displacement. Third, sea- level rises are expected to destroy the low-lying coastal areas that are home to millions of people, who will have to relocate permanently.
  The Indian Ocean Tsunami in 2004 indicated that women and children were very vulnerable in these situations. The World Disaster Report recognises the widespread consensus that women and girls are at higher risk of sexual violence, sexual exploitation and abuse, trafficking and

domestic violence in disasters.[18] The children and women are most vulnerable in psychological stress after disasters with lack of privacy, collapse of regular routines and livelihood, frustration and violence.[19]

- **Loss of biodiversity and nutrition:** Loss of biodiversity can compound insecurity because many rural women in different parts of world depend on forest products for income, traditional medicinal use, and nutritional supplements in times of food shortages. Thus, loss of biodiversity challenges the nutrition, health and livelihood of women and their communities.[20] Nutritional status partly determines the ability of natural disasters.[21] Women are more prone to nutritional deficiencies because the nutritional needs, when they are pregnant or breastfeeding. For example, in South Asia and Southeast Asia, 45-60 per cent of women of reproductive age are underweight and 80 per cent of pregnant women have iron deficiencies. For girls and women, poor nutritional status is associated with an increased prevalence of anaemia, pregnancy and delivery problems.[22]

- **Food security and livelihood:** In the time of disaster and environmental stress, women and girls are generally expected to care for the sick.[23] Women and girls may also face barriers to accessing health-care services due to poor control over economy. A lower education status implies more constraints for women to access health information or early warning systems as they are developed. It also means that girls and women have decreased access and opportunities in the labour market, increased health risks associated with pregnancy and child birth and less control over their personal lives.

- **Urban health:** In rural areas, conflict and divorce, and unemployment forces increasing numbers of women to live in urban areas. The rising rate of female-headed households in urban/pre-urban areas results in a shift of urban sex ratios and feminisation of urban poverty, exposure of dwelling and managing on their own the disproportionate daily burden of infrastructural needs such as waste management, fuel, water and sanitation make urban female heads of households particularly vulnerable to natural disasters.[24]

Environmental and ecological changes on world health

Climate change is also induced by human activities due to excess greenhouse gas emissions. The international policy and local policy makers should take action for eradication of primary health problems arising from such global environmental and socio demographic changes. The world trade organisation should give greater priority for adverse health and environmental effects of international free trade.[25] The WHO framework convention on Tobacco control and alert and response in the emergence of infectious diseases[26] as well as the United Nations Environmental Programme Montreal protocol to protect the Ozone Layer.[27]

The environmental and ecological changes on a global scale will influence world's health. First, the new strains of influenza virus will affect and is increasing in the rural villages of Southeast Asia and East Asia.[28]

Second, the decline in available seafood protein is a threat to health and reflects the unprecedented combination of ocean warming, acidification, deoxygenating, destruction of coastal fish nurseries and overfishing.[29] Third, sometimes, the health risks are posed by the deprivation, displacement and conflict due to the shortage of fresh water. For example, majority population in Bangladesh, Vietnam, Egypt and Iraq Live downstream on great river flows which are threatened by the loss of glacier mass and snowpack due to global warming.[30]

Finally, to maintain food supplies and adequate nutrition present a major challenge for increasing world population. The global food production is also reduced due to land degradation, water shortages and climate change. There are growing pressures on agriculture that are food production, distribution and consumption. These global and systematic changes impact population health.

The global environmental hazards to human health include climate change, stratospheric ozone depletion, changes in ecosystem due to loss of biodiversity, changes in hydrological systems and the supplies of freshwater, land-degradation and urbanisation. WHO provides health expertise into the UN Conventions on Climate Change.

Stratospheric ozone depletion, UV radiation & health

For several decades, it has been recognised that the release of chlorofluoro carbons and other atmospheric pollutant depletes stratospheric ozone, which is causing skin cancer and cataracts. The reorganisation of direct effects on human health was a major stimulus to the Montreal Protocol, which acts to reduce emissions of pollutants that weaken the ozone layer. The WHO and international agreement is proving to reduce it in long term.

Ecosystem products and services for health

Human health depends upon ecosystem products and services such as availability of fresh water, food and fuel sources, which are requisite for good human health and productive livelihoods. Human health impacts on the changes of ecosystem services affect livelihood, income, local migration and may cause political conflict.

Human interventions are altering the capacity of ecosystems to provide their goods such as fresh water, food, pharmaceutical products and services such as purification of air, water and soil. Ecosystem disruption can impact on health in a variety of ways and through complex pathways. The Millennium Ecosystem Assessment (MA) was called for by the UN Secretary General in 2000, to assess the consequences of ecosystem change upon human well-being.

Urbanisation and health
Urbanisation is process of global scale changing the social and environmental aspects of human life. Urbanisation is a result of population migration from rural to urban. As urban populations grow, the quality of global and local ecosystems and the urban environment will play an important role in public health with provision of safe water and sanitation and injury prevention between urban poverty, environment and health.

Biodiversity and health
Biodiversity provides many ecosystem services. Climate is an integral part of ecosystem. Marine biodiversity is affected by ocean acidification due to carbon in the atmosphere. Biodiversity plays a crucial role in human nutrition through its influence on world food production. Human activities are disturbing both the structure and functions of ecosystems and biodiversity. Major infectious diseases are increasing due to deforestation, land use change, water management e.g. dam construction, irrigation and uncontrolled urbanisation.

Land degradation & desertification
Land degradation and desertification can affect human health through complex path ways, as land is degraded and at some places deserts expand. Food and food production is reduced, water sources dry up and people are distressed to move to other places. The potential impacts of desertification on health include:
- High threat of malnutrition from reduced food and water supplies.
- Water and food borne diseases due to lack of clean water.
- Respiratory diseases caused by atmospheric and air pollution.
- The spread of infectious diseases as population migrate.

Water services for health
Fresh water is essential to maintain human health. Threats to fresh water resources mean threats to human health. This fresh water is found in rivers, lakes and underground which is increasingly threatened by land use, deforestation, climate change, growing population and industry. The water is polluted in urban areas. Protecting fresh water ecosystems means protecting the health.

Implications for environmental health and equity in climatic developing countries
Climate change is an emerging threat to global public health. Emissions of the greenhouse gases (GHGs) are determined by consumption patterns in cities of the developed world. The IPCC reported that in 1990, buildings were responsible for 20 per cent of global emissions and agricultural production and waste contributed 18 per cent of GHG emissions with total emissions estimated to increase by about 50 per cent by 2020.[31] GHG emissions and health protection in developing countries are likely to become increasingly prominent in policy development. There is a need for more active input from the health sector to ensure that development and health policies contribute to a preventive approach to local and global environmental sustainability, urban population health and health equity.

Urban health vulnerabilities to global climate change
Between 1990 and 2001, the temperature is likely to rise by 1.4 - 5.8°C with associated changes in the hydrological cycle causing a range of health impacts.[32] In 1900, the 15 per cent of world population lived in urban areas, but now it is 50 per cent of world population who prefer to live in urban areas.33 900 million people estimated that 1/3rd of global urban population, 70 per cent of urban developing-country population now live in low incomes, poor housing and ecosystem degradation and this projection will increase 2 billion by 2020.[33] The heat waves, floods and storms, communicable diseases and air pollution affect the human health and pollute the environment.

Temperatures. Heat waves can cause dramatic impacts on urban health. In Europe (2003), in summer, high temperature was recorded by human induced climate change.[34] The heat waves in island, measured up to 5-11°C warmer than the rural areas.[35]

Sea-level rise and floods
Many cities of the world are located in the coastal areas and they may be

more vulnerable by the sea-level rise and they are more exposed by the windstorms and floods. Heavy rains often result around Caracas, Venezuela in 1999 and Mumbai, India in July 2005. Economic development offers the opportunity for improvement in housing quality and flood protection.[36]

Infectious diseases

Many waterborne and vector borne infectious diseases are commonly influenced by climate conditions in the cities. The vector borne virus dengue has increased in tropical developing regions globally in the past few decades due to unplanned urbanisation, producing Aedes Mosquitoes.[37] The global environmental changes may favour the emergence of new infectious diseases which may spread faster within and between cities due to travel links and higher rates of person-to-person contact.[38]

Air pollution

The ozone is affected by atmospheric conditions and harbingers a warmer days. Greenhouse gas from industrialised countries' emits into atmosphere and creates lots of diseases on the earth. In the time of flood, there is muddy water which pollute the atmosphere and spread many diseases through the air. According to WHO, air pollution is a significant factor for health conditions including respiratory infections, heart disease and lung cancer., The health effects caused by air pollution may include difficulty in breathing, wheezing, coughing, asthma and critical in respiratory and cardiac conditions.

Improvement of urban health from global climate change

Urban health is important. The effect on health of urbanisation is two edged. On the one hand, there are the benefits of ready access to health care, sanitation and secure nutrition and on the other hand, there are the evils of overcrowding pollution, social deprivation, crime and stress related illness. There are major opportunities for improving health status with health equity avoiding global climate threats. The global health burdens are connected with energy and transport policies.[39] Air pollution in urban causes approximately 8,00,000 deaths per year worldwide.[40] Few studies suggest that implementing available technologies to reduce fossil fuel use could bring major health benefits. The indoor air pollution from solid fuel use that kills over 1.5 million each year and is still widespread among the poorest population in cities of developing countries.[41] 'Healthy Urbanisation' should consider the wide range of technological and planning options which could increase a wealthy population and how to implement the technology and planning for GHG emissions and for health.

The percentage of urban trips by motorised private transport as opposed to public transport, walking or cycling (which is typically three to five times more energy efficient) ranges from 89 per cent in the USA, 50 per cent in the Western Europe, 42 per cent in high income Asia and 16 per cent in China.[42] The direct interactions between urban housing, climate change and health goes beyond indoor air pollution. Urban planning affects energy consumption and associated emissions through the building and transport sectors. This offers a scope for actions at the individual level or at the municipal or national level which have the capacity to enhance health, reduce consumption and provide economic benefits.

The world is rapidly urbanising with significant changes in our living standards, lifestyles, social behaviour and health, says Dr. Jacob Kumarsen, Director of WHO. The World Health Organisation has chosen the theme of "Urbanisation and Health" for World Health Day on April 7, 2010, in recognition of the effect urbanisation has on collective health globally and on every individual. Its goal is to draw worldwide attention to the theme of urbanisation and health and to involve governments, international organisations, business and civil society in a shared effort to put health at the heart of urban policy. The cities can meet the challenges for safe, rewarding and healthy atmosphere, which is given below:

- Promote urban planning for healthy behaviours and safety.
- Improve urban living conditions including adequate shelter and sanitation for all.
- Involve communities in local decision making.
- Ensure cities are accessible and age-friendly.
- Make urban areas resilient to emergencies and disasters.

WHO and global health community's function on climate change

- The effects of climate change on health are given attention by policy makers and the public. Human beings are already exposed to short and long term health risks of climate variability and change. Today, millions die due to climate-sensitive diseases and health conditions include malnutrition which causes over 3.5 million deaths per year, diarrhoeal diseases, which kill over 1.8 million and malaria, which kills almost 1 million.
- In May 2008, the World Health Assembly (WHA) passed a resolution on climate change and health, drawing attention to the threat posed by climate change to the achievement of the Millennium Development Goals and health equity. The resolution calls on WHO to strengthen

its work in raising awareness about the health implications of climate change, support capacity building and research in health protection from climate change in countries and urge action by the health sector.

- WHO is committed to strengthen, its scientific, normative and policy development functions its operational programmes (e.g. disease, water, sanitation and hygiene) and its support to ministries of health and other health actors throughout the world.

WHO action plan on climate change

Many policies have adopted to reduce greenhouse gas emissions and produce major health co-benefits. WHO pledges to carry out the following specific actions, to be reported both to the 124th session of the WHO executive Board in January 2009.

Strengthening of health systems

WHO emphasises the need to strengthen health systems to enable countries to deal with both gradual changes and sudden shock. It strengthens key functions and forward planning posed by climate change. It gives stress on the public health interventions such as control of neglected tropical diseases and actions to improve the environmental and social determinants of health, from provision of clean water and sanitation to enhance the welfare of women.

Outreach and advocacy

WHO will work towards greater awareness of the health implications of climate change amongst policy-makers and the general public. One such example is awareness-raising activities, such as the 2008 World Health Day (WHD) campaign on "Health protection from climate change". WHO will continue to draw the attention of the public and policy makers to the serious health risks presented by climate change, to global health and the achievement of the health related Millennium Development Goals through reports, presentations and participation in international climate change meetings and processes. The mitigation and adaption strategies will improve health and reduce vulnerability in the future.

Monitoring, surveillance & forecasting

These systems reduce vulnerability and improve health in the future. WHO is functioned for the monitoring of vulnerability to climate change and encourages member states to strengthen the capacity of their health systems for monitoring and minimising the public health impacts of climate change through adequate preventive measures, preparedness, timely response and effective management of natural disasters.

Health development
WHO is worked by the support of donor countries in vulnerable regions to facilitate better estimate of the scale and nature of the health problems associated with climate change and using strategies for best practice to implement and strengthen effective, preventive and adaptive interventions. WHO consults with member states on WHO's technical support for implications of climate change in health and health systems.

Partnerships
WHO will strengthen its close cooperation with countries and agencies inside and outside of the UN system, governments, civil society and member states. WHO has kept collaborations with health research organisations and with the United Nations (e.g. FAO, WMO, UNDP, and UNEP) and other international and national agencies that are involved in mitigation and adopting to climate change. Specifically, WHO is committed to engage actively in the United Nations Framework on Climate change (UNFCCC).

Research and knowledge
WHO will continue to support national and regional assessments, burden of disease studies, identification of adaptation strategies for the health sector, guidelines and training materials, research and reviews on early warning systems. WHO will lead the definition of a global research agenda on climate change and by convening a consultation process and international meeting that brings together from the health community, researchers, research funders and UN agencies.

Health co-benefits and risks of public health adaptation strategies to climate change
Public health is threatened by climate change and the people in all countries are seriously suffered. Fossil Fuels are burned extensively by human beings and they have been identified as key drivers of climate change since the mid-20th century.[43] A range of health impacts have been identified such as increased cardiopulmonary, infectious, allergic, mental illnesses and weather related injuries.[s]

The **Table 1** shows health impacts of climate change.

Table 1: Health impacts of climate change

Climate change effects	Examples of related health risks
Temperature changes	Cardiovascular, Cerebro-vascular and respiratory diseases
Extreme weather changes in precipitation	Injuries, dioarrhoeal diseases, malnutrition, respiratory infections, depression, anxiety
Air pollution	Respiratory, cardiovascular disease
Pollen Production	Allergic diseases
Microbial contamination & transmission	Malaria, dengue fever, Schistosomiasis, Lyme disease, hantavirus pulmonary Leptospirosis syndrome,
Reduced crop yield	Malnutrition
Displaced population	Poverty, depression, anxiety, malnutrition

The prime health impacts relate to increase in air pollution, water and food borne contamination, spread of disease, carrying vectors and extreme weather events. Some population groups of seniors, the socially disadvantaged, aboriginal people, the chronically ill and disabled people are more vulnerable to the health impacts.[44] The developing countries may face the food insecurity, socio-economic dislocation, political instability and conflict.[45]

Adaptation strategies have been identified about health risks from climate change and there are five key health adaptation strategies which have been identified by public health authorities and researchers which is given below in **Table 2.**[46]

Table 2: Adaptation strategies

Health risks	Adaptation strategies
Heat related	Urban planning and increasing in shading, establish hot-weather response plans and early warning systems. Buddy system to check on neighbours during heat waves. Improve land use planning and environmental management.
Extreme weather related	Enhance quantitative data on short term and long term health impacts on extreme weather events maintain and improve disaster management program for local public health facilities to provide rapid health needs.
Air pollution	Identify vulnerable population and increase capacity for hospital care and physician clinics.
Food and water borne disease	Maintain and upgrade water treatment, sewage and sanitation facilities, Surveillance of water and food borne diseases, Make available new drugs and treatments.
Vector and rodent borne diseases	Travel, importation and quantitative laws, Surveillance of vector population, Develop and make available new vaccines.

The WHO has called for the adoption of both GHG mitigation measures and adaptation strategies to address climate change. Many adaptation and mitigation strategies with reduced effects of climate change have been identified by international organisations such as International Panel on Climate Change (IPCC) and WHO[47] and by national and sub-national Health agencies.

Health co-benefits occur when adaptation strategies produce health advantages and original health targeted to be improved, if proper design and implementation of health adaptation strategies can maximise co-benefits and reduce negative consequences.[48] There is a comprehensive review of the research into the health co-benefits and risks of climate change with adaptation strategies and the adaptation costs and benefits are scattered in terms of sect oral and regional coverage and does not directly address health co-benefits.[49] The research is needed on economic costs to implement the adaptation strategies and expected benefits from sea-level rise and impacts on coastal areas.[50]

Adaptation strategies work for health co-benefits

How to work and take step for improvement of the health co-benefits from adaptation strategies are given below:

- Health co-benefits from adaptation strategies increase social capital. Health co-benefits identified the adaptation strategies to improve social capital. 'Social Capital' refers to "the ability of actors to secure benefits by virtue of membership in social networks or other social structures.[51] A low level of social capital or social isolation is regarded as contributing to the vulnerability of population groups who are excluded from access to resources needed to protect health.[52]

- Risks to health from adaptations that increase social capital. Social networks may reinforce misperceptions about possible health effects of heat waves and this could act as a barrier to adaptation. Strong bonding networks could effect on vulnerability to the health impacts of climate change and social capital suggesting that it is more complex than simply positive for climate change vulnerability.[53]

- In order to increase health co-benefits from adaptation strategies what is needed is to improve urban design and planning. Adaptation strategies are designed to decrease extreme heat events and air pollution in cities. Strategies are increased or improved shade and green spaces, smarter road design, development of walkable neighbourhoods, the creation and maintenance of bike paths and improvement to the public transportation.[54] Strategies to reduce heat stress reduce air pollution and better urban planning may reduce GHG emissions, if designed appropriately.[55] These adaptation actions may convey long term health co-benefits by lessening the rate of climate change and its associated health impacts.

- Indirect health co-benefits improve through increased public health adaptation to climate change that support and strengthen the resources, infrastructures, information systems and processes. Adaptation on health impacts in most countries, climate change related investment to prepare the health sectors cannot directly be attributed to a warming world.[48] The benefits of health adaptation strategy and climate change development are often directly transferable to community and regional activities to improve population health. These public health functions constitute for public health agencies at different levels- population health assessment, health surveillance, health promotion, health protection, disease and injury prevention.[56] By adaptation strategies for climate change impacts, communities will improve their public health systems and public health infrastructure.

Adaptation refers to changes in "processes, practices or structures

to moderate or offset potential damages or to take advantage of opportunities associated with changes in climate" and involves adjustments to decrease the vulnerability of communities and regions to the impacts of climate change and variability.[57]

Adaptation and sustainable development strategies to health

Adaptation strategies are essential within and between nations to reduce the health impacts of climate change and narrow health inequities. Poverty, inadequate public health infrastructure and other factors create problem for adaptive capacity in developing countries. The most cost effective and urgently needed adaptation strategy for the developing world against the global warming crisis is the rebuilding and maintaining public health infrastructure.[58] Equity has been identified by WHO as a major factor in a community's adaptive capacity and developing countries contain 90 per cent of the world's disease burden and 10 per cent of the world's health resources.[59]

Adaptive and sustainable development strategies are formulated in collaboration between the health sector and all other sectors to ensure the long-term health of all communities of the world. For example, the fossil fuels are projected in increasing greenhouse gas emissions from

1999 to 2020.[60] Harrison (2000) states that, the Millennium Development Goals Report is to ensure Environmental sustainability with a target to "integrate the principles of sustainable development into country policies and programmes and reverse the loss of environmental resources". Environmentally sustainable strategies are to preserve forests, rivers and other ecosystems. Greenhouse gas eradication is possible through the implementation of a variety of new technologies and practices.[58] However, implementation of these new and improved technologies and methods are limited by economic, political, institutional, financial and behavioural barriers.[61]

The IPCC defines adaptation as the, "Adjustment in natural or human systems to a new or changing environment. Adaptation to climate change refers to adjustment in natural or human systems to a new or changing environment. Adaptation to climate change refers to adjustment in natural or human systems in response to actual or expected climatic stimuli or their effects which moderates harm or exploits beneficial opportunities. Various types of adaptation can be distinguished, including anticipatory and reactive adaptation, private and public adaptation and autonomous and planned adaptation."

The United Nations Framework Convention on Climate Change (UNFCC) refers that, "All parties shall take climate change considerations into account, to the extent feasible, in their relevant social, economic and environmental policies and actions and employ appropriate methods, for example, impact assessments, formulated and determined nationally, with a view to minimising adverse effects on the economy on public health and on the quality of the environment of projects or measures undertaken by them to mitigate or adapt to climate change."

Mitchell and Tanner (2006) defined adaptation as an understanding of how individuals, groups and natural systems can prepare for and respond to climate or their environment. Health adaption approaches include:-

- Monitoring emerging health risks;
- Planning urban adaptation strategies such as planting trees to minimise heat, build up in cities and manage storm water or promoting the use of cool roofs to reduce energy needs and improve air quality.
- Preparing emergency response plans, which include providing cooling centres for extreme heat events.
- Improving public communication during specific health risks such as extreme heat events or low air quality days.

Adaptations to climate changes refer to adjustment in ecological, social or economic systems. National action programmes and plans include environmental safety, desertification, food management, protection of the ozone layer, ecological public education and the UNFCCC are to design and implement adaptation measures to improve and develop systematic observation network to enhance public awareness. Adaptation to the adverse effects of climate change is vital in order to reduce the impacts of climate change that are happening now and to increase resilience to future impacts.[62]

The adaptive measures include public health response in changing patterns of disease transmission and natural disasters such as setting up effective surveillance and response systems, formulating integrated measures for rapid response after natural disasters and strengthening institutional capacity.

Climatic health mitigation

Climate mitigation is any action taken to permanently eliminate or reduce the long-term risk and hazards of climate change to human life, and property. IPCC defines, "An anthropogenic intervention to reduce the sources or enhance the sinks of greenhouse gases". In the United Nations Framework Convention to Climate Change (UNFCCC), three conditions are made towards the goal of greenhouse gas stabilisation in the atmosphere:

• That it should take place within a time-frame sufficient to allow ecosystems to adapt naturally to climate change;
• That food production is not threatened; and
• That economic development should proceed in a sustainable manner.

According to UNFCCC, the mitigation process is to limit climate change by reducing the emissions of GHGs whereas adaptation aims to alleviate the adverse impacts through a wide range of actions. The mitigation strategies mostly involve identification and selection of actions to reduce GHG emissions. The emissions are occurred by the human activities and it is necessary that such strategies should promote sustainable and equitable development while reducing the concentration of GHGs. Industrialised countries have only 25 per cent of the world's population, but are responsible for most of the current and past global GHG emissions including 75 per cent of carbon dioxide emissions.[63] The reduction and eventual elimination of greenhouse gas emissions is an increasingly prominent issue in the United States and the effects of climate change become more evident. Almost all scientists agree that greenhouse gas emissions from human activities are adding to the overall warming of the climate.

The main goal is to reduce GHG emissions which means to limit carbon-dioxide emissions. Carbon-dioxide is emitted to the atmosphere by three main sources- energy production and use, industrial activities and land use changes. The Intergovernmental Panel on Climate Change (IPCC)'s special report on Renewable Energy sources and climate change mitigation noted the significant potential of renewable energy to mitigate climate change and to provide wider benefits. Several studies indicate, how greenhouse gas mitigation measures in the electricity generation can benefit health due to changes of air pollution emissions. The potential 'health co-benefits' will be from greenhouse gas emission reduction policies in the transport, built environment, electricity generation and agriculture. Health benefits would result from less polluted urban air, increased levels of physical activity, better insulated housing and reduced intake of high energy , high fat -animal -foods.[64]

Cycling, walking and rapid transit systems are associated with a wide range of potential health benefits that climate assessment needs to consider more systematically. Health benefits may include; physical activity from walking and cycling, which can help prevent heart diseases, some cancers, type 2 diabetes and some obesity-related risks; lower urban air pollution concentrations; lower rates of traffic injury risks and less noise stress. Transport systems that prioritise active transport and rapid transit systems, along with better urban land use, also can help improve access for vulnerable groups, including children, the elderly, people with disabilities and lower wage earners, enhancing health equity. This new WHO report, part of the health in the Green Economy series, considers the evidence regarding health co-benefits and risks of climate change mitigation strategies for transport are reviewed by the Intergovernmental Panel on climate Change.[65]

The mitigation measures may include activities like building energy- efficient facilities use of natural ventilation and day light-on-site rain water capture and treatment, appropriate waste/sewage treatment, improving access to health care by mass transport and expanded use of tele-health. For example, rain water harvesting is one conservative measure widely promoted in the WHO South-East Asia Region.[66]

Millennium Development Goals (MDGs) on health

At the millennium summit in September 2000, the largest gathering of world leaders in history adopted the UN Millennium Declaration, committing their nations to a new global partnership to reduce extreme poverty and setting out a series of time bound targets, with a deadline of 2015 that have become known as the Millennium Development Goals.

The Millennium Development Goals targets for addressing extreme poverty, income poverty, hunger, disease, lack of adequate shelter, education and environmental sustainability. They are also basic rights. The rights of each person on the planet are to health, education, shelter and security. Its goals are:-

- Eradicate extreme hunger and poverty
- Achieve Universal primary education
- Promote Gender Equality and Empower Women
- Reduce child mortality
- Improve maternal health
- Combat HIV/AIDS, Malaria and other diseases
- Ensure Environmental sustainability
- Develop a global partnership for development

To achieve the Millennium Development Goals (MDGs) enormous progress has been made. Global poverty continues to decline, more children are attending primary school, children deaths have dropped dramatically, safe drinking water has been greatly expanded and targeted investments in fighting malaria, and AIDS and tuberculosis have saved millions.

Secretary-General Ban Ki-Moon has established a 'UN system task team' to coordinate preparations for beyond 2015. In July 2012, Secretary-General announced the 27 members of a High-level panel to advice on the global development framework beyond 2015. The work of the panel will reflect new development challenges. The post-2015 agenda will reflect new development challenges. The outcome document "The future we want" called for the creation of an intergovernmental open working group on Sustainable Development Goals to develop a proposal for consideration by the 68th session of the General Assembly.

The eight Millennium Development Goals build on agreements made on UN's conferences in the 1990s represent commitments to reduce poverty and hunger and to tackle ill-health, gender inequality, lack of education, lack of access to clean water and environmental degradation.

WHO's work on the MDGs:-
- WHO supports national and regional efforts to achieve the MDGs through normative technical work in a cross cutting way or by providing Management Support.

- WHO works with other organisations of the United Nations Systems' to improve the quality of country health data and aims to build capacity in countries to collect, analyse and act for MDG target.
- WHO Monitors for public health achievement goals at country level include immunisation coverage for new antigens, prevalence of interventions against these diseases.
- WHO strengthens technical collaboration with countries. WHO supports the cooperation strategies in meeting MDG targets.
- WHO and the World Bank coordinated the high level forum on the health MDGs.

International public health policy on climate change
- Threats to health are cited to justify actions to mitigate or adapt to climate change. A comprehensive strategy is urgently needed for both mitigation and adaptation, because the health community has a duty to counter emerging threats and increased attention to climate change offers opportunities to focus on the most disadvantage population's current needs. A global problem requires a strategy of international dimensions that can translate into regional and local actions. It impacts in one location such as infectious disease epidemics or population displacements caused by droughts or rising sea-level, quickly spread across national borders. Coordinated investments in preventive measures contribute to 'the global public good' of reducing the risk of health emergencies. Adaptation to climate change is essentially in public health protection. The challenges of rapid environmental, demographic and social changes for greater emphasis on disease prevention, providing a better balance with the current focus on curative and reactive measures. Climate change and other environmental stresses should help to re-focus political and financial commitments to implement these measures and preventive actions.
- The environmental health threat is the decline in global fresh water resources, extraction and contamination. Climate change is expected to worsen this decline in water quality and quantity in particularly dry regions such as the eastern Mediterranean and North Africa. Scaling of water and sanitation services and providing point-of-use disinfection would reduce the current burden of disease and the health impacts decreasing water supplies. Governments should protect health by strengthening and enforcing their regulatory frameworks to ensure the safe use of new water resources that will

become important. Waste water and grey water in agriculture and aquaculture.[67]

- Effective surveillance and response systems are essential in managing any infectious disease, but they are important under conditions of rapid change. Climate change also strengthens the case for reinforcing response systems for infectious disease out breaks, including pre-defined action plans and maintenance of the control resources and personnel capacity necessary to mount effective responses.

- The impacts of current and future natural disasters could be reduced by the health sector defining integrated measures that address the root causes of vulnerability and planning for effective responses after such events. Post- flooding health effects can be reduced by adequately planned and funded health-sector responses, including interventions to control outbreaks of vector-borne and water-related diseases.

- Since 2000, WHO has worked with the world meteorological organisation and the UN Environment programme to raise awareness of climate change implications for the health sectors of highly vulnerable regions. The necessary next step is to inform and support national health actors in taking concrete actions to protect health. In 2006, the WHO approaches new project to project health in changing climates.[68] UN Development programme and global environmental facility initiative is being implemented in seven vulnerable countries to identify priority health risks from climate change given local environmental socio-economic and health contexts. This initiative is important preparation for the wider action that will be necessary to meet the world-wide challenge of adapting to health risks from climate change.

- "Safe investment" is another adaptation funds method for climate change within the health sector. Climate change has the capacity to suppress agricultural yields, with the greatest risks in Africa,[69] where malnutrition is already the largest single contributor to disease burdens.[70]

- The public health community needs a true preventive strategy to ensure the maintenance and development of healthy environments for local to global levels. In the long term, sustainable development and protection of ecosystem services are fundamentally necessary for human health.[71] Improving access to public and active transport would greatly reduce Carbon Dioxide emissions, while also cutting the 8, 00,000 annual global deaths from outdoor air pollution, the 1.2 million annual deaths from traffic accidents and the 1.9 million deaths from physical inactivity.[70] Changing the poorest communities' domestic energy technologies could reduce the 1.5 million annual indoor air pollution deaths.[72] The critical next stage for the health community is more direct and engagement with the environment and economic sectors to ensure that health-supporting technological and development choices are implemented.

References
1. *Campbell - Lendrum, D et al., Environmental burden of series, 2007.*
2. *Oke T. R (1973): city size and urban heat island, Atmospheric Environment.*
3. *The world health report 2004 changing history: WHO, 2004.*
4. *JA Patz, D Campbell - Lendrum, T Holloway, JA Foley. Impact of regional climate change on human health, Nature 2005, 438: 310-7*
5. *Albritto DL, Meiro- Filho LG, Technical summary, in climate change, 2001.*
6. *Patz, JA et al., Effects of environmental change on emerging parasitic diseases, 2000.*
7. *WHO, World Health Report 2002: Reducing risks, promoting healthy life, WHO, Geneva, 2002.*
8. *Castello et al., 2009.*
9. *Pachauri and Reisinger, 2007.*
10. *Based on data from the United Kingdom Government met office, Had CRUT 3, annual time series. Hadley Research centre, 2008.*
11. *Robine JM et al., Death toll, exceeded 70,000 in Europe during the summer of 2003, Les competes Rendus/Series Biologies, 331, 2008, pp. 171-78.*
12. *Arnell NW, Climate change and global water resources. Global environmental change- Human and policy dimensions 2004, 14, pp. 31-52.*
13. *Climate change 2007. Impacts, adaptation and vulnerability, Geneva IPCC, 2007.*
14. *Zhou XN et al., American journal of tropical medicine and Hygine, 78, 2008, pp. 188 -194. 15. Smit KR, Balkrishnan K. Mitigate climate, http://www.thecommonwealth. org/files/190381/file name/4 kirksmith, 2009pdf.*
16. *World population prospects, the 2010 revision, New York, http://esa.un.org/UNpd/ Wpp/index. htm.*
17. *Haines A, Cassels A.2004.*
18. *IFRC. World disaster report, International Federation of Red Cross and Red Crescent societies, Geneva,2007.*
19. *Bartleft, 2008.*
20. *Boffa, 1999, Food and Agricultural organisation. Roe. D et al., 2006, International Institute for Environmental and Development. Arnold JEM, 2008, managing ecosystems to enhance the food security of the rural poor.*
21. *Cannon T.2002, Gender and Climate hazards in Bangladesh.*
22. *FAO, 2002, The state of food insecurity in the world. Food and Agriculture organisation. 23. Brody A. et al., 2008, Gender and climate change.*

24. *Chant S.2007, Gender, Cities and the Millennium Development Goals in the global south.*

25. *Drager N, Fidler DP, foreign policy, trade and health, 2007.*

26. *WHO Framework convention on Tobacco control, 2003.*

27. *UNEP Montreal Protocol, 1987.*

28. *Weiss R, Mc Michael A. Social and environmental risk factors in the emergence of infectious diseases, 2004.*

29. *Pauly D. Watson R, Alder J, Global trends in world fisheries, 2005.*

30. *Moores, Climate change, water and China's national interest, China Security 2009.*

31. *IPCC, climate change 2001: Synthesis Report Cambridge: WMO/UNEP, 2001.*

32. *Mc Michael AJ, Githeko A. Human Health, climate change, 2001.*

33. *Wall DH, Rabbinge R, Gallopin G, et al., Implications for achieving the Millennium Development Goals, 2005.*

34. *Stott PA, Stone DA, Allen MR Nature, 2004. Human contribution to the European heatwaves of 2003.*

35. *Aniello C, Morgan k, Busbey A, Newland L, Mapping microurban heat islands, 1995.*

36. *Pielke RA, Landsea C, Mayfield M, Laver J, Pasch R, 2005, Hurricanes and global warming.*

37. *Gulbler Dj, Meltzer M, Adv virus Res, 1999; Review impact of Dengue fever on the developing world*

38. *Daszak P, Cunningham AA, Hyatt AD, Actatrop, Zool, Feb 23; Review anthropogenic environmental change and the emergence of infectious diseases in wild life.*

39. *Ezzati M, Lopez a, Murray C, (eds.), Comparative quantification of Health risks, Geneva, World Health Organisation, 2004.*

40. *WHO, The World Health Report 2002, Geneva.*

41. *WHO, fuel for life: Household Energy and Health, Geneva, 2006,*

42. *Newman P, ken worthy J. sustainability and cities 1999.*
    Ken worthy J. Transport Energy use and Greenhouse gases in urban passenger Transport system, 2003.

43. *Hegerl GC, Zwiers FW, Braconnot P. et al., 2007: understanding and attributing climate change.*

44. *Haines A, Patz JA, JAMA, 2004. Review Health effects of climate change. Seguin J, editor, 2008. Human health in a changing climate.*
    Friels, Bowenk, Campbell Lendrum D, Frumkin H, Mc Michael AJ, Rasanathan k, Annu Rev, Public Health, 2011.

45. *Costello A, et al., 2009, Managing the health effects of climate change.*

46. *Confalonieri et al., 2007, climate change impacts, adaptation and vulnerabilities.*
    Keim ME-preventing disasters: public health vulnerability reduction as a sustainable adaptation to climate change, 2011.
    Cash D. 2011. Climate change adaptation report. Beston Executive office of energy and Environmental Affairs and Adaptation Advisory Committee.

47. *Mc Michael AJ, et al., 2003, climate change and human health: risks and responses, WHO, Geneva.*

48. *Ebi et al., vulnerable.*

49. *Adger et al., 2007, climate change impacts adaptation and vulnerability.*

50. *Nicholls RJ, Tol RSJ. Impacts and responses to sea level rise.*

Pay the price: the economic impacts of climate change for Canada, Ottawa: QNRTEE, 2011. 51. *Portes A 1998. Social capital: its origin and applications in modern sociology.*

52. *Cutter et al., 2003; social vulnerability to environmental hazards.*
53. *Wolf J. et al., 2010. Social capital, individual responses to heat waves and climate change adaptation.*
54. *Stone et al., 2010 urban form and extreme heat events and air pollution in cities.*
    Younger et al., 2008. The built environment, climate change and health opportunities for co- benefits.
    Wood cock et al., 2009. Public health benefits of strategies to reduce greenhouse gas emissions: urban land transport.
55. *WHO, 2009, protecting Health from climate change.*
56. *Frank et al., 2011. The future of public health in Canada.*
    Frumkin et al., 2008. Review climate change: the public health response.
57. *IPCC, 2001b.*
58. *IPCC, 2001.*
59. *WHO, 2000.*
60. *Haines et al., 2006.* 61. *Ambuj et al., 2006.*
62. *UNFCCC, 2011. Adaptation. http://UNFCCC.int/adaptation/items/4159.php-26 Sep, 2011.*
63. *Davidson OR. 2011- Strategies to mitigate climate change in a sustainable development framework.*
64. *Haines A. et al., (2009), public health benefits of strategies to reduce GHG emissions; overview and implications for policy makers.*
65. *IPCC, 2007.*
66. *WHO, Zoll. Regional office for South East Asia; sustainable development and healthy environment; water, sanitation and healthy.*
67. *WHO guidelines for the safe use of waste water, excreta and grey water-3rd edition, WHO, Geneva, 2006.*
68. *Climate change and health, WHO, Geneva, 2006.*
69. *M Parry et al., 2005, global food supply and risk of hunger.*
70. *M Ezzati et al., 2004: comparative qualification of health risks.*
71. *Millennium Ecosystem Assessment, WHO, Geneva, 2005.*
72. *Fuel for life; household energy and health, WHO, Geneva, WHO, 2006.*

5. CLIMATE CHANGE, NATURAL DISASTERS AND UN

Introduction

Climate change means the alternation of the world's climate which is causing through fossil fuel burning, clearing forests and other practices that increase the concentration of greenhouse gases (GHG) in the atmosphere.[1] According to the UNFCCC that climate change is the change that can be attributed, "directly or indirectly to human activity that alters the composition of the global atmosphere and which is in addition to natural climate variability observed over comparable time periods".[2] The Intergovernmental Panel on Climate Change (IPCC) defines "climate change" as a change in the state of the climate that can be identified by changes in the mean and the variability of its properties and that persists for an extended period typically decades or longer.[3]

In the past two decades, the global climate is changing due to changes in climate variability and weather extremes. The climate variability and extreme events, much of the science presented from the Intergovernmental Panel on Climate Change (IPCC), the World Meteorological Organisation (WMO) and the United Nations Environment programme (UNEP) established this panel in 1988, to respond to policy makers for the best scientific, technical and socio- economic information on climate change primarily in peer-reviewed scientific literature.

Climate change will affect disaster risks in two ways, first, increase in weather and climate hazards and second, through increase in the vulnerability of communities to natural hazards, particularly through ecosystem degradation, reductions in water and food availability, and changes to livelihoods. Over the period 1991-2005, 3,470 million people were affected by disasters, 9,60,000 people died and economic losses were US $1,193 billion.[4] Over the last two decades (1988-2007), 76 per cent of all disaster events were hydrological, meteorological or climatological in nature; these accounted for 45 per cent of the deaths and 79 per cent of the economic losses caused by natural hazards.

According to IPCC, many long-term precipitation trends (1900-2005) have been observed, including significant increases in eastern parts of North and South America, Northern Europe and Northern and Central Asia, and more dry conditions in the Sahel and Southern Africa, throughout the Mediterranean region and in the parts of Southern Asia. The frequency of heavy precipitation events has increased over most land areas, which is consistent with global warming and the observed increases of atmospheric water vapour.[5]

Natural disasters can reduce development progress, the effectiveness of aid investments and slow progress towards the achievement of the Millennium Development Goals (MDGs). There is evidence that natural disasters disproportionately affect developing countries between 1991 and 2005, more than 90 per cent of natural disaster deaths and 98 per cent of people affected by natural disasters were from developing countries.[6] Disasters are increasing in number and size every year due to rapid population growth, urbanisation and climate change. The natural hazard events such as earthquakes, tsunamis, cyclones and floods become natural disasters.

The process of climate change and the natural disasters will lead the scale and complexity of human mobility and displacement.[7] The United Nations High Commissioner for Refugees (UNHCR) is a leading agency of the United Nations responsible for and possessing the area of forced displacement. It is projected that climate change will over time trigger larger and more complex movements of population, both within and across borders, and has the potential to render some people stateless. UNHCR would encourage more reflection on the humanitarian and displacement challenges that climate change will generate and most of the displacement provoked by climate change manifested, for example, through natural disasters, could remain internal in nature for developing the legal framework for the protection of Internally Displaced Persons (IDPs).[8]

The 27th International conference on the Red Cross and Red Crescent of 1999 adopted the Action plan that "the International federation, while drawing upon existing research and the competence of relevant International bodies, will undertake a study to assess the future impact of climatic changes upon the frequency and severity of disasters and the implications for humanitarian response and preparedness". The International Federation was assisted in the preparation of the study by the Red Cross/Red Crescent centre on climate change and Disaster Preparedness of the Netherlands Red Cross (Climate centre). In June

2002, the Netherlands Red Cross established the climate centre in order to raise awareness, develop risk reduction policy and programmes in relation to climate change and disaster preparedness and to advocate the dialogue on climate change and the humanitarian consequences with policy makers on all levels. The third Assessment Report (TAR) of IPCC "climate change 2001" provides the most comprehensive assessment of the current scientific knowledge on climate change.

It is emergency, the increasing frequency and severity of catastrophes that strike human kind and how to prevent and mitigate the effects of disaster. Effective mitigation measures and the benefits of vulnerability reduction greatly exceed the costs. Cost benefits analyses suggest that appropriate investments in prevention could reduce the burden of disasters.

The UNESCO (United Nations Educational, Social and Cultural Organisation) stresses the merit of mitigation and preventive measures through scientific understanding and technology. This approach depends on imparting information effectively, involving local communities, making disaster prevention part of education and raising the awareness of the public. In June 2006 at UNESCO Headquarters, with the United Nations International Strategy for Disaster Reduction, a world campaign had the privilege on education, entitled 'Disaster Risk Reduction Begins at School'. The campaign aims to promote disaster reduction education in school curricula and to improve school safety by encouraging the application of strict construction standards, through knowledge, education, information and public awareness. This emphasises the positive effects of education on disaster risk management which is experienced by the countries. The Fourth Assessment Report was completed in 2007 and is included more information on extreme weather events as well as on adaptation to climate change. The climate change science describes the human greenhouse gas emissions and the main projections for the future, particularly in terms of global temperature, sea level rise, temperature and precipitation.

The poor in climate change
The poor accept higher levels of risk due to limited opportunities and resources to their income and live and work in informal settlements located in high risk areas. Overall, approximately 72 per cent of African's urban population lives in informal settlements, where investment in drainage infrastructure that can reduce flood risk is often lacking and existing infrastructure is inadequately maintained.[5] Among the poor, disabled,

elderly, orphans, windows and other vulnerable and marginalised groups are more likely to be affected by weather related events. In many cases, women are more affected than men due to their lower mobility and cultural sensitivities that may prevent them from seeking livelihood opportunities away from high-risk areas or to use shelters during extreme events. As a result, for example, some 91 per cent of fatalities in Bangladesh after cyclone Gorky were of women.[10]

Climate change could affect poverty targets directly as well as indirectly by curbing economic growth. Recent modelling studies indicate on global poverty about 10 million additional poor under climate change scenarios by 2055, assuming steady annual economic growth of 2.2 per cent.[11] The economic growth is also sensitive to temperature rise which could significantly increase the number of poor.[12] Overseas Development Institute (ODI) also indicates significant number of poor living in hazard prone countries by 2030.[13] These global studies also suggest that an immediate reduction of greenhouse gases would only have a significant impact on poverty beyond 2100. This is due to the longevity of many greenhouse gases in the atmosphere and inertia in the climate system, under-crossing the urgent need to implement resilience- or adaptation- measures targeted towards the poor.[14]

Rural areas are expected to have the highest number of poor, poor population in urban areas are expected to suffer proportionally more under projected extreme dry events due to their vulnerability to food price increase. An estimated 16 per cent increase in poverty is expected in urban areas compared to a 12 per cent increase amongst rural population.[15] Already, prolonged droughts, land degradation, development patterns and conflict in the Sahel and Horn of Africa have displaced population into more marginal land. Under extreme dry events, a highly vulnerable country like Zambia could see an additional 4.6 per cent of its population impoverished by the end of the century.[16]

Another key challenge relates to the fact that many countries with the highest projected future poverty risk are also the ones with the lowest level of current risk preparedness.[17] The recent ODI report concludes that, without concerted action, some 325 million people could be living in 45 countries most exposed to hazards by 2030, highlighting the close links between poverty, hazards and risk governance, and the need to integrate social protection into development strategies.[18]

The poor are already resilient both by nature as well as by necessity; however, they need further funding, information and support to escape poverty traps and to better cope with weather-related disasters because poverty and vulnerability are so closely intertwined, climate and disaster resilient development must be central to the global goal of ending poverty and promoting shared prosperity.

Climate change projections - affect key sectors

The projections of future climate patterns are based on computer- based models of the climate system that incorporate the important factors of atmosphere and ocean including greenhouse gases from socio-economic scenarios for the coming decades. The IPCC has examined different models on the basis of the evidence and estimated that by 2100.[19]

- The global average surface warming will increase by 1.1 - 6.4°C.
- The sea level will rise between 18 and 59cm.
- The oceans will become more acidic.
- Hot extremes, heat waves and heavy precipitation events will continue to become more frequent.
- There will be more precipitation at higher latitudes and will be less precipitation in most sub-tropical land areas.
- Tropical cyclones such as typhoons and hurricanes will become more intense, with larger peak wind speeds and more heavy precipitation associated with ongoing increase of tropical sea surface temperatures.

The IPCC Fourth Assessment Report of the Working Group II "Impacts, Adaptation and vulnerability" describes the effects of climate change including extreme events. The effects of key sectors are as follows:

Water

Drought affected areas will become widely distributed. Heavier precipitation events are very likely to increase frequency leading to higher flood risks. By mid-century, water availability will decrease in mid-latitudes, in the dry tropics and in other regions supplied by melted water from mountain ranges. More than one-sixth of the world's population is currently dependent on melted water from mountain ranges.

Food

Some mid-latitudes and high-latitudes areas will initially benefit from higher agricultural production, for many others at lower latitudes, especially in seasonally dry and tropical regions, the increase in temperature and frequency of droughts and floods are likely to affect crop production negatively, which could increase the number of people at risk from hunger and increased levels of displacement and migration.

Industry, settlement and society
The most vulnerable industries, settlements and societies are located in coastal areas and river flood plains and whose economics are closely linked with climate sensitive resources. Extreme weather events become more intense or more frequent; the economic and social costs of those events will increase.

Health
The projected changes in climate are likely to alter the health status of millions of people, including through increased deaths, disease and injury due to heat waves, floods, storms, fires and droughts. Increased malnutrition, diarrhoeal disease and malaria in some areas will increase vulnerability to extreme public health and development goals will be threatened by longer term damage to health systems from disasters.

Climate change and its impacts can be lessened by reducing emissions of greenhouse gases such as CO_2 and methane. In 1992, the UN Framework Convention on Climate Change (UNFCCC) was established to try and stabilise greenhouse gas concentrations sufficiently to prevent dangerous anthropogenic interference with the climate system. As scientific evidence for climate change grew stronger during the 1990s, parties to the UNFCCC created the Kyoto Protocol, which includes mandatory reductions of greenhouse gas emissions for developed countries. However, projected climate change and its effects cannot be prevented entirely.

Therefore, the IPCC scientists' advice a combined strategy of reducing greenhouse gas emissions and adaptation to the impacts of climate change.[20] Adaptation can reduce the adverse effects of climate change. Adaptation will work best if it is integrated into policies, which deal with current climate- related risks in the context of ongoing sustainable development and disaster risk reduction.[21]

Climate change occurring disasters and risk factors
The IPCC and international authority on the science of climate change observed and concluded that the climate change is already happening due to increase in the global average temperature by 0.6°C over the last century, which largely can be linked to increase of greenhouse gas concentrations in the atmosphere. Due to human activities, a growing number of physical

and biological responses such as the melting of glaciers in most regions of the world and change in the behaviour and distribution of species are being detected.[22] The global average temperature will increase by 1.4 to 5.8°C by the end of 21st century in comparison to 1990 levels and this magnitude change for at least 1000 years.[23] This temperature is associated with local and regional changes in climatic conditions. Climate change impacts on exposure to hydro-meteorological hazards such as storms, floods and droughts which influence the risk factors on vulnerability to environmental hazards occurring disasters. The following changes for managing disasters:-

Changes frequency of climatic extremes
Disasters are associated with extreme events. Climate change will lead to higher maximum temperatures and heat waves over almost all land areas.[24] Heat waves can impact even in developed countries in 2003, when a heat wave with temperatures rising above 40°C[25] was linked to more than 35,000 excess deaths in France, Italy, Netherlands, Portugal, Spain, Germany and the United Kingdom.[26] Human induced greenhouse gas emissions have at least doubled the risk of this heat wave occurring.[27]

Climate change is causing a higher probability of floods, landslides and soil erosion with associated damages and for most mid-latitude continental interiors increased summer drying for the risks of droughts and the wind increased cyclones, hurricanes and typhoons.[28] If the period of 1990-1999 was compared to 1950-1959, the economic losses occurred were 14 times and increased the disasters due to many factors, including improved information and reporting, uncontrolled urbanisation in hazard prone areas and environmental degradation, increased weather related hazards and the geological hazards such as earthquakes, volcano eruptions reflects long time-scale variations.[29]

Changes in climatic conditions and variability
Due to devastating climatic extremes, climate change occurs; climatic conditions and climate variability increase risk factors and the ability to cope with the aftermath of climate extreme events and other natural hazards. For example, the impact of the affected production from an agricultural area if occurs drought or a flood on livelihoods represents a more difficult challenge for climate sensitive livelihoods, if remedial measures are not taken. Rural livelihoods and food security in Africa are particularly threatened by climate change as shown in a comprehensive assessment by IIASA.[30]

Emerging threats and problems

Most impacts of climate change are existing threats and many events may be threshold events. For example, such changes in combination with sea-level rise are likely to increase the vulnerability of islands nations, where coral reefs sustain fisheries and provide protection from storm surges. Such threshold events may help to maintain the health of ecosystems by reducing compounding impacts. Glacial melt may lead to glacial lake outburst floods with destructive down-stream which effects the water in natural dams. The spread of climate sensitive diseases occurs in regions where these diseases did not occur before the appearance of species in areas.

Many changes can be anticipated and planned through a combination of scientific research, monitoring and foresight planning, if the thresholds are known. The climate system is a complex one between the ocean, atmosphere and terrestrial ecosystems. The more rapidly the climate system is forced by increasing levels in greenhouse gases. Therefore, reducing greenhouse gas emissions and other impacts on the environment represents a precautionary measure, which in the long-term will reduce the risk of surprises with potential adverse consequences.

Hazards are part of nature but often turn into disasters as a result of human actions or inactions. Severe flooding may be exacerbated by deforestation and building in natural flood plains. This is the threat of global climate change and rising sea levels as a result of increased greenhouse gas concentrations in the atmosphere which the Intergovernmental Panel on Climate Change (IPCC) says is caused by human activity. The world's 10 largest cities, among them eight are located on coast, within 100 kms of a sea coast where people live permanently, if predictions of disastrous change in weather patterns by the IPCC come true will be worst affected.

The IPCC predicts that climate change will increase droughts, heat waves and fires in some areas and tropical storms and higher precipitation will increase floods, landslides and mud slides. Natural disasters are unjust because they strike hardest at some of the world's poorest countries like Honduras, Guatemala and Nicaragua that lost more than their entire gross domestic product in hurricane Mitch in 1998 and are still struggling to recover.

Such disasters are among the biggest obstacles in achieving the UN's Millennium Development Goals for poverty reduction. The scope of socio- economic impacts of natural disasters is bringing slowly about disaster relief and disaster prevention with risk reduction and the international community realises the vulnerability to disasters is linked to poverty and vice versa. UNESCO is closely involved in raising public awareness and improving education about the natural disasters, helping vulnerable populations to cope with risk.

Climate change, disaster and migration
The International Organisation for Migration (IOM) effort's to support vulnerable affected by environmental hazards through Disaster Risk Reduction (DRR) and Climate Change Adaptation (CCA) active to sustainable development. The IOM's programmes around the world have the effectiveness of DRR and CCA for reducing risk and vulnerability and for improved management of migration, particularly in times of crisis.[32] The IOM argues that migration and environmental migration are cross cutting issues that need to be effectively managed and fully recognised and mainstreamed into sustainable development strategies at all levels and in DRR and CCA strategic frameworks, including the Hydro Framework for Action and the United Nations Framework Convention on Climate Change (UNFCCC) process.

Environmental migration in a context of vulnerability
The IOM's strategic response to the challenges posed by increased human mobility in a context of heightened risk and vulnerability as a result of changing climate and other structural factors such as poverty, demographic pressure, weakening ecosystems and unmanaged urbanisation has been very tough. There has always been a fundamental interdependency between migration and the environment, but in today's environment, the climate change adds new complexity to this nexus.[32]
- Natural hazards, including geological, hydrological and climatic events when destroying lives, homes and livelihoods can lead to sudden large scale movements of people to the nearest safe place. In 2008 alone, over 46 million people were displaced by sudden-onset disasters, of which 20 million were climate-related disasters.[33]
- Environmentally induced displacements, although most visible because of media attention, IOM considers that most environmental migration is and will increasingly be taking place in the context of slow onset processes.
- Migratory responses to environmental degradation, especially at early and intermediate stages, are often temporary in nature and serve as an adaptation strategy to environmental change.

- At advanced stages of environmental degradation, when livelihoods can no longer be sustained, communities often resort to permanent migration to less affected areas or to urban and semi-rural centres.[34] In the worst cases, the third country may be needed.
- Inadequately managed environmental migration can negatively affect the stability of environments and communities by increasing pressures on land and water resources. Dhaka's slum population, estimated at 3.4 million, is receiving as many as 4, 00,000 migrants a year, most of them poor, arriving from rural and coastal areas experiencing increasing environmental hardship.[35]
- Whether a survival or adaptation strategy migration is not an option open to everyone.

As migration requires resources, the most vulnerable people are often not able to move, it depends on incomes, social networks, local patterns of gender relations and the perceived alternatives to moving.

The application of multidimensional vulnerability assessments including human security assessments has shown that human vulnerability to environmental factors is compounded by a broad range of issues, including demographic pressure, poor urban governance, declining ecosystems, poverty, conflicts and vulnerable rural livelihoods.[36] Climate change is already undermining the livelihoods and security of many people. Over the last two decades the number of recorded natural disasters has doubled from some 200 to over 400 per year. Nine out of every 10 natural disasters today are climate related.[37] The Norwegian Refugee Council recently indicated that as many as 20 million people may have been displaced by climate induced sudden-onset of natural disasters in 2008 alone.[38]

Due to rise in temperature land becomes less productive, and the process of urbanisation will accelerate, generating additional competition for scarce resources and public services in cities across the globe. The vector-borne diseases will increase as a result of climate change. Increased social tension and political conflict with cost of food and energy will be frequent. In regions affected by the long-term consequences of climate change, people will also move in large numbers. Some will move to more hospitable areas in home countries while others will seek to leave their own country and enter other states. Since new forms and patterns of movement are emerging, the concepts traditionally used to categorise different types of movement are becoming increasingly blurred.[39] New legal frameworks may need to be negotiated. The representative of the Secretary-General on the Human Rights of Internally Displaced Persons, Walter Kalin has identified five climate changes that directly or indirectly cause human displacement. They are:-

- Hydro-meteorological disasters such as flooding, hurricanes, typhoons, cyclones, mudslides etc.
- Zones designated by governments as being too high-risk and dangerous for human habitation.
- Environmental degradation and slow onset disaster such as reduction of water availability, desertification, recurrent flooding, salinisation of coastal zones etc.
- The case of 'sinking' of small island states and
- Violent conflict triggered by a decrease in essential resources such as water, land, and food owing to climate change.

Hydro-meteorological disasters or environmental degradation cause internal displacement, as states have primary responsibility for their citizens, national and local authorities have a vital role to play and IDPs should receive protection and assistance in accordance with the 1998 Guiding Principles on Internal Displacement.

People cross international border that are affected by severe disasters, but they would not normally qualify as refugees in the international protection within the existing international refugee framework, nor classify as migrants. The persons who have a need for international protection under international law, but this need may be turned into an entitlement. The low-lying small island states by rising sea levels may prompt internal relocation as well as migration. The international refugee law would not automatically apply. The question of statelessness is directly implicated.[40] The human displacement is a decrease in vital resources such as water, land, food which trigger armed conflict and violence. Those fleeing to other countries could qualify as refugees or be afforded protection under regional refugee law instruments[41] or 'complementary forms of protection' under the relevant international law instruments or under the national law of receiving states.[42]

Implications for UNHCR to displacement

- The movements prompted by climate change could fall within the traditional refugee law framework, within international or regional refugee instruments or complementary forms of protection as well as within UNHCR's mandate.[43]

- Climate-related issues are projected to become more direct and common driver of conflicts in energy sources, fertile land and fresh water. The demand for protection and assistance under the refugee framework will grow. An analysis of this particular aspect of the protection regime and a search for ways to strengthen it in UNHCR's climate change strategy.

- The implications for UNHCR relates to the potentially most dramatic manifestation of climate change, that of the 'sinking island' where the inhabitants of island states such as the Maldives and Tuvalu leave their own country as a result of rising sea levels and the flooding of low-lying areas.

- UNHCR's role conflict-induced internal displacement would be triggered as a result of Cluster Approach, UNHCR has assumed global leadership of the protection cluster[44] and co-leads the global Camp Coordination and Camp Management cluster (CCCM)[45] with the International Organisation for Migration (IOM) and the Emergency Shelter Cluster[46] with the International Federation of Red Cross and Red Crescent Societies (IFRC).

- UNHCR's involvement with people who have been displaced within their own country as a result of natural disasters. UNHCR had an established presence and programme in a country that was struck by such a disaster, the office offered its support to the authorities as a sign of solidarity and as a contribution to broader international and UN relief efforts.

- Any new approach must be rights-based, since experience during the 2004 Indian Ocean Tsunami and other recent disasters have confirmed that such emergencies generate new threats to the human rights of affected populations. In terms of preventing and responding to such threats, UNHCR considers the Inter Agency Standing Committee Operational Guidelines on the Protection of Persons Affected by Natural Disasters and the related with valuable resources to address the special needs and vulnerabilities of persons forcibly displaced by the effects of natural disasters.[47]

- Climate change will affect the delivery of operations, in view of its implications in the areas of water, sanitation, agriculture, environmental protection and health. UNHCR's mandate is clear and uncontested such as in longstanding refugee situations.[48]

- There is likely to improve relationship between governments and partner organisations and increasing dependency on pre-existing partnerships with climate change-affected countries and a need to strengthen, where possible, dialogues between relevant governments and UNHCR.

1951 UN refugee convention

Environmental refugees' or 'climate refugees' concept is used for people who are obliged to leave their usual place of residence as a result of long-term climate change or sudden natural disasters. UNHCR has serious reservations with respect to the terminology and notion of environmental refugees or climate refugees.

The phrase 'refugee' is a legal term. A person who has been determined a refugee will have satisfied the criteria under the 1951 Refugee Convention, the 1969 OAU convention or UNHCR's mandate. UNHCR's opinion that the use of such terminology could potentially undermine the international legal regime for the protection of refugees whose rights and obligations are quite clearly defined and understood. Environmental factors can contribute to prompting cross-border movements and the environmental degradation is migrating people from a country of origin and seeking international protection for the grant of refugee status under international refugee law. UNHCR does recognise that there are indeed certain groups of migrants, falling outside of the scope of international protection, which are in need of humanitarian and other forms of assistance.[49]

Some states and NGO's have suggested that the 1951 refugee convention should simply be amended and extended to include people who have been displaced across borders as a result of long-term climate change or sudden natural disasters. UNHCR considers that any initiative to modify this definition would risk a renegotiation of the 1951 Refugee Convention which would not be justified by actual needs.

Disaster management and risk reduction of climate change

Most disasters or more correctly hazards that lead to disasters cannot be prevented but their effects can be mitigated. Hazards may be natural in origin, but it is the way in which societies have developed that causes them to become disasters.[50] A planning to reduce the impact of disasters is not new. The international community has made substantial effort to reduce the impact on people and livelihoods of disasters with both natural and technological triggers. Many techniques to prepare to reduce potential losses from and to respond and adapt to, hazards have been developed.[51] Disasters planning are a necessary step and is needed to realise the Millennium Development Goals (MDGs) and sustainable development.

These are the elements of a consensus that were reaffirmed at the World Conference on Disaster Reduction (WCDR) in Kobe, Japan January 18-22, 2005. At that meeting climate change was recognised as posing an immediate and long-term threat to the achievement of the MDGs and sustainable human development and as such, should be an integral part of disaster planning.[52]

The impacts of disasters, the UN General Assembly declared 1990-99 the International Decade for Natural Disaster Reduction (IDNDR) and organised the world conference on Natural Disaster Reduction in Yokohoma strategy and plan of Action for a safer world which stressed that every country has the primary responsibility to protect its people, infrastructure and national, social and economic and ecological assets from the impact of natural disasters. The strategy emphasised the urgent need to move from a mainly reactive approach to disaster mitigation to a new paradigm based on a more comprehensive approach that includes preventive measures.

Disaster risk management implies addressing the underlying social, economic and environmental vulnerabilities and reduce the probability of disaster occurring. Disaster risk management tries to address hazard risks as an integral part of development. It is based on a continuous assessment of vulnerabilities and risks and involves many actors and stakeholders such as governments, technical experts and local communities.[53]

Multiple hazards
Climate change is a source of multiple hazards that threaten long-term development actions by the international community, the consensus and planning approaches that have linked development and disaster should extend to climate change. Disaster risk management would take a comprehensive multi-hazard focus which includes various types of geological and hydro-meteorological hazards to which a particular country or region is exposed.

Disaster prevention and contingency plans can build on scientific knowledge of exposure to certain type of hazards and their frequency of occurrence. For example, geological hazards are often localised such as volcanic eruptions and earthquakes. Other hazards are associated with events of semi-periodic occurrence such as El Nino events occurring every 2-7 years. The risk of disaster occurring within a specific region and take remedial measures such as the implementation of forecasting and early warning systems, education initiatives, promoting appropriate infrastructure and natural resource management. Several countries are also prone to earthquakes, volcanic eruptions, droughts, cold spells, landslides and avalanches. The Disaster Risk Index (DRI), which is built on UNDP's experience with the Human Development Index, a global data based on natural disasters is maintained by the Centre for Research on the Epidemiology of Disasters in Brussels.[54]

Approaches to disaster management

The disaster management is to reduce the risk posed by actual and potential hazards.[55] Hazards can be broadly grouped into three areas: natural; technological; and complex emergencies.

Natural and technological hazards

The principal focus of planning for natural and technological hazards is risk assessment and reduction. The plans for natural and technological disasters need to protect society from hazards. This approach to risk reduction and civil protection has been developed through legislation, the defining of institutional responsibilities and the allocation of financial resources coupled with local responses and community involvement.[56] Disaster planning is based on risk assessment and lessons learned which are codified into a set of risk management and emergency plans designed to enable effective and efficient policies and practices. This approach to risk management can be effective in areas prone to natural hazards such as flood plains, storm corridors and active zones. To work effectively this holistic approach to planning requires accountable, democratic government institutions, financial support, political will and the trust of civil society. In Less Developed Countries (LDCs) such an approach to risk and disaster management also exists and it involves commissions and institutions at the national, sub-national, regional and municipal level, which have proliferated since the beginning of the International Decade for Natural Disaster Reduction (IDNDR, 1990-99).

Humanitarian and complex emergencies

Due to rapidly changing environments triggered by violent conflict, government agencies responsible for social protection may not be able to gain access to civilian population and IDPs'. The response to these cases by the international community- United Nations agencies, international organisations and international non-governmental organisations are to complement government efforts to bring relief to those affected.

In complex emergencies, planning needs assessment, delivery of goods and services to meet requirements. Human demand for water, food, shelter, sanitation, healthcare, security and for children's education, job training and counselling is balanced against available resources.[57]

Humanitarian interventions deal with immediate relief whereas longer term recovery and development are the remit of other agencies. The humanitarian and development sectors have different agendas and operating modes in human well-being. The humanitarian sector often can be typified as neutral and state- avoiding whereas the development sector relies on the state as a partner.[58] The humanitarian sector focussed on needs and rights does raise concerns about appropriateness, as the humanitarian system is largely ignorant of the views of the affected people as to the assistance being provided.[59] With the prospect of increasing frequency of climate change-related disasters, both rapid onset such as floods and slow onset such as drought and famine, maintaining and increasing communication will be a challenge.

Changing exposures and vulnerability

A pro-active disaster risk management approach needs to recognise the interaction of globalisation processes, population and demographic trends, economic development and trade patterns, urbanisation and other factors, which impact on the exposure and are vulnerable to hazards. The local and global environmental issues change risk patterns and increase vulnerabilities to various disaster risks. Climate change will alter the disaster risk associated with hydrological hazards by affecting climatic extremes through changes in average climatic conditions and climate variability.

Climate change can be described as both a complex and protracted hazard with multifaceted and multidimensional hazard that has short, medium and long-term aspects. Population movements in response to climate change may also result in new exposure to hazards. Climate displaced persons may suffer complex emergencies and strife as they flee from clan, tribal and national boundaries. An earthquake during a drought may come at a time when reservoirs and water pressure are too low to combat fires adequately.[60]

The UK department for International Development argues that climate change increases the urgency to integrate risk management into development interventions and points out that the impacts of climate change-related disasters are multifaceted. Not only they can lead to loss of life and destruction of homes, infrastructure and livelihoods, but they can also cause significant financial damage, which can impede or compromise development. The losses caused by Hurricane Mitch to Honduras and Nicaragua in 1998 are more than the combined Gross Domestic Product (GDP) of both countries, setting development behind 20 years.[61] LDCs are more vulnerable to climate-related disasters and will be the most poorly equipped to deal with the adverse impacts of climate change.[62]

Climate change increases the frequency and intensity of climate related hazards. Poverty and social impacts will impact humans through a variety of direct change in climate variables and indirect pathways such as pests and diseases; degradation of natural resources; food price and employment risks; displacement; conflicts and negative spirals.[63] Climate change also has implications for the urban poor and for rural-urban change. Most informal urban settlements are built illegally and without formal planning. Limited availability of water, high child and infant mortality and high disease burden such as malaria, tuberculosis, diarrhoeal etc. are common characteristics of such informal settlements.[64]

Climate change impacts on rural livelihoods, migration from rural to urban areas is increasingly likely to shape the foundation of adaptation strategy of rural poor. For both the poor and non-poor, migration can be accumulative strategy.[65] Hence migration is not an option for all, especially the chronically poor or specific vulnerable or excluded people, who may face discrimination and severely limited mobility. Poor people in Africa often faces high risks, use informal and ineffective means to protect themselves against risks in the context of very low coverage of government and market-based instruments.[66] The innovative approaches to social protection and DRR will be needed to bolster local resilience, support livelihood diversification strategies and reinforce people's coping strategies.

Risk management and climate change
Climate related hazards include severe droughts, floods and storms, human health hazards and complex biological impacts on the productivity and stability of livelihoods that depend on natural resources. More destructive cyclonic storms may become more common.[67] Long-term sea level rise will have major impacts on low lying land. Extreme temperatures will heighten the problems of drought prone areas to many communities. Climate-related risks and the adaptation measures build and reinforce resilience for vulnerability to climate variability and extremes in dealing with climate related risks.[68]

According to UNFCCC and the 1997 Kyoto Protocol that climate change brings many potential hazards such as rising sea levels, increased storms and flood frequency, the spread of infectious diseases, declines in biodiversity and reduced availability of food and water.

There is distinction between risk management in More Developed Countries (MDCs) and Least Development Countries (LDCs). Civil protection in the UK has evolved from a long tradition through institutional structures. Thus, the UK government is institutionally capable of adopting a proactive approach to a number of long-term problems and all hazards and approach to risk management concentrates up to 10-15 years with established institutions and capacities. It may be possible to accommodate longer time horizons for climate change 50-100 years in many other MDC governments.[69] But in LDCs, it is different from externally capacity-building programs for disaster risk management which include an institutional strengthening component, but in general these efforts are often focussed narrowly on the creation of disaster specific legislation, administrative arrangements and institutional structures.[70]

The rational approach would be to invest in building capacity and resilience, for example, the millions of US dollars were spent by donors on famine relief in Niger during 2005 (A drought situation) and limited donor for Senegal's proposal was to build a 'green wall' against the encroaching Sahara Desert- what China has been investing in as it protects Beijing and the 2008 Olympics from the Gobi desert to the North.[71] Disaster management has evolved from a relief and response approach to a risk management approach with greater focus on reducing vulnerabilities, aimed at mitigation and prevention and poor finance.

There are a large number of bodies- governmental, non-governmental, public and private involved in disaster management and direct interest in disaster risk reduction including the humanitarian and developmental sectors. The UNFCCC is a legal entity established by treaty in 1992. Decision-making is the responsibility of the Conference of Parties (COP). Protocols agreed by the COP are binding on UNFCCC signatories. COP is a policy-making and implementing body with a focus on mitigation, reduction and adaptation. The UNFCCC could use resources and mechanisms available through the Special Climate Change Fund. COP-9 supports 'capacity building, including institutional capacity for preventive measures, planning, preparedness and management of disasters relating to climate change'.[72] In this particular matter, the UNFCCC has the resources, while the UN/ISDR has the network and capacity.

Since the 1990s, IOM (International Organisation for Migration) and other International organisations have been called by states to assist and protect displaced population affected by natural disasters that overwhelmed national and local capacities. Disaster Risk Management (DRM) is the core of humanitarian emergency response that includes life-saving activities implemented in the course or in the immediate of a disaster. Timely and efficient assistance and protection depend on emergency, preparedness increasing the capacity of the humanitarian community to identify potential crises. To effectively intervene in emergency and post-crisis situations, IOM has developed a comprehensive framework with the objective of displacement through the solutions and the creation of sustainable development. The Emergency response is to save lives and provide protection and moving towards recovery, reconstruction, mitigation and preparedness.

The IOM (International Organisation for Migration) has developed its capacities in terms of DRM to provide assistance and protection to displaced population. This has happened mostly within the Inter-Agency Standing Committee (IASC)[73] where IOM has built its emergency policy framework and its operational procedures of other humanitarian actors from Non-Governmental Organisations (NGOs), the United Nations and the International Red Cross and Red Crescent movement communities. Since the 2005 humanitarian reform, IOM's recognised role with displaced population, its leadership and commitment are shown in its role as Global Cluster Lead for Camp Coordination and Camp Management (CCCM) in natural disasters and its active engagement in the Logistics, Emergency Shelter, Protection, Health and Early Recovery Clusters. Most recently, IOM has taken a leading role in response to the two major natural disasters that occurred in 2010: the devastating earthquake in Haiti (CCCM Global Cluster Lead) and the floods in Pakistan that left more than 1.9 million households in need of shelter (Emergency Shelter Cluster Lead).

For environmental degradation and natural disasters, the sustainable development is a key objective that can be achieved through Disaster Risk Management (DRM) which respond to disasters; Disaster Risk Reduction (DRR) which reduce risk and exposure to hazards; and Climate Change Adaptation (CCA) which address the effects of a changing climate. DRM, DRR and CCA have an important role to play in contributing to the MDGs and to country specific development strategies with an emphasis on sustainable development. IOM is the leading organisation in promoting linkages between migration and development at the policy, research and operational levels.[74] IOM has been working on migration and include the development agenda which managed population movements. It can play an essential role in the recovery process by injecting much needed cash into a devastated economy and reducing the dependence on humanitarian aid.[75]

Ecosystem and disaster risk reduction

Ecosystems are integrated as human ecological systems that work together to provide the range of goods and other benefits necessary to support life, livelihoods and human well-being. The environment, development and disasters are linked and multi-dimensional role of environment in the context of disasters and how environment disaster linkages in turn are affected by, can also shape development process and outcomes.[76]

Disaster can have adverse consequences on the environment and on ecosystems which could have immediate to long-term effects on the population whose life, health, livelihoods and well-being depend on environment and ecosystems. Environmental impacts may include:

- Direct damage to natural resources and infrastructure, affecting ecosystem functions;
- Acute emergencies from the uncontrolled, unplanned or accidental release of hazardous from industries; and
- Indirect damage as a result of post-disaster relief and recovery operations that fail to take ecosystems and ecosystem services.

Environmental conditions can be a major driver of disaster risk as highlighted by the 2009 Global Assessment Report.[77] Degraded ecosystems can aggravate the impact of natural hazards that affect the magnitude, frequency and timing of these hazards. This has been evidenced in areas like Haiti, where very high rates of deforestation have led to increased susceptibility to floods and landslides during hurricanes and heavy rainfall events.[78] Environmental degradation also contributes to risk by increasing socio economic vulnerability to hazard impacts, as the capacity of damaged ecosystems to meet people's needs for food and other product is reduced.[79] Poor communities are particularly affected, as their livelihoods depend heavily on natural resources and ecosystem services.[80] Appropriate management of ecosystems can play a critical role in reducing vulnerability and enhancing resilience of local communities, as healthy socio-ecological systems are better able to prevent, absorb and recover from disasters.[81] Human-induced climate change will also significantly compromise ecosystems' structures and functions, weakening natural resilience against hazards.[82] Disaster and post-disaster recovery interventions can adversely impact ecosystems and thus jeopardise the resource base needed for long-term development including achievement of the Millennium Development Goals (MDGs).

It is argued that ecosystems contribute to reducing disaster risk in two important ways. First, ecosystems such as wetlands, forests and coastal systems, can reduce physical exposure to natural hazards by serving as natural protective barriers in mitigating hazard impacts, well managed ecosystems can provide natural protection against common natural hazards such as landslides, flooding, avalanches, storm surges, wildfires and drought.[83] In the post 2004 Indian ocean tsunami, numerous coastal reforestation projects were initiated in Asia to restore affected areas and to provide protection against coastal hazards, especially the more frequent events such as storms and cyclones. For example, Indonesia announced plans to reforest 6, 00,000 hectares of depleted mangrove forest in five years, and governments of Sri Lanka and Thailand launched big programmes to rehabilitate mangrove areas for coastal protection.[84] The Intergovernmental Oceanographic Commission (IOC) recommends that the potential of a variety of coastal systems-coral reefs, and coastal vegetation- should be harnessed for coastal protection and acknowledges the importance and cost-effectiveness of natural infrastructure in mitigating lower magnitude coastal hazards and sustaining multiple uses of the coastal zone.[85]

The second way, in which ecosystems can lessen disaster risk, is by reducing social-economic vulnerability to hazard impacts. Ecosystems also sustain human livelihoods and provide essential goods such as food, fibre, medicines and construction materials which are equally important for strengthening human security and resilience against disasters. In Mexico, the World Bank is undertaking a large scale coastal wetland and mangrove swamp restoration project to address coastal protection against hurricanes, saltwater intrusion due to sea-level rise as well as water supply and food production to communities.[86] Well-managed ecosystems are considered more resilient to the impacts of extreme events and are able to recover more effectively than degraded ecosystems.[87] However, it is important to recognise that ecosystems also have limits in providing physical protection against hazards. For example, forests do not seem to protect against large-scale flooding from severe events such as tropical cyclones or tsunami.[88]

Ecosystems management provides the unifying base for promoting Disaster Risk Reduction (DRR) and Climate Change Adaptation (CCA) with the overall goals of achieving sustainable development, human well-being and livelihood security. Ecosystem management initiatives could be enhanced by including disaster risk and climate change considerations, while DRR, climate change adaptation and development planning need to recognise the potential of harnessing ecosystem services and also address vulnerability linked to ecosystem degradation.

Figure 1, given below, indicates the ecosystem based disaster risk reduction, a more sustainable approach to DRR and climate change adaptation (CCA).[89]

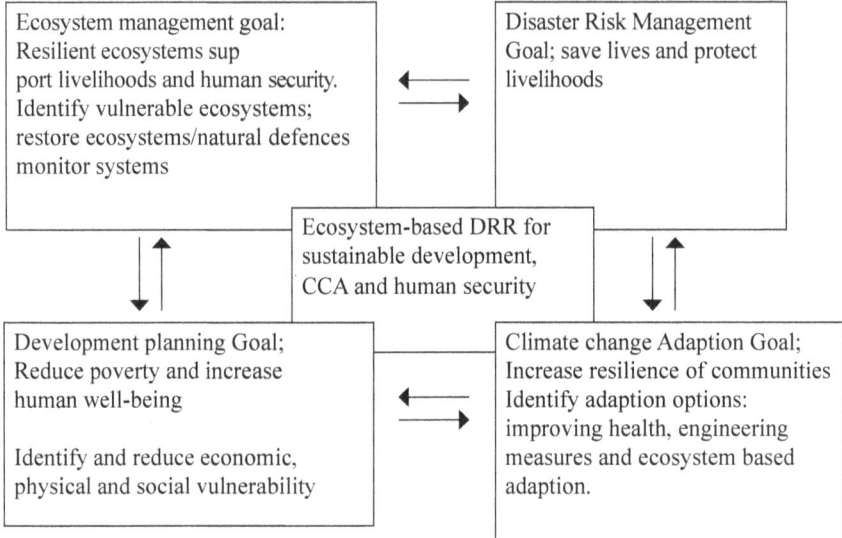

Implications for disaster risk

Since 1999, the 27th international conference of the Red Cross and Red Crescent decided to study climate change in relation to humanitarian response and preparedness, and scientific concerns in the IPCC Third Assessment Report 2001. The past few decades have seen a reduction in the number of people killed by natural disasters, the number of people affected and socio-economic losses. 600 million people were affected by hydro-meteorological disasters- triple the decade's average.[90] The rise in losses and people affected reflects a growing vulnerability to natural hazards, and in particular to weather and climate related hazards, which dominate the disaster statistics.[91]

Weather is the set of meteorological conditions such as wind, rain, snow, sunshine and temperature at a particular time and place. The term 'climate' describes the long-term characteristics of the weather at a place. The ecosystems, agriculture, livelihoods and settlements of a region are very dependent on its climate. The weather conditions, taking account of the average conditions as well as the variability of these conditions. The statistics of extreme conditions such as severe storms or unusually hot seasons are part of the climatic variability. Some slowly changing climatic phenomena for whole seasons or even years: the best known of these are the EI Nino phenomenon.[92] The natural disasters such as hurricane Katrina indicates, mankind is vulnerable to extreme weather events even in wealthy nations.

This growing vulnerability is intimately tied to development patterns: environmentally unsound practices, global environmental changes, population growth, urbanisation, social injustice, poverty and short-term economic vision are producing vulnerable societies.[93] Climate change will directly affect the work of many governments and national societies. In the work of disaster preparedness and response, national societies and other humanitarian organisations are dealing with risks on a daily basis. Climate change not only raises the risks, but also increases the uncertainties.

In some places, climate change impacts may appear to be less important than other issues facing the country and its national society. Sub-Saharan Africa, for example, is gripped by the devastating HIV/ AIDS, which clearly requires priority attention. It is important that disaster preparedness and health programmes pay attention to local knowledge about risks and vulnerabilities. National societies' volunteer networks and the community-based nature of vulnerability and capacity assessments make the International Federation well qualified to carry out such risks.

The global threats posed by climate change are discussed further below in relation to the four core areas of the International Federation's strategy 2010:

Promotion of humanitarian values

Extreme weather events compound other development problems that reduce the coping capacity of those affected. The poorest, the most socially or economically marginalised, the weakest and most ill, are also those people most vulnerable to the impacts of climate change.

Disaster response

All over the world, extreme weather events may become more frequent, intense and long-lasting - which in turn may lead to more disasters. Climate scientists project a widespread increase in the risk of flooding for tens of millions of people due to heavier rainfall and sea level rise.[94] Droughts, heat waves and other weather- related extremes are likely to further stretch the disaster response capacities of national societies.

Disaster preparedness

The key strategy in dealing with the uncertainties of climate change is to enhance existing activities, which minimise current disaster risks. This means two things; improving the disaster preparedness efforts of governments, the national societies and other humanitarian organisations and integrating disaster risk reduction into development strategies. Misguided development is increasing people's vulnerability to extreme weather events through poor land use, deforestation, uncontrolled population growth and urbanisation, social injustice, poverty and economic short-termism.

The international federation also has a role to convince other development actors of the need to integrate disaster risk reduction in their programmes. Some projected impacts of climate change are certain proactive risk reduction measures now. For example, mountain glaciers and ice caps are melting across the world with major implications for the communities who live downstream.

In the Himalayas, the risks posed by Glacial Lake Outburst Floods (GLOFs) must be planned for and reduced. Governments and national societies operating in small island states and along low lying coastlines must plan for effects of sea level rise to protect from erosion and storm surges. The mangrove reforestation projected by Red Cross volunteers in Vietnam to protect coastal dykes from the destructive power of high waves, saving lives and improving livelihoods.

Health and care in the community

As many part of the world are likely to get warmer, disease-bearing mosquitoes and tsetse flies may increase their range, spreading malaria, dengue and Leishmaniasis.[95] Robert T. Watson, formerly chair person of the IPCC warned that, "projected changes in climate could lead to an increase in the number of people at risk of malaria of the order of tens of millions annually"[96] More flooding will increase the risk of water borne diseases such as cholera, dysentery and hookworm. Countries already suffering health and sanitation problems and heat waves will result in additional heat stress mortality and increases in droughts and extreme events would stress on water resources and flood security.

Mitigation and adaptation through disaster risk reduction

UNFCCC is to address the root cause by reducing greenhouse gas emissions from human activity and will require radical changes in the way of fossil fuel use, industry operations, urban development and land use. Mitigation is defined by the IPCC as "an anthropogenic intervention to reduce the anthropogenic forcing of the climate system; it includes strategies to reduce greenhouse gas sources and emissions and enhancing greenhouse gas sinks".[97] Examples of mitigation actions include more efficient systems, developing new low-energy technologies for industry and transport reducing and switching consumption of energy such as solar and wind power. Natural carbon sinks such as forests, vegetation and soils can be managed to absorb carbon dioxide and technologies are being developed to capture carbon dioxide at industrial sources and to inject it into permanent storage deep underground.

Adaptation is defined by the IPCC as "the adjustment in natural or human systems in response to actual or expected climatic stimuli or their effects, which moderates harm or exploits beneficial opportunities.[98] Adaptations include preparing risk assessments, protecting ecosystems, improving agricultural methods, managing water resources, building settlements in safe zone, developing early warning systems, instituting better building designs, improving insurance coverage and developing social safety nets. These measures are linked to sustainable development as they reduce the risk to lives and livelihoods and increase the resilience of communities to all hazards. Adaptation and mitigation should be considered jointly, as some adaptation measures can contribute to reduce greenhouse gas emissions while mitigation measures can be planned to help reduce disaster risks.

Poorer developing countries are especially vulnerable to climate change because of their geographic exposure, low incomes and greater reliance on climate sensitive sectors, particularly agriculture. People exposed to the most severe climate- related hazards due to their limited adaptive capacity. This poses multiple threats to economic growth, wider poverty reduction, and the achievement of the Millennium Development Goals.[99] Stern (2008) argues that social protection could become one of the priority sectors for adaptation in developing countries.[100] The relationship between climate change adaptation, Disaster Risk Reduction (DRR) and social protection, the Institute of Development Studies (IDS) researchers have developed the concept of 'adaptive social protection'.

Social protection policy needs to learn from DRR and adaptation approaches to ensure programmes to continue and to effectively support livelihoods and protect the poor and excluded from risks in the face of climate change. Social protection, DRR and climate change adaptation seek to take integrated, multi-sectoral approaches to mitigate risks faced by poor people and they can contribute to poverty reduction and help move people into productive livelihoods.

Hyogo frame work policy
"Disaster risk reduction" can be defined as "action taken to reduce the risk of disasters and the adverse impacts of natural hazards through systematic efforts to analyse and manage the causes of disasters including through avoidance of hazards, reduced social and economic vulnerability to hazards and improved preparedness for adverse events.[101]

The Hyogo Framework for Action provides the foundation for the implementation of disaster risk reduction at the world Conference on Disaster Reduction in January 2005, in Kobe, Japan, with the support of 168 Governments to "promote the integration of risk reduction associated with existing climate variability and future climate change into strategies for the reduction of disaster risk and adaptation to climate change between 2005 to 2015".[102]

The Hyogo Framework elaborated five priorities for action in reducing disaster risks, these are given below:

• Ensure that disaster risk reduction is a national and local priority with a strong institutional basis for implementation.

This priority includes: encouraging a core ministry with a broad mandate including finance, economics or planning, to be responsible for mainstreaming climate change adaptation policies and activities; organising a national high- level policy dialogue to prepare a national adaptation strategy that links with disaster risk reduction strategies; formalising collaboration and the coordination of climate-related risk reduction activities through a multi-sector mechanism such as a national platform for disaster risk reduction; and developing mechanisms to actively engage women, communities and local governments in the assessment of vulnerability and impacts and formulation of local adaptation activities.

• Identify, assess and monitor disaster risks and enhance early warning.

Important steps under this priority include developing and disseminating high quality information about climate hazards and their likely future changes; conducting assessments of vulnerability and especially vulnerable groups; preparing briefing for policy makers and sector leaders; reviewing the effectiveness of early warning systems; implementing procedures to ensure warnings reach vulnerable groups; and undertaking public information programmes to help people understand the risks they face and how to respond to warnings.

- Use knowledge, innovation and education to build a culture of safety and resilience at all levels.[103]

This principle applies equally to adaptation and disaster risk reduction. Specific steps should include collating and disseminating good practices; undertaking public information programmes on local and personal actions that contribute to safety and resilience; publish community successes; training the media on climate related issues; developing education curricula on climate adaptation and risk reduction; supporting research programmes on resilience; and improving mechanisms for knowledge transfer from science to application for risk management in climate sensitive sectors.

- Reduce the underlying risk factors.

Many environmental and societal factors that create risks from natural hazards measures' can include incorporating climate risk-related considerations in development planning processes and macro-economic projections; requiring the use of climate risk-related information in city planning, land use planning, water management and environmental and natural resource management; strengthening and maintaining protective works such as coastal wave barriers, river levees, flood ways and flood ponds; requiring routine assessment and reporting of climate risks in infrastructure projects, building designs and other engineering practices; developing risk transfer mechanisms and social safety nets; supporting programmes for diversification of livelihoods; and instituting adaptation activities in plans for recovery from specific disasters.

- Strengthen disaster preparedness for effective response at all levels.

Resilience building and early warning systems contribute to this priority. Other specific actions can include revising preparedness plans and contingency plans to account for the projected changes in existing hazards and new hazards not experienced before; building evacuation mechanisms and shelter facilities; and developing specific preparedness plans for areas where settlements and livelihoods are under threat of permanent change.

Action needs to be taken by all actors including governments, international organisations, the business community and NGOs for reducing the risks associated with climate change. The International Federation through the National Societies has made a great contribution to the efforts, particularly in the core areas of strategy 2010. In particular, this study recommends the following seven steps in risk reduction.

Preliminary climate risk assessment
Governments and National Societies should make a preliminary assessment of the projected impacts of climate change and the implications for their role and activities. This assessment needs to include both scientific inputs and community consultations to learn local people, at the risks and how a changing climate would affect everyday lives. An assessment of climate change-related risks could form part of a broader Vulnerability and Capacity Assessment (VCA).

Assess priorities and plan follow up
Such an assessment could raise important concerns that would need to be prioritised. Follow-up activities could be initiated by the government with the National society and in partnership with other national or regional organisations.

Raise awareness
The preliminary climate risk assessment should lead to a programme to raise awareness about climate change and possible impacts on vulnerable people. If climate change is identified as a priority, the next step would be to integrate climate change into ongoing education activities with local communities. In the National Society Context, this could be done through First Aid Programmes, community-based disaster preparedness and risk reduction, health and care in the community or during VCAs (Vulnerability and Capacity Assessment).

Establish and enhance partnerships
The preliminary climate risk assessment will involve various experts such as scientists, meteorologists, etc. These contacts are maintained and strengthened on future impacts of climate change and possible adaptation strategies. Scientific organisations could learn from the International Federations and National Societies, field experience of disaster risk reduction.

Highlight climate-related vulnerability with other actors
People's vulnerability to climate change needs to be kept on the agenda between National Societies, governments and other actors. This could

involve injecting a humanitarian perspective into development issues such as the management of coastal zones and natural resources, policy development for heat waves in urban areas, or land use planning in flood-prone areas. National societies could also help integrate disaster risk reduction into development strategies.

Document and share experiences and information
The impacts of climate change are uncertain and unpredictable. Governments and National societies should find innovative approaches to deal with new uncertainties. Lessons in disaster preparedness and risk reduction should be documented and shared between National societies within the International Federation and with other organisations involved in adapting to climate change.

Advocacy shapes the global response to climate change
Climate change is a global issue with local impacts. As the world's largest humanitarian network, the International Federation is uniquely placed to relate the vulnerabilities and capacities of exposed communities of international humanitarian and development policy. This makes the international federation potentially a key player in contributing to the local, regional and international responses to climate change.

Adaptive social protection
The framework for social protection measures to strengthen poor people's resilience to disaster risks that acknowledge the changing and unpredictable nature of climate related impacts. This concept of adaptive social protection is characterised by a number of features that include:
- An emphasis on transforming productive livelihoods as well as protecting and adapting to changing climate conditions rather than simply reinforcing coping mechanisms.
- Understanding the ground structural root causes of poverty for particular people, permitting more effective targeting of vulnerability to multiply shocks and stresses.
- Incorporation of rights-based rationale for action, stressing equity and justice dimensions of chronic poverty and climate change adaptation based on economic efficiency.
- An enhanced role for research from both the natural and social sciences to inform the development and targeting of social protection policies and measures in the context of the burden of both geophysical hazards and changing climate-related hazards.

- A longer term perspective for social protection policies that takes into account the changing nature of shocks and stresses.

UNESCO's role and activities of international organisations
UNESCO forms part of a network of UN agencies, inter-governmental groups and non-governmental or civil society organisations that re-team together as part of the international strategy for Disaster Reduction. They are:

International Strategy for Disaster Reduction (ISDR)
The ISDR promotes disaster reduction as an integral component of sustainable development. It is a global framework in which countries, institutions and individuals can cooperate and is coordinated within the United Nations by an Inter-Agency Secretariat located in Geneva.[104]

United Nations Development Programme (UNDP)
The UNDP has operational responsibilities at national level for natural disaster mitigation, prevention and preparedness. It works to ensure that disaster risk considerations are factored into national and regional development programmes, and that countries take advantage of disaster recovery to mitigate future risks and vulnerabilities.[105]

United Nations Environment Programme (UNEP)
UNEP provides leadership and encourages partnership in caring for the environment by inspiring, informing and enabling nations and people to improve their quality of life without compromising that of future generations. Through its Global Resource Information Database, UNEP provides world-wide environmental information and early warning about environmental hazards.[106]

United Nations Children's Fund (UNICEF)
UNICEF generally works on warning, prevention, preparedness and recovery activities for the care of children and women in disaster-prone areas. Since the Indian Ocean tsunami, UNICEF, in conjunction with governments and a wide array of partners, has built temporary schools, rehabilitated water systems, and organised family care for children who lost their parents and has kept children healthy through immunisation and other health initiatives.[107]

UN Habitat

The Disaster Management Programme of UN Habitat is based on the idea that the rehabilitation of social and economic conditions following disasters or conflicts offer a unique opportunity to rethink past development practices, improve the sustainability of human settlements and prepare communities to prevent against future risks and threats.[108]

World Meteorological Organisation (WMO)

The WMO coordinates global scientific activity to provide the advance warnings that save lives and reduce damage to property and environment. The organisation deals with hazards related to weather, climate and water, which account for nearly 90 per cent of all natural disasters. WMO also contributes to reducing the impacts of forest fire, volcanic ash and human-induced disasters such as those associated with chemical and nuclear accidents.[109]

World Health Organisation (WHO)

The WHO deals with disaster preparedness connected with health. Its purpose is to reduce avoidable loss of life and the burden of disease and disability in disaster-affected countries. It works with other international organisations and non-government organisations as well as local authorities and civil society in responding to health emergencies.[110]

Food and Agriculture Organisation (FAO)

The FAO's objective is to strengthen the capacity of communities to prepare for natural disasters. It deals with immediate food issues, and plays an important role in reversing degradation and reducing vulnerability to hazards. This is complemented by a special program for food security. The UN world food programme is focussed on emergency and post-disaster food relief and support for rehabilitation.[111]

International Atomic Energy Agency (IAEA)

The agency is concerned with the zoning of nuclear power plants in areas prone to seismic activity, and it has been actively concerned with the design of reactors that can withstand the most severe natural disasters. A core element of the IAEA's work is to help countries to upgrade nuclear safety and to prepare for and respond to emergencies.[112]

The World Bank Group

The group promotes disaster risk management as a priority for poverty reduction, linked to environmental management. It concentrates on reconstruction measures that strengthen resilience to future disaster and identify innovations in risk transfer and financing.[113]

Other Organisations
A number of inter-governmental and non-governmental organisations are engaged in disaster reduction. They include: the International Federation of Red Cross and Red Crescent Societies, the Council of Europe, the Asian Disaster Reduction Centre, the Asian Disaster Preparedness Centre, Caribbean Disaster Emergency Response Agency, the International Council for Science, the International Social Science Council, ActionAid International, the International Council of Engineering and Technology, Register of Engineers for Disaster Relief, the International Consortium of Landslides and the International Council on Monuments and Sites.

About 94 per cent of natural disasters result from four major causes- earth quakes, storms, floods and droughts. Following are descriptions of the principal kinds of risks and the part UNESCO is playing in reducing them.

Earthquakes
UNESCO has supported the establishment of international, regional and national centres for the recording, exchange and analysis of seismological data. It helps train engineers and seismology in the former Yugoslav, Republic of Macedonia, Britain, Japan, Peru and Iran. UNESCO and the US Geological survey have been jointly involved since 1993 for Reducing Earthquake Losses in the Enlarged Mediterranean Region; likewise, UNESCO since 2001 has collaborated with the US Geological survey and earth science organisations on the programme for reducing earthquake losses in South Asia.

Regional disaster reduction programmes have been carried out with the help of UNESCO field offices. For example, UNESCO Tehran office provides advice from the experience in the construction of the ancient city of Bam and the reduction of similar risks in Iran, one of the countries most exposed to earthquake hazard, UNESCO seeks to mitigate disaster by supporting the development and implementation of building codes for collapsing buildings that kill people.

Earthquakes also provide scientists with a living laboratory; a considerable knowledge about the behaviour of earthquakes which is obtained as a result of many post- disaster reconnaissance missions conducted by UNESCO.

Tsunami

Due to earthquakes, the ocean floor, volcanic eruptions, submarine landslides and meteorite impacts can touch the monstrous waves known as Tsunamis. The giant Tsunami of December 26, 2004 caught countries around the Indian Ocean unprepared. Scientists were immediately aware of the massive earthquake on the sea floor but there was no way of sounding the alarm. More than 2, 40,000 people were killed or missed in the Indian Ocean disaster. Within a month of the calamity, governments at the Kobe meeting agreed to set up a Tsunami warning system for the India Ocean and an initial network is in place since July 2006 with help from the pacific Tsunami warning centre in Hawaii and the Japanese Meteorological Agency.

The member states of the Inter-governmental oceanographic commission of UNESCO decided at their General Assembly in June 2005 to coordinate the establishment of a global warning system for ocean-related hazards in close cooperation with other UN bodies. A Tsunami warning system for the North East Atlantic and Mediterranean is scheduled to be in place by the end of 2007. Planning is also underway for an early warning system for ocean- related hazards in the Caribbean. The tsunami warning systems will eventually form part of a global warning and mitigation network being coordinated by the inter-governmental oceanographic commission. This will be contributing to the Global Earth Observation System of Systems, a world wide effort involving 60 countries, the European commission and 43 international organisations.

Floods

Flood occurs when rain water or snows melt exceeds the capacity of the soil to absorb it or rivers to carry it away. As the leading UN agency for water related issues, UNESCO with the world meteorological organisation inaugurated the international flood initiative during the 2005 world conference on Disaster Reduction in Spain. The organisation participates in all aspects of water resource research through the International Hydrological Programme.

UNESCO's office in the Indonesian capital has contributed to a flood mitigation project in a district subject to recurrent severe flooding. Under the project, community representatives were trained to communicate and show how people could lessen the impact of future inundations, UNESCO's aim to create communities that are well- informed and adequately prepared for such events.

Tropical cyclones

Tropical cyclones are also known as typhoons or hurricanes depending on geographical location. A cyclone accompanied by an exceptionally high storm surge swept over the coastal wetlands of Bangladesh in 1970, killing 3,00,000 people. A future rise in sea levels associated with global warming occurring bigger storm surges and shows more vulnerability to Tsunamis. Increasing population density in coastal regions has increased human vulnerability to cyclones. According to the Inter-governmental Panel on Climate Change, "Many coastal areas will experience increased levels of flooding, accelerated erosion, loss of wetland and sea-water intrusion in fresh water sources."

The inter-governmental oceanographic commission of UNESCO has been involved since the mid-1980s in a long-term project on the management and planning for coastline change in the Caribbean. The IOC is also one of the sponsors and hosts the secretariat of the global ocean observing system, one of the ultimate aims of which is to mitigate damage from natural hazards and population.

Droughts and desertification

Changing climate patterns are also responsible for the increasing drought in many parts of the world. Particularly in Africa, it contributes to the spread of deserts and land degradation. Desertification already threatens over one-third of the earth's land surface, directly affects the lives of 250 million people and threatens another 1.2 billion in 110 countries. Tens of millions of those affected in sub-Saharan Africa are expected to swell migratory pressures toward northern Africa and Europe.

UNESCO has been involved in dry lands research since 1950s and emphasises the need for a developmental rather than a crisis management approach. In 2006, UNESCO organised an international scientific conference in Tunis at which delegates stressed the need for interdependence and conservation of cultural and biological diversity and the integrated management of water resources.

UNESCO's awareness in global disaster reduction

- UNESCO aims to have such lessons made part of school programmes everywhere. In 2006, it hosted the launch of a campaign entitled "Disaster Risk Reduction Begins at School" to increase awareness of disaster reduction. It also campaigns to create schools building capable of resisting natural hazards such as earthquakes by developing and applying appropriate construction codes.

- UNESCO plays a pivotal role in a 'cluster' of international, government and non-government organisations pledged to build 'a culture of resilient communities' based on knowledge, innovation and education. It emphasises the integrated education about disaster reduction into school curricula to make school out of hazards.
- Awareness is not only taught in the class room, but it is passed on from generation to generation and this knowledge is essential from local to traditional.
- UNESCO is closely involved in risk assessment and rescue operations to protect monuments and urban historic centres, sites, museums and archives in cooperation with other international conservation institutions.
- UNESCO has implemented numerous projects to safeguard culture sites and objects in the wake of disasters including the 1950 earthquake in Cuzco, Peru or the 1996 flooding in Florence and Venice, and the temple of Prambanan, Indonesia severely damaged by earthquakes in 2003 and 2006, respectively.
- UNESCO has published several manuals and guidelines on protecting cultural sites, including a policy document in 2006 entitled, "A strategy for Reduction Risks at World Heritage Properties". Cultural and Natural heritage and intangible artistic skills are important contribution to sustainable development which includes the mitigation of disasters.

There are practical examples of adaptation and disaster risk reduction in various sectors. These are given below:-

Agricultural and food security
The measures include altering crop strains to enhance their drought and pest resistance, changing planting times and cropping patterns and altering land topography to improve water uptake and reduce wind erosion. Burkina Faso is one country which is researching new drought resistant millet and sorghum for decreased rainfall regimes.[114]

Water sector
Adaptation measures include actions on both water supply and water risks such as protecting water supply infrastructure and traditional water supply sources, developing food ponds, water harvesting, improved irrigation, desalination, non-water-based sanitation and improved watershed and trans-boundary water resource management. Integrated

water resource management (IWRM) provides the accepted framework for such actions.

Health Sector
Measures include early warning systems, systematic action on water-and-vector- borne diseases to raise public awareness of watershed protection, vector control, and safe water-and food-handling regulations; the enforcement of relevant regulations; and support for education, research and development on climate related health risks.

Awareness and education
Measures include curriculum development for schools, supply of information to community groups and women's networks, radio and television programmes, public poster campaigns and leadership by national figures and celebrities. Awareness-raising for strategic inter mediatise such as teachers, journalists and politicians and support to technical experts and groups are also important.

Environmental management
Healthy ecosystems provide significant benefits for resilience, livelihoods, risk reduction and adaptive capacity. Measures include strengthening of environmental management in areas of greatest risk from weather hazards; protecting ecosystems such as coral reefs or mangrove forests.

Early warning systems
Measures include improving existing systems to cover the changed hazard circumstances, instituting specific means to disseminate warnings to affected people in a timely, useful and understandable way and providing advice on appropriate actions to take upon receiving warnings.

Development planning
Adaptation and disaster risk reduction measures can be made a formal part of development processes and budgets and programmed into relevant sector projects, for example, in the design of settlements, infrastructure, coastal zone development, forest use etc. In order to achieve sustainable land management, avoid hazardous areas and build safe schools, hospitals and other public facilities.

Cost effective and disaster risk reduction through adaptation
According to World Bank, "Accelerated changes in demographic and

economic trends have disturbed the balance between ecosystems, increasing the risk of human suffering and losses. Today's populated areas- cities and agricultural zones- constitute an increasingly valuable asset base-potential human, and social and economic losses from natural disasters grow year by year, independently of Nature's forces. Increased vulnerability requires that natural disaster management be at the heart of economic and social development policy of disaster-prone countries." The economic losses are increasing day to day. In the 1990s it losses three times higher than in the 1980s, almost 9 times higher than in the 1960s and 15 times higher than in the 1950s.

According to UNESCO the number of extreme events is increasing as a result of global climate change and population pressures, disaster relief rather than risk prevention is generally considered by aid donors and non-government organisations. Relief is action-oriented and easy to quantify and more difficult to put figures on disaster risk reduction. In the case of Mozambique, one of the world's poorest nations, it suffers from periodic cyclone damage and flooding. After a warning in 2000 due to heavy rainfall, Mozambique asked aid donors for US $2.7 million to enable it to carry out immediate preparedness and mitigation activities, but received less than half this amount. Subsequently, Mozambique made three successive appeals totalling US $160 million for emergency assistance, all of which were met and a further US $456.48 million dollars was pledged at an International Reconstruction Conference in Rome later that year.

Rivers across the region burst their banks, washing away all the economic progress that Mozambique has made since its long civil war. Aid workers said the flood waters, which submerged vast areas of land and destroyed much of the country's infrastructure, caused more destruction than the civil war.

Investment in disaster prevention and mitigation is entirely consistent with planning targets for sustainable development. Natural hazards cannot be eliminated, but unnatural hazards- those caused by man's actions- can be minimised provided that communities are informed and resilient and ecosystems are allowed to perform in disaster reduction is an integral part of sustainable development and the global fight against poverty.

Disaster risk reduction offers cost-effective approaches to reduce the negative impacts of flooding, landslides, heat waves, temperature extremes, droughts and intense storms. Some examples include:-
- China spent US $3.15 billion on flood control between 1960 and 2000, which is estimated to have averted losses of about US $12 billion.

- The Rio de Janeiro flood reconstruction and prevention project in Brazil yielded an internal rate of return exceeding 50 per cent.
- The disaster mitigation and preparedness programmes in Andhra Pradesh, India yielded a benefit/cost ratio of 13.38.
- A mangrove-planting project in Vietnam aimed at protecting coastal population from typhoons and storms yielded an estimated benefit/cost ratio of 52 over the period 1994 to 2001.
- Property-owners in the US and Gulf States who implemented hurricane protection methods employed at nearly 500 locations avoided US $500 million in property losses from Hurricane Katrina, after customer investments of only US $2.5 million. These customers sustained eight times less damage than those who chose not to implement the protection measures.[115]

The rising population and assets in naturally at-risk areas remains the most important driver of growing disaster risk.[116] This includes rapidly expanded settlements in low-lying coastal areas and flood plains, inadequate spatial planning and regulation enforcement and lack of compliance or weak building standards.

Consequently, the world's 136 largest cities could be facing annual flood losses of US $1 trillion by 2050.[117] The extreme weather events associated with temperature, precipitation and sea level rise include warmer spells and heat waves increased heavy rainfall events since the 1960s.[118] Tropical cyclones and droughts are also likely to increase, although these projections vary by region and are subject to low confidence levels.[119]

Clearly attributing disaster costs to climate change remains extremely difficult due to loss and damage of local climate change, and current scientific efforts have focussed on climate change to particular hazard intensities, but they remain limited.[120] Climate change, poorly planned development, poverty and environmental degradation influence the risk of a climate event becoming a disaster. The loss and damage factors need to be managed collectively. Climate and disaster resilient development provides an opportunity with DRM in Risk Transfer, disaster preparedness and resilient reconstruction.

Achieving climate and disaster resilient development requires the international community and national governments to promote approaches that link climate and disaster resilience to broader development paths. Climate and disaster resilient development is consistent with the Doha decision on loss and damage which promotes the integration of climate risk management into development planning. For example, gradual adaptation in crop production and consumption patterns could reduce long-term losses in per capita. In Brazil, modelled population mobility across municipalities reduced climate change impacts on poverty by 63 per cent.[121]

Early warning systems have been proven to save countless lives worldwide and typically yield benefits that are 4-36 times higher than initial costs.[122] For example, Bangladesh and India show the benefits of prevention in terms of lives saved. Cyclone Phailin, which hit Odisha and Andhra Pradesh in October 2013 resulted in 40 deaths, compared to the 10,000 during the event in 1999.

Climate and disaster resilient development involves additional upfront costs which cannot be neglected, these cost are 'building back better' during disaster reconstruction; upgrading hydro-meteorological systems; risk assessments; and establishing and maintaining risk financing instruments. Another important upfront investment is institutional strengthening and improved coordination, which can take time to develop.

Financing decisions for climate and disaster resilient development may need simple guidelines such as the indicative sliding scale previously used by the Global Environment Facility (GEF). This could take into consideration the country's level of development and the financing needed. For example, Zambia to justify a 30 per cent level of top-up financing for local development plans that incorporated climate resilience.[123] The losses caused by weather-related disasters from 1980-2010, arguing that attribution of disasters to climate change as opposed to local vulnerability, remains a very difficult challenge. They also suggest focussing on promoting climate and disaster resilient development, while recognising that it has a higher initial cost.

Improving weather forecasts and early warning systems must be effectively linked to action on the ground, to save both lives and property. Preparedness activities must include strengthening the capacity of local organisations to plan for and respond to the effects of disasters. In the case of cyclones, local authority can use early warnings to evacuate large number of people to safer location. Long lead times enable people to protect property and infrastructure; reservoir operators can reduce water gradually to accommodate incoming flood waters. Early warning can also provide information on the occurrence of a public health hazard and enable a more efficient response to seasonal drought and food insecurity. Therefore, effective systems require a combination of government leadership, multiagency coordination to ensure effective responses based on pre-agreed operating procedures and community participation.[124]

The World Development Report 2014 provides five insights on managing global risks including climate and disaster risks to development[125]:-

*   Taking on risks is necessary to pursue opportunities for development. The risk of inaction may well be the worst option of all.
*   To confront risk successfully, it is essential to shift from unplanned and ad hoc responses when crises occur to proactive, systematic and integrated risk management.
*   Identifying risks is not enough: the trade-offs and obstacles to risk management must also be identified, prioritised and addressed through private and public action.
*   For risks beyond the means of individuals to handle alone, risk management requires shared action and responsibility at different levels of society, from the household to the international community.
*   Governments have a critical role to play in managing systemic risks, providing an enabling environment for shared action and responsibility, and channelling direct support to vulnerable people.

Disaster Risk Reduction and UNFCCC

The Bali action plan's directions for adaptation is that, "Risk management and risk reduction strategies including risk sharing and transfer mechanisms such as insurance; disaster reduction strategies and means to address loss and damage associated with climate change impacts in developing countries that are particularly vulnerable to the adverse effects of climate change."[126] Bali Action Plan is highly relevant to reducing disaster risk, particularly vulnerability assessments, capacity building and response strategies as well as integration of actions into sectoral and national planning. The integrate disaster risk reduction and adaptation into national development strategies has emerged as a key conclusion from international policy forums, in particular the "Stockholm plan of action for Integrating Disaster Risk and climate change Impacts in Poverty Reduction".[127] In support of the Bali Action Plan and based on consultation with ISDR system partners and UNFCCC parties, the UNISDR has identified and promoted the following three areas of action over 2008.[128]

Develop national coordination mechanisms to link disaster risk reduction and adaptation. This can be done through national consultation meetings, climate change and development, formally cross-linking the national platform for disaster risk reduction and national climate change team and encouraging systematic dialogue and information exchange between climate change and disaster reduction bodies, focal points and experts.

- Conduct a baseline assessment on the status of disaster risk reduction and adaptation efforts. This involves efforts by countries to collect and summarise national risk information, including socio-economic data concerning vulnerability and institutional capacities, together with reviews of relevant existing policies, particularly development strategies and sector plans, Hyogo Framework implementation, adaptation programmes and national risk transfer mechanisms.
- Prepare adaptation plans drawing on the Hyogo Framework; Based on the assessment of needs and gaps include the joint development of a disaster reduction plan and an adaptation plan. It should capitalise on National Adaptation Plans of Action where present and other adaptation initiatives and should use the concepts and language of the Hyogo Framework where appropriate, ideally with action on all five of the Hyogo Framework's priorities, to ensure a comprehensive, integrated and systematic approach to adaptation.

At COP 16, parties developed the Cancun Adaptations Framework, which helped identify areas of activities that qualified as 'adaptation' and which later may be considered for climate finance support. On November 15, 2011, the Global Migration Group adopted a joint statement on the impact of climate change on migration- the first common position on this issue to be adopted at the level of the United Nations. It recognises that migration and displacement resulting from environmental degradation and climate change is a global challenge that requires urgent international action. On this basis, it puts forward strong recommendations to strengthen the human rights and improve the livelihoods of those affected, to assist the Least Developed Countries (LDCs) in integrating climate change and migration into national development strategies.

Policy makers to take a holistic approach to tackle adverse impacts of climate change and a set of recommendations developed by the German Marshall Fund Study Team on climate change and migration. The group of international organisations provided key insights in elaboration of the Nansen principles with the recommendations made at the International Dialogue on Migration on "Climate change environmental degradation and migration", hosted by IOM in 2011. Some of these recommendations include:-

- Supporting the collection, analysis and sharing of more useful primary data on population movements, particularly on internal, short-term and cyclical migration and displacement which is essential for informing better policy-making for adaptation. Taking these important population data with hydro- meteorological data using geographic information systems and mapping technology will contribute to effective policy formulation at national, urban and local levels.

- Identifying guiding principles, effective practices and institutional frameworks to help governments in developing appropriate comprehensive policies, laws and targeted programmes to address both internal and external displacement.
- Building capacities in terms of policy, institutional, administrative and legal frameworks as well as operational and technical capacities.
- Migration can be part of strategies of vulnerable populations to adopt to climate change with sustainable rural and urban development and social protection mechanisms that ensure access to adequate and nutritious food as well as land or other assets. Anywhere necessary, governments will be better able to help people move in safely and with dignity.
- Supporting disaster risk reduction and conflict mediation strategies while strengthening humanitarian responses. Investments should be made today in resilience building strategies designed to prompt uncontrolled crisis situations as well as in more effective humanitarian responses to natural hazards and conflict.
- Without the resources to migrate, who may be more vulnerable to forced displacement and less able to use mobility in an adaptive way?

The Cancun Adaptation Framework should take other key issues such as population growth and location, age structure and conflict sensitivities into consideration as each of which are essential to an adaptation framework that protects and empower vulnerable people in a changing climate and ensures people's human rights, health and well-being. Most significant impacts of climate change will come in places where vulnerability is high, where population is changing rapidly and in areas of heightened conflict.

Strengthening disaster risk reduction in climate change adaptation *at* COP 18, Doha, and Qatar 2012

Previous COPs have adopted a number of decisions linking disaster risk reduction to climate change adaptation. These decisions include the Adaptation Committee and the Loss And Damage work programme that were detailed in COP 17, in Durban, South Africa in 2011 as part of the actions envisioned in the Cancun Adaptation Framework adopted in Cancun, Mexico 2010 at COP 16. Furthermore, both the 'Bali Action Plan' under the Ad hoc Working Group on Long-term Cooperative Action under the convention (AWG-LCA) and the 'Nairobi Work Programme' on impacts of vulnerability and adaptation to climate change have considered and supported stronger efforts to reduce the risks of disasters.

The key messages on climate change adaptation and disaster risk reduction are given below:

- Climate change will worsen the impact of disasters. Weather and climate related disasters are already a major and costly concern.
- Adaptation and disaster risk reduction are closely linked. Adapting to the impacts of climate change and reducing risk to disasters are priorities that are best addressed in an integrated manner.
- Reducing the risks of disasters lead to sustainable development. Unsustainable development is the main factor in growing disaster risks. Climate change adaptation and disaster risk reduction are part of the sustainable development agenda as outlined in the Rio+ 20 outcome document 'The Future we want'.
- Risk reduction tools are ready for adaptation use. There are many well- proven tools and methods that can be applied and contribute to the acceleration of climate change adaptation such as risk assessments, environmental protection, early warning systems and insurance.
- Climate change adaptation is reflected in a post-2015 framework for disaster risk reduction. Any future global blueprint for disaster risk reduction provides opportunity to guide adaptation action.

Accordind to the IPCC Fourth Assessment Report and IPCC Special Report Managing the Risk of Extreme Events and Disasters to Advance Climate Change Adaptations (SREX),[129] climate related hazards are likely to increase in frequency, intensity, spatial extent and duration as a result of changing climate. Furthermore, the UN Global Assessment Report (GAR) on Disaster Risk Reduction in 2011 estimated that over 80 per cent of economic losses are attributed to weather- related disasters.[130] Climate change is altering the face of disaster risk, not only through rise in sea-level and temperatures, but also through increased socio- economic vulnerability resulting from water stresses, impacts on agriculture, ecosystems and health.

Disaster risk reduction and adaptation agendas at COP 18 are outlined below:

- The five workshops organised since COP 17 in Durban under the UNFCCC on the loss and damage and work programme, provided opportunities to understand better how up to-date risk assessments and models contribute to estimates of potential climate change impacts that in turn support development of informed-policy decisions.
- Urge that work on the loss and damage work programme promote local and national owned disaster loss data and related risk assessments for more effective planning and prioritising of adaptation actions.
- Development of National Adaptation Plan (NAPs) can provide useful mechanisms to address climate change and disaster risk in national and local sustainable development planning. The guidelines for the NAPs initiated in COP 17 provide an opportunity on disaster risk reduction capacities, tools and institutions can assist developing and least developed country parties to formulate and implement adaptation plans.
- Recommend that NAPs include risk assessments and will help to prioritise actions that reduce the risk of extreme climate events in national adaptation plans. Risk assessments will also identify gaps in adaptive capacities at national and local levels.
- Recognise that there are national and local institutions engaged in the implementation of the Hyogo Framework for Action 2005-2015: Building the Resilience of Nations and Communities to Disaster with expertise and functioning of effective mechanisms to address the risk of natural hazards in national and local development planning.
- Use national and local reporting on progress against the implementation of the Hyogo Framework for Action to further guide adaptation planning.
- Recommend that the Guidelines promote the integration of climate change adaptation and disaster risk reduction national plans, which have proven effective in a number of countries, for example in the Pacific Island States.
- The work plan of the Adaptation Committee can further strengthen the linkages between climate change adaptation and disaster risk reduction and allow more systematic approaches to ensure methodologies offered by disaster risk reduction practices address the negative impacts of climate change.
- Recommend that the Adaptation Committee explore the establishment of a sub-group to identify and reflect upon existing mechanisms, institutions and policy frameworks- including the Hyogo Framework for Action and its future arrangements.

- Promote the use of existing regional approaches, strategies and policy frameworks to reduce disaster and climate change impacts, including trans-boundary losses and damages, into the Adaptation Committee both on assessing and addressing the adverse effects of climate change.

References

1. Greenhouse gases (GHGs) are "those gases constituents of the atmosphere, both natural and anthropogenic, that absorb and emit radiation at specific wavelengths within the spectrum of thermal infrared radiation emitted by the Earth's surface, the atmosphere itself and by clouds". The primary greenhouse gases include H2O, CO2, N2O, CH4 and O3. IPCC fourth assessment report, working group I; Glossary of terms: http://ipccwg1.ucar.edu/wg1/Report/ AR4WG1_print_Annexes. pdf.
2. UNFCCCArticle1, Definitions: http://unfccc.int/essential_background/convention/ background/ items/1349.php.
3. IPCC fourth assessment report, working group I, glossary of terms: http://ipcc-wg1. ucar.edu/ wg1/report/ar4wg1_print_annexes.pdf.
4. Hydrological, meteorological and climatological disasters include storm, flood, wet-mass movement, extreme temperature, drought and wildfire. Data downloaded from EM - DAT database, centre for research on the Epidemiology of Disasters (CRED), http://www.emdat.be/
5. IPCC fourth assessment report, working group I report: http://195.70.10.65/ipcc report/ar4- wg1.htm.
6. OFDA/CRED International Disasters Database EM - DAT.
7. UNFCCC, available at: http://www.unhcr.org/refworld/docid/3boof2770.html.
8. 1998 guiding principles on internal displacement, available at http://www2.ohchr. org/english/issues/idp/standards.htm.
9. UNISDR, (2009b). Risk and poverty in a changing climate. Global Assessment Report On Disaster Risk Reduction, United Nations office for Disaster Risk Reduction, Geneva.
10. World Bank (2012c). The Sendai Report: Managing Disaster Risks for a resilient future. World Bank, GFDRR and government of Japan, Washington DC.
11. Skoufias, E., ed. 2012. The poverty and welfare impacts of climate change: Quantifying the effects, identifying the adaptation strategies, World Bank, Washington, DC.
12. Dell et al., "Temperature and income: Reconciling new cross-sectional and panel estimates", American Economic Review, 99 (2), 2009, pp. 198-204.
13. Shepherd et al., (2013). The geography of poverty, disasters and climate extremes in 2030. ODI, met office Hadley Centre, RMS Publicati, Exeter.
14. IPCC, 2013. Climate change 2013: the Physical Science Basis. Summary for policymakers. Working Group I contribution to the IPCC Fifth Assessment-report/ ar4/wg1/ar4-wg1-spm.pdf. World Bank (2012d). Turn down the Heat. A report for the World Bank by the Potsdam institute for climate impact research and climate analytics, Washington, DC.
15. Ahmed et al., "Climate volatility on poverty vulnerability in developing countries", Environmental Research Letters, 4 (3), 2009, at http://iopscience.iop.org/1748-9326/4/3/034004/fulltext/.

16. *Ahmed et al., 2009.*

17. *World Bank. (2013b). Managing risk for development. World development report 2014. Washington, DC.*

18. *Shepherd et al., 2013.*

19. *IPCC Fourth Assessment Report, Working Group I, summary for policymakers: http://195.70.10.65/pdf/assessment-report/ar4/wg1/ar4-wg1-spm.pdf.*

20. *IPCC TAR Synthesis report, 2001; summary for policymakers: adaptation is a necessary strategy at all scales to complement climate change mitigation efforts.*

21. *Poverty and climate change: Reducing the vulnerability of the poor through adaptation, June 2003, World Bank. Org/povcc.*

22. *IPCC 2001a. Climate change 2001: The scientific basis. Contribution of the Working Group I to The Third Assessment Report of the IPCC (eds. Houghton, J.T, Ding, Y., Griggs, D.J., Noguer, M., Vander Linden, P.J., Dai, X., Maskell, K. and Johnson, C.A) Cambridge University Press, Cambridge.*

    IPCC, 2001b. Climate Change 2001: Impacts, Adaptation and Vulnerability. Contribution of the Working Group II to the Third Assessment Report of the IPCC (eds. Mc Carthy, J.J., Canziani, O.F. Leary; N.A., Dokken, D.J and White, K.S), Cambridge University Press, Cambridge.

23. *IPCC, 2001a. I*

24. *PCC, 2001.*

25. *WMO (World Meteorological Organisation) 2004, WMO statement on the Status of the Global Climate in 2003. WMO - No.966. Geneva. pp. 11.*

26. *Koppe, C.S., Kovatas, G. Jendritzky, B. Menne. Heat-waves: risks and responses, World Health Organisation. Global Change and Health, series no 2. Regional office for Europe, Copenhagen, 2004.*

27. *Stott, P.A., Stone, D.A. and Allen, M.R. 2003. Human Contribution to the Europe Heat Wave in 2003. Nature 432, pp. 610-614.*

28. *IPCC, 2001.*

29. *UN/ISDR, 2004: (United Nations inter-agency secretariat of the international strategy for Disaster Reduction) living with risk. A global review of disaster reduction initiatives. Geneva.*

30. *Fischer, G., Shah M., Van Velthuizen, H. and Nachtergaele, F.O. 2001. Global Agro-Ecological Assessment for Agriculture in the 21st Century.*

31. *For an overview of IOM's activities that address natural disasters and the consequences of climate change, see the compendium of IOM's activities in migration, climate change and the environment, IOM, Geneva, 2009, at http://www.iom.int/template/ gmaps/migration_ environmental/. The full report can be down loaded at: www.iom. int/jahia/webdav/shared/ shared/mainsite/activities/env-degradation/compendium_ climate_change.pdf.*

32. *On the complex nexus linking migration, climate change and environment: Assessing the evidence, IOM, 2009, available at index.php? Main page=product-info and products-id=539.*

33. *Figures presented in the study monitoring disaster displacement in the context of climate change, IDMC/OCHA, 2009, available at www.internaldisplacement. org/8025708F004CFA06/ (httppublications)/451D224B41C04246C12576390031 FF63? Open Document.*

34. *The expression 'semi-rural' is used to describe outlying areas adjacent to suburbs,*

which often combine both rural and urban characteristics.

35. *World Bank, Dhaka: Improving living conditions for the urban poor. Bangladesh Development Series. Paper no. 17, World Bank, Dhaka, 2007.*

36. *UNISDR, Risk and poverty in a changing climate: in rest today for a safer tomorrow 2009 Global Assessment Report on Disaster Risk Reduction, United Nations, Geneva, 2009, available at www. preventionweb.net/english/hyogo/gar/report/index. php?id=1130&pid:pih:2.*

37. *Sir John Holmes, under-Secretary-General for Humanitarian Affairs and Emergency, Relief Co-ordinator, Opening Remarks at the Dubai International Humanitarian Aid and Development Conference and Exhibition 'DIHAD 2008 Conference' & April 2008, available at http://www.reliefweb.int/rw/rwb.nsf/db900sid/YSAR-7DHL88? Open document.*

38. *Ever land, S., 20 Million Climate Displaced in 2008; Norwegian Refugee Council, June 8, 2009, available at http://www.nrc.no/?did=9407544.*

39. *See 'The Climate Change-Displacement Nexus'Presented by Prof. Walter Kalin, Representative of the Secretary-General on the Human Rights of Internally Displaced Persons, Panel on Disaster Risk Reduction and Preparedness: addressing the Humanitarian Consequences of Natural Disasters, ECOSOC Humanitarian Affairs Segment, July 16, 2008, available at http://www.brookings.edu/speeches/2008/0716 climate change Kalin.aspx.*

40. *An analysis of legal aspects of forced displacement occurring in the context of climate change by UNHCR to 6th session of the ad hoc working group on long-term cooperative action (AWG- LCA 6) under UNFCCC. AWG-LCA met in June 2009 in Bonn, for talks which represented on important opportunity for states parties to the UNFCCC to negotiate on a draft in Copenhagen, December 7-18, 2009. Both UNHCR submission papers to the UNFCCC are available on UNHCR's climate change web page, at http://www.unhcr.org/cgibin/texis/vtx/ search/?page=&comid=4a2d26df6&cid=49aea9390&keyworfs=UNFCCC.*

41. *See for example the broader definitions of refugee contained in the OAU convention governing the specific aspects of refugee problems in Africa 1969, at article 1(2), http://www.unhcr.org/refworld/docid/3ae6b36018.html and the Cartagena declaration on refugees 1984, at conclusion 3, available at http://www.unhcr.org/ refworld/docid/4538838e 10.html.*

42. *The rights under the 1966 international covenant on economic, social and cultural rights (ICESCR), at http://www.unhcr.org/refworld/ docid/4538838e10.html.*

43. *1951 Refugee Convention, the mandate includes: victims of manmade disasters and persons of concern of the High Commissioner: see ECOSOC resolution 2011 (LXI) of August 2, 1976, available at http://www.unhcr.org/refworld/docid/3ae69ef418.html.*

44. *See, http://www.humanitarianreform.org/humanitarianreform/default.aspx? tabid=79.*

45. *See, http://www.humanitarianreform.org/humanitarianreform/default.aspx? tabid=672.*

46. *See, http://www.humanitarianreform.org/humanitarianreform/default.aspx? tabid=77.*

47. *IASC Operational Guidelines on the protection of persons affected by natural disasters and the related pilot manual, Brookings-Bern Project on International Displacement, March 2008, available at http://www.unhcr.org/refworld/docid/49a2b8f72.html.*

48. *UNFCCC submission paper, 'protecting the health of vulnerable people from the humanitarian consequences of climate change and climate related disasters of Ad Hoc Working Group on Long-Term Cooperative Action, June 1-12, 2009 by WHO with UNHCR and other agencies, available at http://www.unhcr.org/refworld/docid/4a2d189e1a.html.*

49. *Mixed migration and associated gaps in the international protection regime featured prominently within the High Commissioner's dialogue to protection challenges, held on December11 and 12, 2007, the chairman's summary of the dialogue is available at the following link: http://www.unhcr.org/refworld/docid/479744c42.html.*

50. *Maskrey A. (ed.) (1993) Los disasters no son naturals. La Red/ Intermediate Technology Development Group (ITDG), Lima. http://www.desenredando.org/public/libros/1993/idnsn/index.html.*
   Hewitt, K. (1996) Regions of Risk: A Geographical Introduction to Disaster. Longman, Harlow. Bhatt, M (2002) Corporate Social Responsibility and Disaster Reduction: local overview of Gujarat, case study for corporate social Responsibility and Disaster Reduction. A Global Overview, DFID-funded study conducted by the Benfield Grieg hazard research centre, University College London. http://www.benfieldhrc.org/siteroot/disaster-studies/csr/csr- gujurat.pdf.
   Wisner, B and J. Adams (2003) Environmental Health in Emergencies and Disasters: A Practical Guide. World Health Organisation, Geneva.
   http://www.who.int/watersanitationhealth/hygiene/emergencies/emergenes 2002/en/

51. *UN/ ISDR, 2004.*

52. *UN/ ISDR, 2005: World conference on Disaster Reduction Kobe, January 18-22, 2005, at http://www.unisdr.org/.*

53. *UN/ISDR, 2004: (United Nations Interagency Secretariat of the International Strategy for Disaster Reduction), A Global Review of Disaster Reduction Initiatives. Geneva.*

54. *UNDP (United Nations Development Programme) 2004, Reducing Disaster Risk: A challenge for Development, UNDP. Bureau for Crisis Prevention and Recovery. New York. pp. 146.*

55. *Alexander, D. (2002b) Principles of Emergency Planning and management. Terra publishing Harpenden.*

56. *O' Brien, G. and P. Read (2005) "Future UK emergency management: new wine, old skin?"*
   Disaster prevention and management, 14 (3), pp. 353-361.
   Alexander, D. (2002a) 'from civil defence to civil protection- and back again' Disaster Prevention and Management, I (3), pp. 209-213.

57. *Wisner and Adams, 2003; UNICEF (United Nations Children's Fund) 2005. Emergency Field Handbook: A Guide for UNICEF staff. UNICEF, New York, NY. http;//www.unicef.org/publications/files/UNICEF-EFH-2005.pdf.*

58. *Harmer A. and J. Macrae (eds.) 2004. Beyond the continuum. The changing role of aid policy n protracted crises. Humanitarian Policy Group (HPG) Research Report 18. Overseas Development Institute (ODI), London, http://www.odi.org.uk/hpg/papers/HPGrepoat.18.pdf.*

59. *Hofimann, C.A., L. Roberts, J. Shoham and P. Harvey (2004) Measuring the impact of Humanitarian aid. A review of current practice. HPG Research Report 17. ODI, London.*

60. *Scawthorn, C. (2000) 'Emergency Water Supply and Disaster Vulnerability', In J.*

Uitto and A. Biswas (eds.) Water for Urban Areas, United Nations University Press, Tokyo, pp. 200-225. 61. DFID (2004b) key sheet 06. Adaptation to climate change: making development disaster- proof. DFID, London.

62. Adger, N. and N. Brooks (2003) country level risk measures of climate-related natural disasters and implications for adaptation to climate change. Working paper 2b. Tyndall centre for climate change research, Norwich, http://www.tyndall.ac.uk/publications/working-papers/wp 26.pdf.

63. Heltberg, R., S. Jorgenson and P. Seigal (2008a). "Addressing human vulnerability to climate change: Towards a 'no regrets' Approach, World Bank, Washington D.C. at http://ssrn.com/ abstract=1158177.

64. Sattherwaite, D. et al., "Adapting to climate change in urban areas: The possibilities and constraints in low-and-middle-income nations"; Human Settlements Discussion Paper Series: Climate Change and Cities1, International Institute of Environment and Development, London 2007.

65. Scott, L, 2008, "Climate variability and climate change. Implications for chronic poverty", working paper 108; www.chronicpoverty.org/pubfiles/108.
Chronic Poverty Research Centre (CPRC), University of Manchester.

66. Helberg, R., S. Jorgenson and P. Seigal (2008b) "climate change: challenges for social protection in Africa", World Bank, Washington, D.C. http://ssrn.com/ abstract=1174774.

67. Emanuel, K. (2005) Divine Wind: The History and Science of Hurricanes, Oxford University Press, New York, NY, 2005.

68. VARG (Vulnerability and Adaptation Resource Group) 2005. Disaster Risk Management in a changing climate: a discussion paper, at http://www.unisdr.org/wcdr/

69. O' Brien. G. and P. Read (2004). Future UK Emergency Management: From discretion to regulation-panacea or long overdue reform? Proceedings of the International Emergency Management Society 11th Annual Conference, May 18-21, 2004, Melbourne, Australia.

70. UNDP (United Nations Development Programme) 2004a. Reducing Disaster Risk: A Challenge for Development. UNDP, Crisis Prevention and Recovery, Disaster Reduction Unit, Geneva, at http://www.undp.org/bcpr/disred/english/publications/rdr.htm.

71. Aloisi, S. 'Senegal mulls "Green Wall" to stop Desert Advance'. Reuters August 1, 2005, at http://forests.org/articles/redear.asp?linked=44784.

72. COP 9 (Conference of Parties 9) 2002. 'The Special Climate Change Fund (SCCF), decision 5/cp-9'. COP 9, Milan, at http://unfccc.int/cooperation-and-support/funding/special-climate-change-fund/items/2602. php.

73. The IASC is the primary mechanism for inter-agency coordination of humanitarian assistance. It is a unique forum involving the key United Nations and non-United Nations humanitarian partners. See, www.humanitarianinfo.org/iasc/pageloader. aspx.

74. For more information on IOM's work on migration and development, see. www.iom.int/jahia/jahia/activities/bytheme/migratiodevelopment/cache/offonce and the related links.

75. IOM Guatemala, Survey on Remittances 2008 and Environment, Working Notebooks on Migration, No. 26, 2008.

76. *Sudmeier - Rieux and Ash (2009). Environmental Guidance Note for Disaster Risk Reduction. IUCN: Gland.*
UNEP/UNISDR (2008). Environmental and disaster risk. Emerging perspectives 2nd edition. UNISDR secretariat: Geneva.

77. *ISDR (2009). Global assessment report on disaster risk reduction. United Nations: Geneva, Switzerland.*

78. *Sudmeier- rieux et al., (2006). Ecosystems, livelihoods and disasters- an integrated approach to disaster risk management. Ecosystem Management Series No.4. IUCN.*

79. *ISDR (2009).*

80. *Millennium Ecosystems Assessment (2005). Ecosystems and human well-being: current state and trends: finding of the condition and trends working group.*

81. *Sudmeier - Rieux and Ash (2009:pp.1.2)*

82. *World Bank (2010); Campbell et al., (2009); Sudmeier-Rieux and Ash (2009); UNEP (2009a); Dolcemascolo (2004). For a review of literature on the links between biodiversity and climate change, see the convention of Biological Diversity, Technical Series 42.*

83. *Pro Act Network (2008), Sudmeier-Rieux and Ash (2009), World Bank (2010);*
Pro act network (2008). The role of Environmental Management and co-engineering in disaster risk reduction and climate change adaptation.

84. *Harakunarak and Aksornkoae "Life-saving belts: post-tsunami re-assessment of mangrove*
Ecosystem values and management in Thailand", Tropical coasts, 12 (1), pp. 48-55.

85. *Inter-governmental Oceanographic Commission (2009). Tsunami risk assessment and mitigation for the Indian Ocean. Knowing your tsunami risk and what to do about it. UNESCO Manuals and Guides 52.*

86. *World Bank (2010).*

87. *World Bank (2010); Sudmeier-Rieux and Ash (2009).*

88. *FAO and CIFOR (2005). Forests and floods drawing in fiction or thriving on facts? Van Dijk et al., (2009). Forest-flood relation still tenuous-comment on 'global evidence that deforestation amplifies flood risk and severity in the developing world'. Global Change Biology. 15, pp. 110-115.*

89. *Sudmeier - Rieux and Ash (2009): p. 11.*

90. *International Federation, World Disasters Report, 2003.*

91. *Ibidem*

92. *IPCC Fourth Assessment Report, Working Group I, Glossary of Terms: http://ipcc-wg1.ucar.edu/wg1/Report/Ar4wg1-Print-Annexes.pdf.*

93. *Living with Risk: A global review of disaster risk reduction initiatives, preliminary version, UNISDR, Geneva, 2002.*

94. *IPCC TAR, Working Group II: Impacts, Adaptation and Vulnerability, Summary for Policy Makers.*

95. *IPCC TAR Working Group II, Impacts, Adaptation and Vulnerability, Chapter 9 (Human Health).*

96. *Presentation at the 6th Conference of Parties to the United Nations Framework Convention on Climate Change, November 13, 2000.*

97. *Greenhouse gas 'Sinks': IPCC Fourth Assessment Report Working Group II, Glossary of Terms: http://195.70.10.65/pdf/glossary/ar4-wg2.pdf.*

98. *IPCC Fourth Assessment Report, Working Group II, Glossary of Terms:*

http://195.70.10.65/pdf/glossary/ar4-wg2.pdf.

99. ADB et al., (2003), "poverty and climate change: reducing the vulnerability of the poor through adaptation", VARG multi development agency paper, United Nations Development Projects (UNDP) United Nations, New York.
    Stern et al (2006), "Stern review on the economics of climate change". www.hm-treasury.gov.uk/sternreview-index.htm,HM Treasury, London and Cambridge University Press.

100. Stern, N. (2008), "Key Elements of A Global Deal on Climate Change London", London School of Economics and Political Science, London.

101. UNISDR (2004): Terminology of Disaster Risk Reduction, at http://www.unisdr.org/eng/library/lib-terminology-eng%20 home.htm.

102. Hyogo Framework for Action 2005 - 2015: Building the resilience of nations and communities to disasters (HFA), see part B, paragraph 4(i)C, at http://www.unisdr.org/hfa/hfa.htm.

103. The term "at all levels" encompasses community, municipal, province, national, regional and international levels.

104. ISDR (International Strategy for Disaster Reduction) at http://www.unisdr.org

105. UNDP (United Nations Development Programme) at http://www.undp.org 106. UNEP (United Nations Environmental Programme) at http://www.unep.org

107. UNICEF (United Nations Children's fund) at http://www.unicef.org

108. UN-Habitat, at http://www.unhabitat.org

109. WMO (World Meteorological Organisation) at http://www.wmo.ch

110. WHO (World Health Organisation) at http://www.fao.org

111. FAO (Food And Agriculture Organisation) at http://www.fao.org

112. IAEA (International Atomic Energy Agency) at http://www.iaea.org

113. The World Bank group, at http://www.World Bank.org

114. Linking Disaster Risk Reduction, Climate Change and Development, Global Platform for Disaster Risk Reduction, Information Note 1, at http://www.preventionweb.net/globalplatform/first-session/docs/media-docs/info-note-I-HL-   dialogue-climate-change.pdf.

115. Mills, E., Locomte, E. (2006): From Risk to Opportunity: How insurers can Proactively and Profitably Manage Climate Change, at http://eetd.lbl.gov/Emills/puBs/PDF/ceres_insurance_climate_repreport_090106.pdf.

116. IPCC 2012. Managing the Risks of Extreme Events and Disasters to Advance Climate Change Adaptation. A Special Report of Working Group I and II of the Intergovernmental Panel on Climate Change. Field, C.B., V. Barros, T.F. Stocker, D. Qin, D. J. Dokken, K. L. Ebi, M. D. Mastrandrea, K. J. Mach, G. K Plattner, S. K. Allen, M. Tignor and P. M. Midgley, eds., Cambridge University Press, Cambridge and New York.

117. Hallegatte, S., Corfee - Morlot, J., Green C., and Nicholls, R. J. 2013. Future Flood Losses in Major Coastal Cities. Nature Climate Change, doi: 10-1038.

118. IPCC, 2013. Climate Change 2013: The Physical Science Basis. Summary for Policy Makers. Working Group I Contribution to the IPCC Fifth Assessment Report.

119. IPCC, 2013.

120. Peterson, T. C., M. P. Hoerling, P. A. Stott and S. C. Herring, eds. 2013. Explaining Extreme Events of 2012 from a Climate Perspective. Special Supplement Bulletin of the American Meteorological Society 94(9).

121. *Skoufias, E., (ed.) 2012. The Poverty and Welfare Impacts of Climate Change: Quantifying the Effects, Identifying the Adaptation Strategies. World Bank, Washington, DC.*

122. *Hallegatte, S. 2012. A cost effective solution to reduce disaster losses in developing countries: Hydro-Meteorological Services, Early Warning and Evacuation. Policy Research Working Paper 6058, Sustainable Development Network, World Bank, Washington, DC.*

   Rogers, D., and V. Tsirkunov, 2013. Weather and Climate Resilience: Effective Preparedness through National Meteorological and Hydrological Services. Direction in Development; World Bank, Washington, DC.

123. *World Bank, 2013c. Zambia strengthening climate resilience (PPCR Phase II) Project, Report Number: 73982-zm, Washington, DC.*

124. *Rogers and Tsirkunov, 2013.*

125. *World Bank 2013b. Managing Risk for Development. World Development Report 2014. Washington, DC.*

126. *At the 13th COP, Bali December 2007. It provides the roadmap toward a new international climate change agreement to be concluded by 2009, and that will ultimately lead to post-2012 international agreement on climate change.*

127. *Stockholm Plan of Action for Integrating Disaster Risks and Climate Change Impacts in Poverty Reduction, at http://www.unisdr.org/eng/partner-netw/wb-isdr/docs/stockholm-plan-of-action.pdf.*

128. *"Sub-Paragraphs 1(c) (ii, iii) of the Bali Action Plan: Back ground and options for Reducing Disaster Risks". Informal paper prepared by secretariat of the international strategy for Disaster Reduction (UN/ISDR) for UN Climate Conference, Bangkok, April 1-5, 2008, p. 6.*

129. *The IPCC produced in 2012, this special report on extreme events "SREX" highlighting inter- linkages between climate change adaptation and disaster risk reduction. It is available on: http://ipcc-wg2.gov/SREX/*

130. *The UN Global Assessment Report on Disaster Risk Reduction is a biennial report of the United Nations coordinated and produced by the UNISDR, at http://www.preventionweb.net/ english/hyogo/gar/.*

6. CLIMATE CHANGE INDUCED DISPLACEMENT AND UN

Introduction

International Community has published the scientific aspects of climate change with the aim, understanding and process. The climate change has created humanitarian problems and challenges.[1] The United Nations High Commissioner for Refugees (UNHCR) encourages the reflection on humanitarian and displacement challenges that climate change occurs. The legal framework for the protection of Internally Displaced Persons (IDPs) has been made for disasters of climate change induced displacement persons in developing countries.[2]

Climate change has wiped-out many schools, security, income and started deteriorating the qualities of life among the people. The Norwegian Refugee Council indicated 20 million people have been displaced by natural disasters in 2008 alone.[3] Land becomes less productive due to temperature rise, the food and energy will be costly and many diseases will be increased due to climate change. Issue of Displacement has become more problematic with unpredictable disasters such as cyclones, floods and mudslide rising due to climate change.

Due to climate change, many serious global problems are created such as disrupted weather patterns, violent storms, rising global temperatures and rising Ocean levels. By this reason climate change has displaced people from a small poor family to rich whole population. These displaced people are called, 'Climate Change Refugees' or 'Environmental Refugees'.

United Nations applies the definition of refugee to only Legal refugee. "A Legal refugee is a person who has fled his or her country due to persecution by the state for reasons of a particular social group or political opinion".[4] The United Nations High Commissioner for Refugees (UNHCR) feels that, "the refugees' convention (1951) is to include climate refugees which may reduce protection for the conventional political refugees."[5] In 1995, Mires and Kent from Green

College, Oxford University, England defined, "persons who no longer gain a secure livelihood in their traditional homelands because of unusual scope".[6]

Another definition on the climate change refugees is that, "people living in an environment where slow motion disasters have occurred or will occur. These disasters are caused by human activity and by the earth's natural cycles of glaciers and warming. Droughts and the rising sea levels are signs of the disease, famine and immigration.[7]

The United Nation Environmental Programme (UNEP) is the only agency which recognises, "the existence of environmental refugees".[8] The term Environmental migrant is used with the similar terms such as forced environmental migrant, environmentally motivated migrant, climate refugee, Environmentally Displaced Person (EDP) disaster refugee, eco-refugee, ecologically displaced person and Environmental-Refugee-To-Be (ERTB).[9] Due to sudden or gradual changes on natural environment these three related impacts of climate change are sea level rise, extreme weather events, drought and water scarcity.[10]

Causes for environmentally displaced persons

There are five factors for environment displacement. These are natural disasters, ecosystem degradation, development, industrial accidents and war.

- **Natural disasters-** Natural disaster means droughts, floods, tropical storms, earthquakes, heavy temperature, sea level rise, famine etc. When the oceans get warmer, as a result, the global climatic change course becomes more powerful as climate causes natural disasters worldwide.[11] In China, the earthquakes killed and injured 90,000 residents.[12] In November 2009, 500 people died and 11,000 people were missing in flood at town of Jeddah in Saudi Arabia.[13]
- **Ecosystem degradation-** Another factor of environmental displacement is the gradual degradation of natural eco-systems such as air pollution, desertification, deforestation, so it causes erosion and potable water availability.[14]
- **Development-** The government's development plan, infrastructure and uncontrolled building, development projects make the native people homeless and displaced for government's profit motive activities. For example, the Narmada Dam project in India will "submerge an area of land greater than the size of New Delhi".[14] Due to this reason thousands of Indians became homeless and suffered disease and hunger.

- **Industrial accidents-** Another factor of environmental displacement is industrial accidents. In the city of Kiev, Ukraine, 1,00,000 people were permanently displaced around a 30 mile radius surrounding the accident. In 1976, 1, 00,000 families became homeless due to high social cost of large scale industrial accidents in Pennsylvania.[15]
- **War-** The last course of environmental displacement and environmental gradation is war. Two prominent examples are the Vietnam War and Saddam Hussein's arson of oilfields of Kuwait which would be a threat to world peace and environment.[16]

Types of displaced persons

The Representative of the Secretary General on the Human Rights of Internally Displaced persons, Walter Kalin, has identified five climate change scenarios that directly or indirectly cause human displacement. How to protect those affected displacement, it is described here.[17]

- Hydro-meteorological disasters (flooding, hurricanes/typhoons/cyclones, mudslides, etc.)
- Government-initiated planned evacuation of areas at high risk of disasters. This is likely to lead to permanent internal displacement.
- Environmental degradation and slow onset disasters, such as reduced water availability, desertification, recurrent flooding and increased salinity in coastal zones. Kalin explains "such deterioration may not necessarily cause displacement, but it may prompt people to consider 'voluntary' migration as a way to adapt to the changing environment and be a reason with better living conditions and income opportunities. However, if areas become uninhabitable over time because of further deterioration, finally leading to complete desertification, permanent flooding of coastal zones or similar situations, population movements will amount to forced displacement and become permanent.18
- Small Island countries are at risk of disappearing because of rising seas. Kalin notes that current International Law provides no protected status for such people and even if they were to be treated as 'stateless'. Small island countries such as Kiribati and Tuvalu emit less than 1 per cent of global greenhouse gases, their small physical size, exposure to natural disasters and climate extremes, very open economics and law adaptive capacity make them particularly susceptible and less resilient to climate change.[19]
- Risk of conflict over essential resources, violent conflict due to decrease of essential resources such as water, land and food. People

displaced by conflict may be eligible for protection as refugees or assistance as IDPs, resource - based conflicts may be particularly challenging at the operational level.

The scenario described above involves different kinds of pressures and impacts which will affect the time, speed and size of movement. Hydro metrological disasters or environmental degradation cause internal displacement. The national and local governments have a vital role to tackle these scenarios. In accordance with the 1998 guiding principle on internal displacement, the IDP should receive protection and assistance. Due to dangerous disasters, people cross the international border as refugees for protection within International refugee framework and their "status remain unclear".[20]

The Low Lying small island states are affected by rising sea levels. The territory is wiped out and the human's life is not saved and they are compelled to remain as international refugees. Another cause of human displacement is a decrease in vital resources e.g. water, land and food production which create conflict and violence due to climate change.

Ecological refugees and the 1951 UN refugees convention
The term ecological refugees include with environmental refugees or climate refugees and categories of displaced persons fleeing various industrial and chemical hazards.[21] Essam El-Hinnawi, defines "environment disruption as any physical, chemical or biological changes in the ecosystem temporarily or permanently unsuitable to support human life."

The UN talks about ecological refugees' that 25 million people who have been forced to leave their homes because in their own environment the degradation of nature, the draught, flooding and the desertification makes living impossible. It is in the hands of man that the building of great dams, the contamination from mining and oil extraction, the devastation caused by the mono-cultures for export and the fevers of wild urbanisation.

The word 'refugee' is a legal term. There are many steps which have been taken for refugees in the 1951 Refugee convention, the 1961 OAU convention or UNHCR's mandate. Some states and NGO's have suggested that the 1951 Refugee convention should be amended to include people who have been displaced across borders as a result of long term climate change or sudden natural disasters.

The term refugee applies to international cross border people. There are difficulties due to climate changes as 'persecution' entails violations of human rights that are serious either of their inherent nature or their repetition.[22]

Although adverse climate impacts such as rising in sea-levels, salination and increase in the frequency and severity of extreme weather events such as storms, cyclone and floods are harmful. The governments of Kiribati and Tuvalu are not responsible for climate change as a whole, nor do they develop policies which increase its negative impacts on particular sectors of the population. The persecutor in such a case is the international community and industrialised countries whose failure to cut greenhouse gas emissions is a predicament, now being faced.[23]

Refugee convention requires such persecution to be on account of an individual's race, religion, nationality, political opinion or membership of a particular social group. Climate change adversely affects some countries, their geography and resources. An argument that such affected social groups' would be difficult to establish because the law requires that the group must be connected by a fundamental, immutable characteristic other than the risk of persecution itself.[24] People fleeing natural disasters and bad economic conditions falls outside the convention.[25]

What is an environmental refugee?
The term environmental refugee was first used by Black (2001:2) Myers' estimates of environmental refugees are driven by three major sources: population growth, sea-level rise and an increase in extreme weather events.[26] He agrees in the forceful displacement of people living in densely populated, low lying regions such as the Nile Delta, the East coast of China and Bangladesh.[27]

Due to natural disasters, drought, flood, famine and other environmental factors, refugees are increasing across the globe. Refugees have increased by the environmental events and stresses from dust bowl to the recent Tsunami. In 1951, the United Nations General Assembly adopted resolution 429 which defined refugees as, "Those having well-founded fear of being persecuted for reasons of race, religion, nationality, membership of a particular social group or political opinion is outside the country of his nationality and is unable, or owing to such fear is unwilling to avail himself of the protection of that country; or who, not having a nationality and being outside the country of his former habitual residence as a result of such events, is unable or, owing to such fear, is unwilling to return to it."

In 1990, the Intergovernmental Panel on Climate Change (IPCC) predicted that, "The millions of human migrations are displaced by the effects of climate change happened due to shore line erosion, coastal flooding and severe drought. The climate change will have increasingly dramatic impacts on ecological and social systems.

The International organisation recognises three types of environmental migrants: environmental emergency migrant, environmental forced migrants and environmental motivated migrants. Environmental migrants: persons who are displaced temporarily but who can return to their original home when the environmental damage has been repaired; persons who are permanently displaced and have resettled elsewhere and persons who migrate from their original home in search of a better quality of life when their original habitant is degraded to such an extent that it does not meet their basic needs.[28]

Essam El-Hinnawi in 1985 defined environmental refugees as: "Those people who have been forced to leave their traditional habitat, temporarily or permanently, because of a marked environmental disruption that jeopardised their existence or seriously affected the quality of their life". UN statistics division defines an environmental refugee as simply, "a person displaced owing to environmental causes, land loss and degradation and natural disasters".

Climate-based displacement may be
* Temporary: a climate event such as a hurricane, flood, storm or Tsunami where people are able to return to their homes.
* Permanent local displacement: groups are displaced locally, but on a permanent basis due to sea level rise, storm, coastal inundation and the lack of clean water.
* Permanent internal displacement where relocation is within the national borders but so far from the group's original location that the move is permanent.
* Permanent regional displacement: where solutions within the national territory and migrants must be received permanently in other countries in the region.
* Permanent inter-continental displacement: an internationally-coordinated relocation is required.
* Temporary regional or international displacement where groups are received temporarily while permanent solutions are planned.

Climate and environment change not only test the international legal frame works and concepts but also test the concept of state control over and management of territory.[29] An estimated "26 million of the 350 million displaced worldwide are considered climate displaced people. Of these, one million each year are estimated to be displaced by weather-related disasters brought on by climate change."[30] The first Assessment Repot of the IPCC in 1990 estimated that by 2050, some 150 million people could be displaced due to climate change.[31]

Climate change induced migration

Environmentally induced migration is a complex and changing human environmental systems.[32] Climate change will be on urban and rural areas, sea- level rise, flooding, storms, droughts and violent hazardous events in regional climates makes people's livelihood systems stressed. It is useful to discuss the implications of global environmental and societal changes in the context of governance systems. The current governance systems address many issues that environmentally-induced migration raises or whether new governance modes are needed.[33]

International legal frameworks and operational mechanisms

Many different international agreements, guiding principles, norms and institutions shape governance of human mobility. These governance includes the 1951 UN convention and 1967 protocol on the status of Refugees (UNHCR). The International Organisation for Migration (IOM) and the migration working group surveyed the limitation of international co-operation on human mobility. These agreements do not involve significant commitments. Worldwide national, regional and international systems exist to address the humanitarian and other aspects related to natural hazards, both rapid - on set and slow onset. The global environmental change, a wide national, local and international rules, norms, treaties and organisations exist, but few have considered the interactions of eco-systems and human mobility.[34]

Governments and humanitarian actors challenge on rapid developing issue

Environmentally induced migration is closer to adaptation. Migration is often a pro-active risk diversification strategy for households facing environmental stresses amidst a range of other risks that must be managed at the same time.[35]

- **Opportunities and challenges** - An opportunity and challenge for governance systems is to create policies and actions that are highly dynamic and non-linear processes. It improves the education and training and livelihoods in communities affected by environmental change.

- **New treaty** - There is new international treaty to address the movement of people displaced by climate change.
  - The 1951 Refugee convention, a protocol to the UN Framework Convention on Climate Change (UNFCCC) for reducing displacement due to climate change.[36]
  - The Mauritius strategy for further implementation of the programme of action for the sustainable development of small island developing states.
  - The pacific islands framework for action on climate change (2006 - 2015).
  - The Niue declaration on climate change 2008 which refers not only to adaptation strategies but also economic and security threats and the possible need for relocation.
  - The UN Secretary-general Ban Ki-Moon announced in 2008 that establishment of an inter-agency climate change centre to support countries in the pacific region.
  - UN Human Rights Council decided to investigate the issues of human rights and climate change.
  - UN's various policy issues relates to climate change and population displacement working group of the UN Interagency Standing Committee (IASC) where UNHCR is directly involved.

There are four primary types of climate-induced displacement. Weather-related disasters, such as hurricanes and flooding; Gradual environmental deterioration and slow onset disasters such as desertification, sinking of coastal zones and submersion of low-lying island states; Increased disaster risk resulting in relocation of people from high risk zones; and social violence attributable to climate change related factors.

An estimate of environmentally induced displacement
Some of the more prominent estimates of environmentally induced displacement are as follows:-
- The International Federation of Red Cross and Red Crescent societies (IFRC) estimated in 2001 for the first time the number of environmental refugees displaced by war.
- UNHCR (2002-12) estimated that 24 million people around the world who had fled because of floods, famine and other environmental factors.
- El-Hinnawi estimates there are already some 30 million environmental refugees, while Director of UNEP Klaus Toepfer

predicts it has already reached more than 50 million and IPCC predicts 150 million environmental refugees by 2050 equivalent to 1.5 per cent of 2050s predicted global population of 10 billion.[37]

- The American statement (1994) observed that 135 million people could be at risk of being displaced as a consequence of severe desertification.
- Myers, in 1993 who predicted 150 million environmental refugees, now believes the impact of global warming could potentially displace 200 million people (Myers 2005).
- The Stern Review, Commissioned by the UK Treasury, agrees 200 million displaced by 2050 (Stern 2006).
- Nicholls (2004) suggested that between 50 and 200 million people could be displaced by climate change by 2080.
- Friends of the Earth (2007-10) predicts climate refugees at 200 million Worldwide and 1 million from Small Island states by 2050.
- UNEP agrees that by 2060, there could be 50 million environment refugees in Africa alone.
- Christian Aid in 2007 postulated that a billion people could be permanently displaced by 2050- 250 million by climate change-related phenomena such as droughts, floods and hurricanes and 645 million by dams and other development projects.

International law for climate -forced -displacement
The report in 2007 of IPCC highlights the climate change and its severe impacts on the environment and human lives urgently need action. The Bali Action Plan recognises adaptation and risk management as important elements to be addressed in the climate change negotiations and agreement. The Bali Action Plan article I (c) addresses the protection of internally displaced persons and persons displaced across international borders.[38]

State parties agreed that there is clear link between the effects of climate change and displacement. They indicate the following reasons for displacement
- Mitigation of climate change by reducing greenhouse gases in accordance with the UNFCCC, its Kyoto Protocol in preventing displacement.
- The UNFCCC's National Adaption Programmes for Actions should systematically address the issues related to displacement.
- The protection of Internally Displaced Persons (IDPs) under human rights law and the guiding principles on International displacement.

- Those persons across international borders, their protection are the responsibility of states under the 1951 convention and international and regional refugee law and human rights law.
- Enhancing states' ability to protect people on their territories fall squarely within the nation of adaptation.
- The states should give priority to vulnerable people, displaced persons or those affected by climate change.

International law addresses to internal displacement in the context of climate change
Under International Law, the states face three level challenges:-
- Addressing the cause: mitigating Climate change.
- Addressing the effects: Reducing risks created by climate change and vulnerabilities.
- Addressing the consequences: Protecting individuals displaced by the effects of climate change.

Protection from the threats of natural hazards: Prevention of displacement evacuations and relocation
The state's first duty is to protect the lives of their population from drastic disasters. The states should take steps to safeguard the life, limb and property which is their jurisdiction against the threats of disasters. The states should take natural hazards in obligations under international human rights law. Some measures include disaster risk mapping, early warning systems, predetermination of humanitarian aid of evacuation routes, prepositioning of humanitarian aid, building capacities of local communities to deal with disasters, evacuation and permanent relocations away from danger zones.

Special protection for affected persons in the time of displacement
The most vulnerable groups of society including the poor, marginalised minorities' female and child headed households, chronically ill persons, persons with disabilities and older people without family support- suffer from the negative effects of natural hazards due to their weakened adaptation capacities.

Restoring the rights of the internally displaced
The natural disasters or environmental degradation can have the potential for displaced persons. Lack of sustainability of durable solutions perpetuates the displacement situation and states risk a violation of international law. Therefore, states must act to make solutions sustainable.

Some elements of sustainability:
- Information on the process, consultation with and participation of the affected communities.
- Safety.
- Recovery of land property upon return, including through settlement of property and land disputes.
- Physical needs.
- Livelihoods.
- Participation.

Climate change and cross-border displacement
The sudden and slow-onset disasters affect people severely not only within state territories but also across international borders and community. The occurrence of disasters in countries raises one important question that is the admissibility of forcible returns of foreigners to their country.

The International protection for refugees
According to the 1951 convention relating to the status of refugees, as modified by the 1967 protocol, in Latin America, the 1984 Cartagena inspired Declaration on Refugees, which has inspired the legislation of many states in the region contains massive violation of human rights and internal conflicts. The 1969 OAU convention governing the specific aspects of refugee problems in Africa includes within the refugee category those persons who are compelled to flee due to events seriously disturbing public order.

Persons displaced across international borders but not qualifying as refugees
Majority of persons leaving their countries are not to qualify as refugees under International Law. In extreme disaster scenarios, the state is unable to advocate with other states on behalf of its citizens in distress. For this reason some advocate for the protection of such persons and have suggested amending the 1951 convention. The International community must support and strengthen states abilities to protect their own citizens, both from the displacement and during displacement.

The principle of prohibitions of returns
The 1951 convention contains in its Article 33(i) a prohibition to return a refugee, "to the frontiers of territories where his life or freedom would be threatened". Thus, it is established that no person, regardless of states or

conduct, may be returned in any manner whatsoever to a country where his or her life or integrity would be at risk.

In future if climate change does cross the threshold as earmarked by IPCC, the number of climate change induced refugees will tend to increase; the refugees will be on their heels for shelter, protection of livelihood and security transcending their territorial boundaries. This will turn to be the greatest humanitarian crisis impinging upon the United Nations to tackle the crisis which remains unprecedented and where states are still clinging to Westphalian tradition of impenetrability of territorial sovereignty. How many states will come forward to accept these environmentally displaced persons coming from other states? The migration will be generally from developing and poor and vulnerable countries to developed regions, areas and states. The question that remains perturbing to the United Nations is whether the developed states will open their borders sacrificing their economy, way of living and law and order. This challenges the capability and potential of the United Nations to tackle the crisis. What remains pivotal to this is the determined willingness of the developed countries to reduce emission of carbon dioxide to the extent that will forbid the climate change from crossing the danger point.

References

1. *Inter-Agency standing committee's Informal Task Force on climate change: - http:// www. humanitarianinfo.org/iasc/page Loader, aspx? page=content-news-news details & news id = 134*
2. *Guiding principles on internal Displacement at http://www2.ohchr.org/english/ issues/idp/ standards.htm*
3. *Elverland, S,"20 Million climate displaced in 2008,"Norwegian Refugee council, June 8, 2009, at http://www.nrc.no/? did= 9407544.*
4. *Policy: - Climate change refugees seek a new international deal (2008) climate news for business, climate change corp. http corp.com/content.asp? Content ID= 5871, cited 2009 June 26.*
5. *ibid. cited June 26, 2009.*
6. *KolmannsKog, V.O., Future floods of refugees. A comment on climate change, conflict and forced migration (2008) Norwegian Refugees council at http://www. nrc. no/arch/- img/9268480. pdf. cited Jun 26, 2009.*
7. *(Cascio, J, Environmental refugee (2005) world changing, change. Your thinking, Time Magazine at http://www. world changing.com/archives/003618.html. cited Aug 10, 2009.*
8. *Reed's, environment and security (2007) Topics/ core issues, climate institute, at http://www. climate. org/topics/environmental security/index.html. cited Jun 30, 2009.*
9. *Boano, C, Zetter, R, and Morris, T, (2008) Environmentally displaced people*

understanding the Linkages between environmental change, Livelihoods and forced migration, Refugee study centre policy Brief NO.1 (RC : Oxford), pg. 4.

10. *Global Governance project (2012). Forum on climate Refugees< Retrieved on May 5, 2012.*

11. *Lambert, J. (2002). Refugees and the environment: The forgotten element of sustainability Brussels, U.K. European Parliament.*

12. *Gore, A (presenter), Guggenheim, D. (Director) (2006). An inconvenient truth: A global warming (Motion picture). United States paramount pictures.*

13. *Flor Cruz, J (2009, May 13) China Marks earthquake anniversary C N N., retrieved from wwww.cnn.com*

14. *Al- Ahmed A. (2009, December, 3). Jeddah flood deaths shame Saudi royals, The Guardian, retrieved from http://www.guardian.co.uk*

15. *Lopez. A (2007). The protection of environmentally displaced persons in international Law. Environmental Law, 37 (2), pp. 365-409.*

16. *Van wormer, K, Besthorn, F.H. & Keefe, T (2007). Human behaviour and the Social environment Mavro Level: groups, communities and organisations, Oxford University Press, New York, NY.*

17. *Kalin, op cit.*

18. *Kalin, note 40 above, 85.see also Mc Adam and soul, note 41 above, on secondary movement.*

19. *N.Mumura et al., 'Small Islands' in IPCC, note 28 above, 692-93.*

20. *Kolin, op cit.*

21. *Westra, Laura, 2009, Environmental Justice and the rights of ecological refugees, Earth Scan, London, UK*

22. *Council directive 2004/83/EC of 29 April 2004 on minimum standards for the qualification and status of third country Nationals or state less persons as refugees or as persons who otherwise need International protection and the content of the protection granted (2004) OJL 304/12.*

23. *IPCC, climate change: The IPCC scientific Assessment, note 8 above, 7 (fn omitted); IPCC, climate change 2007: synthesis Report, note 8 above, 5;6;12;13.*

24. *Goodwin- gIll and M c Adam, note 50 above,>9-80; Applicant A V. minister for Immigration and Ethnic Affairs (1997) 190 C L R 225, 341 (Dawson J).*

25. *Minister for Immigration V. Haji 1brahim (2000) HCA 55; 204 CLRI, para 140.*

26. *Myers, N. 1993 Environmental refugees in a global warmed world, Bioscience, 43 (ii), PP 752- 761.*

27. *Myers, N. and Kent, J, 1995. Environmental exodus: An emergent crisis in the global arena. Washington D.C. Climate Institute.*

28. *El-Hinnawi, E. (1985) Environmental Refugees, Nairobi: UNEP*

29. *Summary of IASC expert meeting on Migration/Displacement and climate change 15th September 2008.*

30. *Global humanitarian forum 2009, The Anatomy of a silent Crisis, Human impact report climate change, Geneva.*

31. *OLi Brown, ' The numbers game' in forced migration review, Vol-31 October 2008.*

32. *Berkes, F, J. Colding, and C. Folke (eds.) 2003. Navigating social - ecological systems: building Resilience for complexity and change, Cambridge University Press, Cambridge, UK*

33. *Pierre, Jand, G.B Peters, 2005 .Governing complex societies. Palgrave McMillan.*

34. *Brown, O. 2008, migration and climate change in international organization for migration. (IOM) Research series no.31 (IOM, Geneva, 2008)*
35. *Berkes,F, J.Colding and C.Folke (eds.).2003.Navigating social-ecological systems:building resilience for complexity and change, Cambridge University Press, Cambridge, U.K*
36. *UN framework convention on climate change (May 1992). Frank Biermann and Ingrid Boas, preparing for a warmer world: Towards a global governance system to protect climate refugees (2010).*
37. *IPCC, (2007) Climate change 2007: Impact, Adaptation and Vulnerability. Working group II contribution to the Intergovernmental panel on climate change 4th Assessment Report. Geneva: IPCC*
38. *(a) Change Migration and Displacement: Who will affected?' Working paper submitted by the informal group on Migration /Displacement and Climate Change of the IASC -31 October 2008 to the UNFCCC Secretariat.*
    (b) "Disaster Risk Reduction Strategies and Risk Management Practice: Critical Elements for Adaptation to Climate Change" Submission to the UNFCCC Adhoc Working Group on Long Term Cooperative Action by The Informal Taskforce on climate change of the Inter-Agency Standing Committee and The International Strategy for Disaster Reduction 11 November 2008

7. ROLE OF UN IN MEETING THE THREATS OF CLIMATE CHANGE

Introduction

In the 19th century, awareness began that carbon dioxide in the Earth's atmosphere could create a 'greenhouse effect' and increase the temperature of the planet. By the middle of the 20th century, it was becoming clear that human action had significantly increased the production of these gases and the process of 'global warming' was accelerating. Today, all scientists agree to stop this process of natural disasters that will change life on earth.

The UN is in the forefront of the effort to save the planet. In 1992, its 'Earth Summit' produced the 'United Nations Framework Convention on Climate Change'. In 1998, the 'World Meteorological Organisation (WMO) and the United Nations Environment Programme (UNEP)' set up the Inter-governmental Panel on Climate Change (IPCC) to provide objective source of scientific information. And the convention's 1997 'Kyoto Protocol' which set emission reduction targets for industrialised countries. The UN has always taken the lead in taking on climate change. In 2007, the 'Nobel Peace Prize' was awarded jointly to former United States vice-President Al Gore and the IPCC for their efforts to build up and disseminate greater knowledge about man-made climate change. The 'Copenhagen Accord' was agreed by Heads of states, Heads of Government, Ministers and other heads of delegation at the UN climate change conference in Copenhagen in December 2009. Paris climate change negotiations in 2015 ended by all the nations of the world brought an end to the stalemate that stood in the way of the successful negotiation on climate change.

Climate change talks in 'Cancun' concluded with a package of decisions to help countries advance towards a low emissions future in December 2010. 'Cancun Agreements' have taken decisions to mitigate adverse impact and protect the world's forests.

In 2011, the world population reached 7 billion. It is expected to grow to 9 billion by 2043, placing high demands on the Earth's resources. In 2011, the UN climate change conference in Durban, South Africa, decided to adopt a universal legal agreement on climate change as soon as possible, but not later than 2015. In December 2012, after two weeks of negotiations at Doha conference, nations moved forward on climate change and extended the Kyoto Protocol.

At the 11th Conference of the Parties (COP) of the United Nations Framework Convention on Climate Change (UNFCCC) in Montreal Canada, 2005, parties agreed to initiate the 'Dialogue on long-term cooperative action to address climate change by enhancing implementation of the convention.[1] It centered on four themes:

* Advancing development goals in a sustainable way;
* Addressing action on adaptation;
* Realising the full potential of technology; and
* Realising the full potential of market-based mechanisms.

Its mandate expired in Vienna. In August 2007, the dialogue led powerfully into the Bali Action Plan and the Ad Hoc Working Group on long-term cooperative Action under the convention that is mandated with its implementation.[2]

The Bali Action Plan picked up adaptation, technology transfer and market based-mechanisms (finance and investment) as pillars for action and advancing development goals that include cross-cutting objective, other elements of the plan and pillar of mitigation.

Environmental changes can threaten global, national and human security. Environmental issues include land degradation, climate change, water quality and quantity and the management and distribution of natural-resources (e.g. oil, forests and minerals). These factors are the direct causes to poverty, migration, small arms and infectious diseases. Experts predict that climate change will trigger enormous physical and social changes like water shortages, natural disasters, decreased agricultural productivity, increased infectious diseases and human migration. Global scientific consensus predicts, the earth's atmosphere is warming rapidly due to human activity. These changes could impact on international security due to competition for natural resources, destabilising weak states and increasing humanitarian crises. The United Nations' Intergovernmental Panel on Climate Change (IPCC) and multilateral body predict that global warming will trigger enormous physical and social changes.

Physical effects
Physical effects of climate change include:
* Higher average surface and ocean temperatures.

- More rainfall globally from increased evaporation.
- More variability in rainfall and temperature with more frequent and severe floods and droughts.
- Rising sea levels from warming water, expanded further by runoff from melting continental ice fields.
- Increased frequency and intensity of extreme weather events such as hurricanes and tornadoes extended ranges and seasons for mosquitoes and other tropical disease carriers.[3]

Socio-economic effects

The adverse socio-economic effects of climate change include:

- Shortfalls in water for drinking and irrigation with concomitant risks of thirst and famine.
- Changes and possible declines in agricultural productivity stemming from altered temperature rainfall or pest patterns.
- Increased rates and geographic scope of malaria and other diseases.
- Associated shifts in economic output and trade patterns.
- Changes and possibly large shifts in human migration patterns.
- Large economic and human losses attributable to extreme weather events, such as hurricanes.

Threats of climate change and security implications

Climate change impacts on quality of natural resources such as fresh water, arable land, coastal territory and marine resources. Some researchers have speculated that these changes could cause armed conflict and violence. Environment and armed conflict is well established. Competition for natural resources such as diamonds, timber, oil, water and narcotics have motivated violence in places such as Kuwait, Colombia and Afghanistan. Global warming is to be the primary cause of any particular armed conflict. Regional climate changes, as with other causes of environmental degradation, could make armed conflicts more likely.

A warmer world will generate more natural disasters and humanitarian crises. Natural disasters are already a major security threat between 1990 and 1999; an estimated 188 million people per year were affected by natural disasters, 6 times more than the 31 million annually affected by armed conflict.[4] Refugee and Internally Displaced Persons (IDPs) are vulnerable not only to the physical and socio-economic effects of disease, malnutrition and loss of income, but they can also become personally insecure and subject to crime, violence and broader militarised conflict. Natural disasters means a country lacks the capability or willingness to help affected population, under-mining the government's legitimacy and increasing popular grievances, that would pose wider security challenges.

The drought, disease and economic stagnation may reach a critical level which becomes failure of the state. The HIV/AIDS that lead to international widespread death from infectious diseases could destabilise vulnerable nations. According to the World Health Organisation (WHO) and the London School of Hygiene and Tropical medicine estimates, there may already be upwards of 1, 60,000 deaths annually from ancillary effects of global warming such as malaria and malnutrition. The study further estimates those numbers could nearly double by 2020.[5]

Vulnerable nations and climate change
There are three types of vulnerable nations, they are- Least developed Nations, Weak states and Undemocratic states.

Least developed nations
There is lack of governance, economic and technical capabilities in poor developing countries or least developed countries. These countries mostly suffer the effects of climate change. They lack the capacity to prevent or react to humanitarian disasters such as widespread flooding. Tropical developing nations face the most severe consequences of climate change including extreme weather events, droughts and disease.

Weak states
Weak states have no capacity to respond to climate change or prevent it from triggering a large scale humanitarian disaster. Drought, crop failure and state failure led to tens of thousands of deaths in Somalia in the 1990s.[6] Thus, failed and failing states-those of weak government institutions, poor border control, repressed populations or marginal economics- stand a higher risk of being destabilised by climate change.

Undemocratic states
Population in undemocratic states will be vulnerable to more numerous and more severe humanitarian crises induced by climate change. 20 years ago, economist Amartya Sen noted that, democracies in which leaders have to be responsive to people can vote them out of power. The 20th century is replete with examples of undemocratic regimes failing to protect populations at risk of drought, floods and other weather-related phenomena.

Climatic change, security and the UN
In the round table conference, 2004 the experts and participants have given the following views about environment and security.

Environment, population, Development and Security
- Environmental problems can constitute security threats both environmental scarcity of natural resources that lead to violent conflicts. Threats of human livelihoods, social and economic inequities create environmental problem as a result there is migration, relative deprivation, tense ethnic divisions, poor governance and declining economic productivity.
- The environmental, security and development communities speak different languages and therefore, they do not adequately communicate, cooperate or coordinate within international organisations, national governments or NGOs.
- The environmental crisis challenges the depth and breadth of policy responses within UN system. The UN needs to coordinate efforts among agencies, remove institutional barriers and mainstream environmental concerns in all of its agencies and policies.
- 'Environmental diplomacy' can promote environmental issues as a pathway to dialogue between parties in tension. 'Peace dividends" can encourage and justify greater monetary investments, while mitigating security problems create more opportunities for development.
- Environmental problems are high on many states, security agendas and threaten a large percentage of the world's population. Most participants agreed that environmental changes endanger human security and some recommended demilitarising and redefining security. The efficacy of redefining security cast doubt on the concept and ability to integrate environmental issues into the security agenda.
- Leadership is critical to recognise UN's response to environmental related conflict into action. Some participants asserted that the UN's senior and mid-level management lacks the expertise and commitment required to integrate environmental issues into its broader security agenda.
- Early warning systems can provide an integrated mechanism to identify when large number of people are suddenly and physically threatened by fatal diseases, homelessness, hunger etc. as a result of environmental change or natural disasters.
- Disaster response could be more effective if viewed as part of a broader environmental security agenda. The UN office for the

coordination of Humanitarian Affairs does not need to mitigate humanitarian crises; it should be linked with development and early warning programs to form an integrated approach to disaster prevention.

- The environmental agenda needs to be refashioned, so that it fits not only the North's security goals but also the South's development goals by promoting sustainable development.

Climate change and security
- Climate change poses a large threat to floods, droughts, rising sea levels, extreme weather events, increased tropical diseases, water scarcity, famines, declines in agricultural productivity and shifts in migration and trade patterns. These changes are likely to be incremental but may also occur suddenly and dramatically.
- The developing world will suffer the most, while the industrialised world is responsible for most of the CO_2 emissions.
- The effects of climate change can threaten security by increasing the severity and frequency of natural disasters, humanitarian crises and destabilising vulnerable nations. Natural disasters, which already kill six times more people than armed conflict have been increasing over the last three decades
- States that are less developed, weak or undemocratic and Small Island States, will suffer most from climate change-induced problems because they have limited adaptive capacity.
- Mitigation is a necessary component of a climate change strategy. Countries can mitigate climate change by moving away from hydrocarbons and abating emissions. All participants conceded that the Kyoto Protocol has been relatively ineffective; most agreed that mitigation must be pursued.
- Adaptation requires strengthening ongoing disaster work and allotting more money for reduction and prevention. Risk reduction could be enhanced by exploring options from the private sectors or by integrating risk management into sustainable development programs. High-risks states need improved early warning systems, vulnerability indices and contingency plans to prepare for disasters and potential conflicts.
- There are a large information gaps between the developed and developing world about climate change; coordinating the message at the highest levels of the UN - using practical applicable statistics-could create a sense of urgency.

- Some UN programs work at cross-purposes when addressing climate change. Participants viewed the MDGs' silence on climate change as a great disadvantage.

Water, conflict and cooperation
- Paucity of clean fresh water, which impedes development and health improvement, can ignite conflict between and within states. Water security is essential to both economic development and political security.
- There are widespread fears of water wars between states. The shared river basins engender tensions and potential for violence in the context of weak governance institutions, unilateral development projects that control water flow across international borders.
- However, water is a common source of conflict at local and individual levels. Within local domains, most water disputes occur due to mismanagement of water resources or the inequitable distribution of benefits.
- Shared management of basins can foster a high level of cooperation and provide a pathway for confidence building and conflict prevention; sharing and codifying data can establish the basis for such cooperation.
- Improving water quality and sanitation can save many of the 2-3 million people who die each year from water-related illnesses.
- UN recognises the importance of water policy and has incorporated it into the MDGs, the Commission on Sustainable Development and many of its agencies. The 13th session of the UN Commission on Sustainable Development focusses on water policy but unlikely to address how this might be incorporated into the UN's conflict and security agenda.

UN and traditional security

Human security
Human security means protecting fundamental freedoms- freedoms that are the essence of life. It means protecting people from severe and widespread threats and situations. It means using processes that are built on people's strength and aspirations. It means creating political, social, environmental, economic, military and cultural systems that together give people the building blocks of survival, livelihood and dignity.[7]

The concept of Human security is in a fundamental way by:

- Moving away from traditional, state-centric conceptions of security that focussed the safety of states from military aggression to security of individuals, their protection and empowerment;
- Drawing attention to a multitude of threats between security, development and human rights; and
- Promoting a new integrated, coordinated and people-centered approach for advancing peace, security and development within and across nations.

International security law at the present stage of development is primarily found in the United Nations collective security system. This is based on the norm of non-use of armed force under Article 2 (4) of the UN charter and the institution of the UN Security Council vested with the primary responsibility for the maintenance of international peace and security under Article 24 of the charter.[8] Collective security provides institutionalised procedures for legalising collective response, designed at least originally to address traditional military oriented threats to the maintenance of international peace and security. Thus collective security is a product of law, based on the delegation of power by sovereign states to a collective entity.[9]

The traditional view of security is defined in military terms, with the primary focus on state protection from threats to national interests. Human security is a human or people-centered and multi-sectoral approach to security, which means the protection of people from critical threats and situations, the empowerment of people to develop their potential, to develop norms, processes and institution that systematically address insecurity.[10] Peter Hough understands the human security approach as both a 'widener' and 'deepener' of the traditional narrow security conception to not only communities and groups but also individual persons or government ministers or private individuals, who can make a securitisation move.[11] Human security may assist in defining new security concerns or redefining the terms of debate surrounding traditional security threats.[12] The Security Council has recently been more active in indicating human security when they refer to children in armed conflict, women and peace and security and the protection of civilians more generally. The Security Council's practice of deploying peacekeeping forces as part of traditional collective security measures with a mandate to protect civilians has caused normative, operational and ethical dilemmas.[13]

Human security differs from traditional security
- State security concentrates on threats directed against the state, mainly in the form of military attacks. Human security draws attention to a wide array of threats faced by individuals and communities. Human security, however, is not intended to displace state security. Human security and state security are mutually reinforcing and dependent on each other. Without human security, state security cannot be attained and vice versa.[14]
- To human development's objective of 'growth with equity', human security adds the important dimension of 'downturn with security' such as conflicts, economic and financial crises, ill health and natural disasters, people are faced with sudden insecurities and deprivations.
- Violations of human rights result in conflicts, displacement and human suffering on a massive scale. Human security makes no distinction between different kinds of human rights-civil, political, economic, social and cultural rights.

There are so many types of Human Security[15] which affected various threats such as Economic Security- poverty, unemployment; Food Security- hunger, famine; Health Security- infectious diseases, unsafe food, malnutrition, lack of access to basic health care; Environmental Security- Environmental degradation, resource depletion, natural disasters, pollution; Personal Security- physical violence, crime, terrorism, domestic violence, child labour; Community Security- Inter-ethnic, religious and other tensions; Political Security- political repression, human rights abuses'.

Human Security emphasises the inter-connection of both threats and responses while addressing these insecurities. . The policy-making implication implies that human insecurity cannot be tackled in isolation through fragmented stand-alone responses. Instead, human security involves comprehensive approaches that stress the need for cooperative and multi-sectoral responses that bring together the agendas of those dealing with security, development and human rights. With human security, the objective must be a stronger and more integrated response from communities and states around the globe.[16]

Protection and empowerment for achieving human security
Protection and empowerment of people are the two building blocks for achieving the goal of human security. They are proposed by the Concept of Human Security (CHS) as the bi-parts of any human security policy framework.

- Protection is defined by the CHS as strategies, set up by states, international agencies, NGOs and the private sector.[17] It refers to the norms, processes and institutions required to protect people from critical and pervasive threats. Protection implies a 'top-down' approach. It recognises that people face threats such as natural disasters, financial crises and conflicts. Therefore, human security requires protecting people in a systematic, comprehensive and preventative way. States have the primary responsibility to implement such a protective structure through international and regional organisations. Civil society, non-governmental actors and private sector also play a vital role for human security.
- Empowerment is defined by the CHS as strategies that enable people to develop their resilience to difficult situations.[17] Empowerment implies a 'bottom up' approach. It aims at developing the capabilities of individuals and communities to make informed choices and to act on their own behalf. Empowering people not only enables them to develop their full potential but it also allows them to find ways and to participate in solutions to ensure human security for themselves and others.

National security

Traditionally, national security is focussed upon the physical protection of a state's territory from military attacks by another state. This is reflected in the inherent right of individual or collective self-defence in Article 51 of the UN charter. The national security has posed challenges to the rules of international law in dealing with non-traditional security threats. National security considerations should be given weight in balancing the need to ensure human rights protection of those who are suspected of being involved in terrorist activities.[18]

The UN collective security system ensures that national security measures were in compliance with the existing rules of international law through the Security Council's adoption of Resolution 1373 which was instrumental in the adoption of extreme border control measures to prioritise national security.[19]

International security

The International Security evolved through the development of a collective security system, under UN. Development lies in the concept of a threat to the peace under Article 39 of the UN charter and the Security Council enlarged the concept of a threat to the peace which is well documented.[20] In 2000, the Security Council discussed the impact of HIV/ AIDS on peace and security in Africa under the Council presidency of US Vice-President, Al Gore.[21] The non-traditional security threats such as public health threats, environmental degradation and climate change, should be accommodated within the purview of the Security Council through expansive mandate for the maintenance of international peace and security.[22] The International institutions are required to operate within the competence defined by the provisions of their constitutive instrument. Thus, the expansion of the concept of international security will entail one of the two consequences: (i) posing challenges to the jurisdictional limits of international institutions and organs e.g. Security Council; (ii) requiring a wider range of mechanisms to respond to diverse security threats than the collective security system.

Peace and security
The Millennium Development Goals (MDGs) have framed or influenced the policies, programmes, projects or activities of governments, international organisations, civil society and the private sector. The discussions on the post-2015 framework create an opportunity to include them are given below:
- violence and fragility have become the largest obstacles to the MDGs.
- The narrow approach of MDGs is problematic given the broadening of the concept of development that has occurred.
- The narrow focus also ignores the interrelations among various aspects such as security, justice and development.
- The post 2015 agenda needs to take a comprehensive approach guided by the millennium declaration which included fundamental values and goals on peace, security and disarmament, development and poverty eradication, human rights, democracy and good governance and protecting the vulnerable.
- The post-2015 framework should include separate goals related to peace and security and a clear, concise and measurable target on violence, which can be measured through indicators on battle-related deaths and international homicide.

MDG and violent conflict
Violent conflict has become the largest obstacle to the MDGs. The consequences of violence on development are significant and long-term. Violent conflict causes death, disease and displacement, destroys physical and social capital, damages the environment, decreases school attendance

and discourages investment.[23] A country that experienced major violence during the period from 1981-2005 had a poverty rate on average 21 per cent higher than a country without violence.[24]

The gap in MDG performance between conflict-affected countries and other developing countries is large and increasing. Fragile and conflict-affected states account for 47 per cent of the population of developing countries, 60 per cent of the undernourished, 61 per cent of impoverished, 77 per cent of children not in primary school and 65 per cent of people without access to safe water and 70 per cent of infant deaths occur in fragile or conflict affected countries.[24]

About 40 per cent of countries have come out of violence within 10 years and 90 per cent of countries that had civil wars in the 21st century in the previous 30 years.[25] Peace building will reduce the risk of violence. Countries need to reduce violence and building resilient institutions and peaceful societies. A focus on justice, human rights, horizontal inequalities, jobs and inclusive politics will reduce the risk of violence.

Development agenda for the post agenda of 2015
The post-2015 agenda needs to take a comprehensive approach, guided by millennium declaration, which included a wide-ranging set of elements.[26] It adopted fundamental values such as the freedom from fear of violence, oppression or injustice and equality. And it identified the following key objectives: peace, security and disarmament; development and poverty eradication; protecting common environment; human rights; democracy and good governance; protecting the vulnerability; meeting the special needs of Africa; and strengthening the United Nations.

In 2005, the Secretary-General, drawing on the UN charter, argued that the idea of freedom that development, security and human rights go hand in hand[27] and in 2005 World Summit outcome shows that Peace and Security, Development and human rights are the three pillars of the United Nations system and the foundations for collective security and well-being. The three pillars are interlinked and mutually reinforcing.[28] Sustainable socio-economic or human development which could include targets on poverty, jobs, food security and nutrition, health, education, energy and the environment.

The post-2015 framework should include indicators to measure health and education inequalities across regions by gender and by income. The post-2015 agenda could include a target on violence. The freedom from violence is a human right but also a development goal.

Peace building goals

The international organisations advocate peace building goals into the post-2015 agenda. They come together in the International Dialogue on peace building and state building. The International Dialogue has proposed the New Deal on Engagement in Fragile states at the 4th High Level Forum in 2011, which consists of five peace building and state building goals. The five goals are:

- Legitimate politics- Foster inclusive political settlements and conflict resolution.
- Security- Establish and strengthen people's security.
- Justice- Address injustices and increase people's access to justice.
- Economic Foundations- Generate employment and improve livelihoods.
- Revenues and Services- Manage revenue and build capacity for accountable and fair service delivery.

These goals are "important foundation to enable progress towards the MDGs". The New Deal outlines an agenda for more effective aid to fragile states based on the five peace building and state building goals, stronger alignment and mutual accountability. The New Deal will be implemented in seven self-nominated pilot countries, they are: Afghanistan, Africa, Congo, Liberia, Sierra, South Sudan and Timor-Leste and was endorsed by 35 countries and six international organisations including United Nations Development Group and the World Bank.

The G7+ members should take an important role in bringing issues around peace, security, justice and governance to the United Nations.

Women, peace and security

Promoting the redistribution of power and the construction of sustainable and democratic political procedures provide opportunities for advancing gender equality, they are women with role of potential peacemakers in re-construction and re-building processes. They advocate that peace and security is essential to peace processes and policy making at all levels. In 2001, donors were committed to the following efforts to:

- Support Women's organisations during conflicts to become involved in dedication, negotiations and attempts to institutionalise the peace process.[29]
- Develop policies and programmes that extend support to women organisations that focus on the conflict situations and encourage women coalitions for peace building across regions and sub-regions such as

in human rights, relief, rehabilitation and peace building. Women for peace works to effectively support and encourage women's initiatives at all levels.[30]

- Encourage capacity building for women in public life. Peace building and Peace making processes should incorporate women as decision-makers at each level and consider their concerns at every stage.
- Support the representation of women in peace processes. Militarisation during the pre-conflict period often marginalises women from decision- making processes.
- Improve women's access to resources during reconstruction, rehabilitation and reconciliation. Many arrangements for public administration and legislation are renegotiated after war and provide opportunities for security or increasing women's legal rights, their control over key resources such as land and access to education and mechanisms for justice.
- Develop special ways of dealing with women, youth and children who have been victims of gender-based violence and abuses as a consequence of conflict.

In 2006, the UN Secretary General's High Level Panel on system-wide was tasked to study the reform of the United Nations System. Civil society advocates pushed that GEAR (Gender Equality Architecture Reform) be part of UN reform. The GEAR campaign is a global initiative of women's human rights and social justice group that proposes the creation of a stronger UN entity for women, gender equality, the empowerment of women and women's human rights in the work of the UN. The GEAR campaign has five global focal points: Women's Environment and Development Organisation (WEDO), Centre for Women's Global Leadership (CWGL), Development Alternatives with Women for a New Era (DAWN), Association of Women's Rights in Development (AWID) and International Planned Parenthood Federation (IPPF). Civil society has always played a vital role in the UN's work on women's rights in every stage of the process at global, regional, national and local levels including in the governing board.

Around the world women have waited for a long time for the United Nations and member states to fulfill the promises made since the first International Women's Year in 1975, the adoption of the Convention on the Elimination of Discrimination Against Women (CEDAW) 30 years ago, as well as in the UN World Conferences in Nairobi (1985) and Beijing (1995).

International Day of peace is on September 21, 2013. The Global Network of Women Peace Builders (GNWP) is launching, "Women speak out for peace: A Global Media Campaign". This wide-reaching campaign would start on September 16 and continue until Peace Day, September 21. This campaign's objectives are following:

- To raise extent awareness and knowledge of UNSCR 1325, 1820 and other supporting Women, Peace and Security (WPS) resolutions and of their practical application at the country and community levels.
- To demand greater accountability from governments, the UN, regional organisations and fellow civil society actors to honour their obligations under the WPS resolutions including but not limited to the development and Implementation of National Action Plans.
- To transform the depiction of women in conflict settings from victims to peace builders, decision-makers and change agents.

This campaign aims to involve as many individuals and organisations as possible. All GNWP members and partners are encouraged to participate. Civil society organisations and other stakeholders such as UN agencies, government officials, activists, students, and young professionals come together to share and fulfill the objectives.[31]

Security system reform

The traditional concept of security, which revolves around the protection of states from military threats, is being redefined in three important respects that provide the basis for the Security System Reform policy agenda:

- The focus of security policy focusses on state stability and regime security to include the well-being of their populations and human rights.
- Security and development are increasingly seen as a public policy and a governance issue. This invites greater public scrutiny of security policy.
- The military is now seen as only one instrument of security policy with traditional legal, social and economic instruments receiving greater attention.

The security system includes the following actors:

- Core Security actors: - armed forces, police, paramilitary forces, presidential guards, intelligence and security services (both military and civilian), coast guards, border guards, customs authorities, reserve or local security units (civil defence forces, national guards, militias).

- Security Management and oversight bodies: - the executive, national security advisory bodies, legislature and legislative select committees, ministries of defence, internal affairs, foreign affairs, customary and traditional authorities, financial management bodies and civil society organisations.
- Justice and Law enforcement Institutions: - Judiciary, Justice Ministries, prisons, criminal investigation and prosecution services, human rights commissions and traditional justice system.
- Non statutory security forces: - Liberation armies, guerrilla armies, private body-guard units, private security companies.

Working principles for effective security system reform
- The core values for SSR to be people centred, locally-owned and based on democratic norms and internally accepted human rights principles and on the rule of law. They should seek to contribute to an environment characterised by freedom from fear.
- SSR should be seen as a framework to structure thinking about how to address diverse security challenges facing populations and states through more integrated development and security system reform policies. Therefore, SSR framework should address both external and internal threats to people's safety, to law and order and to state stability.
- Donor governments should provide their assistance within strategic frameworks that are multi-sectoral. They must be developed jointly with partner governments and civil society and based on an assessment of the security needs of the people and the state. Women's organisations in particular, can play a major role in ensuring that needs assessments capture the security concerns of vulnerable groups. This should involve broad consultation among donor government departments as well as close co- ordination with other donor governments and international organisations. SSR must use and combine the broad range of diplomatic, legal, social, economic, security and political policy instruments available to develop appropriate military and non-military responses to security issues. Security policy will remain narrowly concentrated on agencies that deal with more traditional matters such as defence, intelligence and civil bodies involved in legislative bodies, judicial ministries and civil society actors.
- The security system should be managed according to the same principles of accountability and transparency that apply across the public sector, in particular through greater civil oversight of security

processes. These principles include promoting: (i) the availability of information required by policy makers; transparent and accountable decision making, (ii) a comprehensive approach to public expenditure management; adoption of medium term perspectives for decision making and (iii) a capacity and willingness to shift priorities and reallocate resources to achieve strategic objectives.

The long term objective is to ensure that the security system is effectively integrated into all relevant government-wide budgeting and planning processes.

* As far as possible, SSR processes should address the three core requirements of a well-functioning security system:
  o Developing a national concept of security and the policy and institutional frameworks' states require to handle development and security as distinct but integrated areas of public action.
  o Establishing well-defined policies and strengthening governance of the security institutions that are responsible for formulating executing, managing and monitoring security policy.
  o Building the institutional mechanisms for implementation and capacity throughout the security system, this includes the development of professional security forces both civil authorities and operational tasks.

United Nations Security Council

The end of the 2nd world war gave rise to a new international system and the formation of the United Nations. The organisation contains two main bodies: the General Assembly which comprises all member nations and the smaller Security Council which has the primary responsibility of maintaining international peace and security including the mitigation of and response to international conflicts. The United Nations Security Council (UNSC) focusses on peaceful means of conflict management and resolution through negotiations and the dispatch of UN peace keepers. UNSC Resolutions under chapter VII are binding on all UN members and can include economic sanctions or collective military action.[32]

The UN Security Council consists of 15 member states: five permanent members (P5) and 10 other members of regional representation that rotate on a 2-year basis. The five permanent members represented the five main powers at the end of World War II and include China, France, Russia, the UK and USA. The Security Council is the United Nations' most powerful body, with primary responsibility for the maintenance of international peace and security. It dispatches military operations, imposes sanctions, mandates arm inspections, deploys election monitors and more.

In 1997, UN Secretary General Kofi Annan announced his plan for United Nations reform. The agenda for better management and coordination across the entire UN system as well as stronger human rights promotion and peace keeping operations. In 2002, Annan announced for further reforms to enhance coordination of the organisations in the UN system and greater 'focuses' in the UN's work. In March 2005, Annan presented his most comprehensive reform and policy agenda for terrorism, financing for development, enlarging the Security Council and replacing the Human Rights Commission.[33]

The Economic and Social Council (ECOSOC) is the principal UN body coordinating the economic and social work of the organisation. ECOSOC enjoys little authority in international policy making. The UN charter places ECOSOC under the authority of General Assembly, allowing the body to issue policy recommendations to the UN system and member states. However, other international institutions such as World Bank, the International Monetary Fund (IMF) and the World Trade Organisation (WTO) have assumed leadership in the field of global policy.

Climate change and development

Unchecked climate change impacts on developing countries' prospects for sustainable development, For example, changing rainfall patterns threaten to severely impact agricultural activity in Africa by 2020 reducing the viability of rain-fed agriculture by as much as 50 per cent, in some countries[34] glaciers threaten to world's mightiest rivers into seasonal flows in South Asia and Western South America with severe impacts of Agriculture by 2020.[35] India's agriculture could fall by 30-40 per cent by 2035 at the Ganga, the Brahmaputra and the Ganges has completely lost their head water glaciers in the Himalayas.[36] In the Bangladesh delta, 35 million refugees are estimated due to sea level-rise of some 40 cm by the end of the century.[37]

In the words of the Group of Eight (G8), "mitigation and adaptation strategies should be pursued as part of development and poverty alleviation efforts".[38] The ongoing negotiations and the shape of the final outcome envisioned at the COP- 15, this concept seeks to understand how such outcomes might be realised and fundamental links between climate change and development goes as follows:

• Climate change is taking place; action to address it is imperative.

- Poverty and inequity exist in unacceptable measures; development that addresses them is imperative.
- Economic growth is a means to development- one that will continue to be used in the foreseeable future.
- There are number of potential conflicts between economic growth and action on climate change.
- But there are synergies between climate change action and development: actions aimed at both adaptation and mitigation can contribute to development and unchecked climate change will undermine development goals and economic growth.
- Advancing development goals in a sustainable way should be a central part of efforts to address climate change in all countries.
- The International community should focus in particular to help achieve this in developing and least developed countries.

Advancing development goals in a sustainable manner was one of the four pillars of the dialogue on long-term cooperative action that led to the Bali Action Plan.

Climate change is taking place: action to address it is imperative
In 2007, Fourth Assessment Report (AR4) of IPCC found that warming of the climate system was "unequivocal and that a number of attendant effects were already observable including:
- An increase in global average surface temperature over the last 100 years of 0.740C.
- A decrease in the average extent of mountain glaciers and snow cover in both hemispheres.
- An acceleration in the annual rate of sea level-rise from 1.8 mm between 1961 and 2003 to 3.1mm between 1993 and 2003.
- More intense and longer droughts over wide areas.[39]
The international scientific steering committee, meeting in 2005, United Kingdom explored the temperature thresholds that melting of the Green Land icecap and shutdown of the Atlantic thermohaline circulation warms the North Atlantic countries.[40] The international consensus clears that action is needed and the UNFCCC commits parties to address climate change, by stabilising GHG concentrations so as to "prevent dangerous anthropogenic interference with the climate system".
- The G8 meeting, in Hokkaido, Japan in 2008, reaffirmed their commitment to take, "strong leadership in combating climate change" and endorsed a goal of reducing GHG emissions by 50 per cent by

2050.[41] After the summit G5 (Brazil, China, India, Mexico and South Africa) issued its own statement urging the international community to address the challenge of climate change.[42]

- In Hokkaido, the leaders of major economies, they are Australia, Brazil, Canada, China, the European union, France, Germany, India, Indonesia, Italy, Japan, Korea, Mexico, Russia, South Africa, UK, Northern Ireland and USA, declared climate change as "one of the great global challenges of our time" and pledged to work together to strengthen implementation of the UNFCCC, while recognising that "deep cuts in global emissions will be necessary to achieve the convention's ultimate objective".[43]
- Australia, China, India, Japan, Korea, New Zealand and the ASEAN nations gathered in Cebu, Philippines in January 2007 recognised the urgent need to address global warming and climate change and committed to limit greenhouse gas emission through effective policies and measures.[44]

Poverty and inequity exist in unacceptable measure; development that addresses them is imperative

Poverty and inequity go hand in hand. The richest 5 per cent of the world's population controls almost one half of world income, earning in 15 hours, what the poorest 5 per cent make in a year.[45] In sub-Saharan Africa and South Asia 77 per cent and 59 per cent of the population respectively, do not have home access to electricity- in total over 1.3 billion people.[46] In 2001, global leaders agreed on a set of goals that are Millennium Development Goals (MDGs) to be achieved by 2015. These went far beyond focussing on income poverty to identify development as including goals such as education, health, gender equality, literacy and environmental sustainability.

Economic growth is a means to development- one that will continue to be used in the foreseeable future

Development or increase of human well-being goes beyond simply income poverty with the approach, the United National Development programme.[47] The MDGs may be the clearest international statement that development goes to economic growth. Poverty is defined as a lack of freedom to pursue life's ambitions and development as a process of fostering such freedom.[48]

Economic growth will continue for the foreseeable future and will preserve as a key element of national development efforts. Thus economic growth is important even for development more broadly cast. Economic growth continues to be of primary importance to national governments, whether in developing or developed countries. MDGs have addressed income poverty as the first goal. Even Sen allows for the major enabling role of income as a determinant of one's capabilities throughout his work. Many states are planning for significant stimulus spending with a focus on green growth.

There are a number of potential conflicts between economic growth and action on climate change

Copeland and Taylor (1994) developed a seminal model that can be used to break down the environmental effects of economic growth into three effects: scale, composition ad technique effects.[49] The International Energy Agency's World Energy Model used to derive the projections of energy demand is based on projected economic growth as a key exogenous assumption.

The Organisation of Economic Cooperation and Development (OECD, 1994) is another type of environmental effect.[50] Economic growth can enrich citizens who will demand better environmental regulations and can provide governments with the financial resources.

Synergies between climate change goals and development

There are number of ways in which development efforts can lead to mitigation:

- Efforts to restore forest cover or avoid deforestation/land degradation can have significant pay offs including reduced fuel wood, indoor air pollution from inefficient biomass use and flood control in water sheds.[51]
- Efforts to provide energy to the poor constitute development in their own right.[52]
- Fuel-switching efforts may be aimed at reducing the burden of import costs, improving balance of payments and generating domestic employment.[53]
- Energy diversification and energy security are objectives that have value in and of themselves, helping to ensure smooth and continuous access to energy at affordable rates and shielding countries from the balance of payments impacts of fluctuating global fossil fuel prices.[54]
- There are also a number of ways in which efforts to mitigate GHG emissions can contribute to development:
- Efforts to achieve energy efficiency have enormous potential to reduce GHG emissions.[55]

- Efforts to avoid the emissions associated with deforestation, as in the provision of improved cook stoves or solar cookers to fuel wood users, can yield significant development benefits as well, including reduced indoor air pollution.[56]
- Efforts to capture methane emissions from landfills and livestock operations contribute powerfully to GHG emission reductions.[57]

Mitigation efforts and development objectives are closely and inextricably connected, have strong links to adaptation. The key objective of adaptation measures is to reduce vulnerability to the impacts of climate change, so any successful adaptation efforts will constitute development.

Advancing development goals in a sustainable way
There are a number of important elements in national and international approaches to address climate change, including a focus on adaptation, the engagement of the private sector and the lowering of costs through the inclusion of market mechanisms and a focus on development and new technologies. The foregoing analysis argues strongly that advancing developing goals in a sustainable way should also be central.[58] The argument starts with the urgent need for and the international commitments to, action on both climate change and development.

Advancing development goals sustainably in developing and least developed countries
Economic growth and development are key priorities for policy makers the world over and climate change impacts will be felt everywhere. The international community addresses climate change and there are number of reasons focussed on advancing development goals in both developing and least developed countries.

First, while it is true that economic growth and development are priorities in all countries, the needs in developing and least developed countries are on a different scale altogether than that in the developed world. The Millennium Declaration and the Monterrey Consensus on Financing for Development are strongly focussed on actions to assist developing and least developed countries.[59]

Second, current development paths indicate the greatest need for investment to alter energy paths in developing countries. IEA's baseline projections predict that between 2007 and 2030 non-OECD countries will account for 87 per cent of the world's increase in primary energy demand

and particularly all 97 per cent of the global increase in energy-related CO_2 emissions.[60]

Third, developing and least developed countries are particularly vulnerable to climate change impacts. Developing countries and their poorest population are generally more sensitive to any climate change impacts, given a high dependence on agriculture, strong reliance on ecosystem services, rapid growth and concentration of population and poor health. Developing countries have less capacity to adapt to climate change impacts, having inadequate infrastructure, limited household income and public services.

Fourth, there is a strong argument to be made on equity grounds. The vulnerable countries have contributed least atmospheric concentrations of GHGs.[61] It can be argued that a responsibility among the largest historic contributors to assist in achieving development goals in ways that contribute to adaptation and mitigation goals.[62]

Finally, if developing and least developed countries are to contribute efforts toward mitigation of climate change impacts, they will need the strengthened capacity that comes with development. The UNFCCC holds among its principles that economic development is essential for adopting measures to address climate change.[63]

Bali climate change conference- December-2007, COP 13/CMP 3
The 13th session of the conference of the parties to the UNFCCC and the 3rd session of the conference of the parties serving as the meeting of the parties to the Kyoto Protocol took place in Bali and were hosted by the government of Indonesia. The Bali climate change conference brought together more than 10,000 participants, including representatives of over 180 countries together with observers from inter-governmental and non-governmental organisations and the media.

Governments adopted the Bali Road map, a set of decisions that represented the various tracks that were seen as key to reach a global climate deal. The Bali Road Map includes the Bali Action Plan, which launched a, "new, comprehensive process to enable the full, effective and sustained implementation of the convention through long-term cooperative action now up to and beyond 2012".

Other elements in the Bali Road Map included:
* A decision on deforestation and forest management.
* A decision on technology for developing countries.
* The establishment of the Adaptation Fund Board.
* The review of financial mechanism, going beyond the existing Global Environmental Facility.

The Ad Hoc Working Group on Long-term Cooperative Action (AWG-LCA) was set up to conduct work under the Bali Action Plan. The Ad Hoc Working Group on Further commitments for Annex I parties under the Kyoto Protocol (AWG-KP) was to work in parallel. The central task of the AWG-KP was to decide the emission reduction commitments of industrialised countries after the Kyoto Protocol's first commitment period in 2012.

Decisions of Bali Action Plan
- **1/CP.13** - Bali Action Plan, Indonesia.
- **2/CP.13** - Reducing emissions from deforestation in developing countries.
- **3/CP.13** - Development and transfer of technologies under the Subsidiary Body for Scientific and Technological Advice (SBSTA).
- **4/CP.13** - Development and transfer of technologies and the Subsidiary Body for Implementation (SBI).
- **5/CP.13** - Fourth Assessment Report of the Intergovernmental Panel on Climate Change.
- **6/CP.13** - Fourth review of the financial mechanism.
- **7/CP.13** - Additional guidance to the Global environment facility.
- **8/CP.13** - Extension of the mandate of the least developed countries' Expert Group.
- **9/CP/13** - Amended New Delhi work programmed on Article 6 of the convention.
- **10/CP-13** - Compilation and synthesis of 4th national communications.
- **11/CP 13** - Reporting on global observing systems for climate.
- **12/Cp 13** - Budget performance and the functions and operations of the secretariat.
- **13/CP - 13** - Programme budget for the biennium 2008-2009.
- **14/CP - 13** - Date and venue of the 14th and 15th sessions of the conference of the parties and the calendar of meetings of convention of bodies' resolution.

2009 United Nations climate change conference - COP 15/ CMP 5
United Nations climate change conference in 2009 commonly known as the Copenhagen summit was held at the Bella centre in Copenhagen, Denmark between December 7 and 8. The conference included the 15th Conference of the Parties (COP15) to the United Nations Framework Convention on Climate Change (UNFCCC) and the 5th meeting of the parties (MOP 5) to the Kyoto Protocol. According to the Bali Road Map,

a framework for climate change mitigation beyond 2012 was to be agreed there.[64]

On the Climate Change: Global risks, Challenges and Decisions, a Scientific conference took place in March 2009 at Bella centre. The negotiations began to take a new format when in May 2009 UN Secretary General Ban Ki-Moon attended the World Business summit on climate change in Copenhagen, organised by the Copenhagen Climate Council (COC), where he requested that COC councillors attend New York's climate week at the summit on climate change on September 22, and engage with Heads of government on the topic of climate problem.[65]

The Copenhagen Accord was drafted by the United States, China, India, Brazil and South Africa on December 8, 2009 and judged as a meaningful agreement by the United States government, Copenhagen agreement deals with the key points of the agreement that:-

- The Accord, reached between the US, China, India, Brazil and South Africa, contains no reference to a legally binding agreement as some developing countries and climate activists wanted. There is no deadline for transforming it into a binding deal, though UN Secretary General Ban Ki-Moon said that it needed to be turned into a legally binding treaty next year.

- The text recognises, the need to limit global temperatures rising no more than 2C (3.6F) above pre-industrial levels. The accord does not identify a year by which carbon emissions should peak, a position resisted by some richer developing nations. It outlines a goal of providing $100 billion a year by 2020 to help poor countries cope with the impacts of climate change. The accord says the rich countries will jointly mobilise the $100 billion, drawing on a variety of sources: "public and private, bilateral and multilateral, including alternative sources of finance". A green climate fund will also be established under the deal. It will support projects in developing countries related to mitigation adaptation, 'capacity building' and technology transfer.

- The pledges of rich countries will come under "rigorous, robust and transparent" security under the UN Framework Convention on Climate Change (UNFCCC). In the accord, developing countries will submit national reports on their emissions pledges' under a method "that will ensure that national sovereignty is respected". Pledges on climate migration measures seeking international support will be recorded in a registry.

- The Implementation of the Copenhagen Accord will be reviewed by 2015. This will take place about a year-and-a-half after the next scientific assessment of the global climate by the Intergovernmental Panel on Climate Change (IPCC).

It is recognised that, climate change is one of the greatest challenges of the present day and that actions should be taken to keep any temperature increases to below 2^0C. Many countries and non-governmental organisations were opposed to this agreement, but throughout 2010, 138 countries had either formally signed on the agreement or signalled they would.[66]

To establish a Green Climate Fund to provide financing to projects, programmes, policies and other activities in developing countries, remained in the accord.

•   The Cancun Adaptation Framework included setting up an Adaptation Committee to promote the implementation of stronger, cohesive action on adaptation.

The developed countries agreed to mitigate reduction targets and strengthen and develop low-carbon national plans and strategies in developing countries and should encourage to develop low-carbon national plans and strategies with finance and technology and to implement Reducing Emissions from Deforestation and Forest Degradation (REDD). Governments also agreed to include Carbon Capture and Storage (CCS) in the projects under the Clean Development Mechanism (CDM).

Nationally, Appropriate Mitigation Action (NAMA) refers to set of policies and actions that countries undertake as a part of commitment to reduce greenhouse gas emissions. Different countries may take different nationally appropriate actions on the basis of equity and in accordance with common but differentiated responsibilities and respective capabilities. It also emphasises on financial assistance from developed countries to developing countries to reduce emissions. NAMA is the first used in the Bali Action Plan as part of the Bali Road Map agreed at the United Nations climate change conference in Bali in December 2007 and also formed part of the Copenhagen Accord in United Nations climate change conference in Copenhagen (COP 15) in December 2009.

2010 United Nations climate change conference - COP 16/CMP 6
United Nations climate change conference - COP 16/CMP 6 in Cancun, Mexico was held from November 29 - December 10, 2010.[67] 194 countries represented in this climate change conference, UN Framework Convention on

Climate Change (UNFCCC) Executive Secretary Christiana Figures said in her opening address that, 'governments had revealed a growing convergence and that a balanced set of decisions under both the convention and the Kyoto Protocol could be an achievable outcome in Cancun and a number of politically charged issues need to be resolved in order to reach such an outcome'.

The COP 16 delegates reached an agreement on climate change. They passed Article 6 of the convention, which focusses on climate change education, training, public awareness, public participation, public access to information and international cooperation. Article 6 was the first agreement reached by 194 parties during the COP 16.

The Cancun climate change conference drew almost 12,000 participants including 5,200 government officials, 5,400 representatives of UN bodies and agencies, inter-governmental organisations and non-governmental organisations and 1,270 accredited members of the media. The meeting produced most comprehensive and international response to climate change the world had ever seen to reduce carbon emissions and build a system which made all countries accountable to each other for those reductions. The parties agreed:

- To commit to a maximum temperature rise of 2^0C above pre-industrial levels, and to consider lowering that maximum to 1.5^0C in the near future.
- To make fully operational by 2012 a technology mechanism to boost the innovation development and spread of new climate-friendly technologies.

Climate change adaptation in COP/16/CMP6 - 2010 outcome documents
- Climate change is the top issue on global agenda affecting millions of people worldwide, particularly in developing countries. GHG emissions continue to rise dramatically. The most vulnerable and the poorest people are in Small Island and low lying coastal countries.
- Addressing the global threat of climate change is fundamental for ensuring long-term sustainable development, energy security, availability of food and water resources and disaster risk reduction.
- COP 16/ CMP 6 express the fact that the international negotiations on climate change have not yet satisfactory outcome. The goal of the negotiating process is to adopt new global agreement establishing a fair and effective international framework towards global low-emissions and resilient development; strengthening trust between the countries, increasing the transparency and inclusiveness of the process are critical for its success.

- The governments, local authorities, businesses and civil society actors have to identify for implementation a common response to the challenge of climate change. Parliaments should exercise with constitutional authority and press by the national commitments of developed countries and actions by developing countries to minimise GHG emissions and strengthen mitigation and adaption.
- Multilateral Path is a fair and effective route to resolve global problems.
- The United Nations with its engagement across a wide range of sectors and its universal membership remains the institution that has the scope, expertise and legitimacy to craft and implement, through multilateral framework, effective policies to address the strategic imperative of climate change. The recent adoption in Nagoya, Japan, of a historic new protocol to the convention on Biological Diversity sets an inspirational example for climate change negotiations.
- The agreement will be achieved in particular areas in reducing emissions from deforestation and forest degradation, with a system of measuring, reporting, verifying mitigation, short and long-term finance and finance transparency, a more effective and flexible market mechanism, technology development and transfer and an adoption framework.
- The governments should implement the future climate change agreements with the aim of building green prosperity globally. The Copenhagen Accord did provide a framework for securing a possible future agreement and to stabilise GHG emissions at a level of global temperature to below 2^0C.
- Deep cuts in global emissions are consistent with scientific knowledge and can be implemented without compromising the right to development. The eventual agreement should be based on the principles of accountability and transparency of national action plans.
- The Copenhagen Accord's financial commitment of US \$100 billion by 2020 has not yet been guaranteed within the ongoing negotiations. The establishment by the United Nations of a High-Level Advisory group on climate change financing is a welcome step in this direction to support climate change action in developing countries including through fast-track finance.
- The parliamentary colleagues around the world should take action for approval of national climate related budgets and implementing legislation, emissions reduction and adaptation strategies and gender-focussed programme.

- The Inter-Parliamentary Union to pursue its efforts to mobilise the global Parliamentary Community around the issue of climate change towards low emissions' safer, healthier, cleaner and more prosperous future for all.

2011 United Nations climate change conference - COP 17/CMP 7

The 2011 United Nations climate change conference was held in Durban, South Africa from November 28 - December 11, 2011 to establish a new treaty to limit carbon emissions.[68] The conference agreed to establish a legally binding deal comprising all countries by 2015 which was to take effect in 2020.[69] There was also progress regarding the creation of a Green Climate Fund for which a management framework was adopted. The fund is to distribute US $100 billion per year to help poor countries adapt to climate impacts.[70] Scientists and environmental groups warned to avoid global warming beyond 20C for which more urgent action is needed.[71]

A primary focus of the conference was:

- To secure global climate agreement as the Kyoto Protocol's first commitment period (2008-2012) was about to end.
- To focus on "finalising at least some of the Cancun Agreements", reached at the 2010 conference such as cooperation on clean technology as well as forest protection, adaptation to climate impacts and finance.
- To transfer of funds from rich countries to poor in order to help them protect forests, adapt to climate impacts and green their economies.

The climate change summit was held and talked about humanity's role in climate change among nations, US and China. The decision, that was reached and its implication for the rest of the world and hopefully ensure for survival. At the meeting, there was discussion for about how to best educate population and to deal with something like climate change. The major issue of the summit was, "humanity's role in climate change". The major results from the meeting were to adopt the Kyoto Protocol, lowering trade emissions worldwide and pushing for more environmental friendly energy research in the first world.

The Decisions of COP 17 are:

- Establishment of an Ad Hoc Working Group on the Durban platform for enhanced action.
- Outcome of the work of the Ad Hoc Working group on long term cooperative action under the convention.
- Launching the Green Climate Fund.
- Technology Executive Committee Modalities and procedures.

- National adaptation plans.
- Nairobi work programme on impacts, vulnerability and adaptation to climate change.
- Work programme on less and damage.
- Forum and work programme on the impact of the implementation of response measures.
- Least Developed Countries Fund: support for the implementation of elements of the least developed countries work programme other than national adaptation programmes of action.
- Report of the global environmental facility to the conference of the parties and additional guidance to the global environmental facility.
- Guidance on systems for providing information on how safeguards are addressed and respected and modalities relating to forest reference emission levels.
- Capacity building under the convention.
- Work of the consultative group of experts on national communications.
- Revision of the UNFCCC reporting guidelines on annual inventories for parties.
- Research dialogue on developments in research activities relevant to the needs of the convention.
- Administrative, financial and institutional matters.
- Programme budget for 2012-13.

The UN climate change conference in Durban was a turning point in the climate change negotiations. In Durban, governments clearly recognised the need to draw up the blueprint for a fresh universal, legal agreement to deal with climate change beyond 2020, where all will play their part to the best of their ability and all will be able to reap the benefits of success together. All governments were committed in Durban to a comprehensive plan that would come closer with ultimate objective of the climate change convention to stabilise greenhouse gas concentrations in the atmosphere.

2011 UN climate change conference in Durban implies international community's response to climate change. This conference balanced negotiations in the implementation of the convention and the Kyoto Protocol, the Bali Action Plan and the Cancun Agreements. The outcomes' decision by parties was to adopt a universal legal agreement on climate change as soon as possible and not later than 2015. The President of COP 17/CMP 7 Maite Nkoana Mashabane said, "What we have achieved in Durban will play a central role in saving tomorrow, today."

2012 United Nations climate change conference - COP 18/CMP 8

The 2012 United Nations climate change conference was the 18th Conference of Parties (COP) to the 1992 United Nations Framework Convention on Climate Change (UNFCCC) and the 8th Meeting of the Parties (CMP) to the 1997 Kyoto Protocol. The conference took place from Monday, November 26 to Saturday, December 7, 2012, at the Qatar National Convention in Doha.

The UNFCCC entered into force in 1994, the Conference of Parties (COP) to the UNFCCC has been meeting annually to assess progress in dealing with climate change. The COP is the supreme body of the convention and highest decision making authority. The COP is an association of the countries that are parties to the convention. 195 parties took part in climate change negotiation convention. In 2010, governments agreed that emissions needed to be reduced below 2^0C.

The COP is assisted by two subsidiary bodies. The Subsidiary Body on Scientific and Technological Advice (SBSTA) links scientific, technical and technological assessments. The Subsidiary Body for Implementation (SBI) was created to develop recommendations to assist the COP in reviewing and assessing implementation of the convention and in preparing and implementing its decisions.

Meeting of the parties to the Kyoto Protocol (CMP) meets annually coinciding with the COP. The CMP reviews the implementation of the Kyoto Protocol and takes decisions to promote its effective implementation. The Kyoto Protocol was adopted in 1997 and legally binds developed countries to emission reduction targets. The protocol's first commitment period started in 2008 and ends on December 31, 2012.

In the Doha climate change conference, poor countries have won historic recognition of the plight they face from the ravages of climate change and they will receive funds to repair the "loss and damage" incurred. This is the first time developing countries have received such assurances and the first time the phrase "loss and damage" from climate change has been enshrined in an international legal document.

The COP 17/CMP 7 in Durban November 29, 2011, the state of Qatar and Republic of Korea have agreed to closely cooperate to make the next major UN climate change conference at the end of 2012 a success. UNFCCC Executive Secretary Christiana Figures said, "I congratulate these two countries on their commitment to work together in the lead up to and during the COP. Both countries are leaders in their own ways and can generate strong synergies to put the world on a more climate-safe path. All governments should work together on the next essential climate steps which can be inspired by this collaborative spirit".

The state of Qatar and the Republic of Korea will make joint efforts to globally promote and implement the green growth agenda. The Republic of Korea has championed the concept of green economy which links green growth to sustainable development and poverty eradication. The state of Qatar, as one of the world's main energy exporters, expressed its eagerness in Durban to secure progress in the UN climate change negotiations and support to the endeavours of developing countries including small island developing states, in adapting to the effects of climate change.

The 18th UN climate change conference in Doha in December 2012 was held with the aim of laying road map for a global binding agreement on emissions reduction due to be finalised in 2015.

Negotiations on forests' and climate change mitigation, dealt with under REDD+, took place in two tracks: The Ad Hoc Working Group on Long Term Cooperative Action, that mostly deals with the issues of REDD+ and results- based financing, and the Subsidiary Body on Scientific and Technological Advice (SBSTA) that deals with methodological aspects related to REDD+, providing guidance on key matters such as safeguard information systems, Reference Emission Levels, Forest Monitoring Systems and drivers of deforestation.

Negotiations in Doha were expected to tackle the following topics:
- Rights-based safeguards in information systems,
- The valorisation of traditional knowledge,
- Support for indigenous monitoring systems for REDD+,
- Key threats to indigenous peoples' rights with reference to drivers of deforestation,
- The need to take into account non-carbon values of forests in REDD+ financing.

The SBST negotiations ended without adopting any decisions. Discussions will, therefore, continue throughout 2013, most notably on:
- Modalities for a national forest monitoring system.
- Monitoring, Reporting and Verification.
- Provision of information on how safeguards are addressed and respected.
- Issues related to drivers of deforestation.
- Issues related to non-carbon benefits include biodiversity conservation, protection of livelihoods, governance reforms, land tenure reform and respect of indigenous peoples' rights.

Forest peoples' programme will continue to monitor UNFCCC negotiations in 2013 to provide consolidation of commitments on indigenous peoples' rights in climate related activities in forests.

United Nations Conference on sustainable development, Rio + 20, June 20-22, 2012

The United Nations conference on sustainable development is being organised in pursuance of General Assembly Resolution. The conference will take place in Brazil on June 20-22, 2012 to mark the 20th anniversary of the 1992 United Nations Conference on Environment and Development in Rio de Janeiro.

At the Rio + 20 conference, world leaders, thousands of participants from governments, private sector, NGOs and the groups will come together to shape how can reduce poverty, advance social equity and ensure environmental protection. The conference focussed on two themes: (a) a green economy in context of sustainable development poverty eradication, and (b) the institutional framework for sustainable development. The preparations for Rio + 20 have highlighted seven areas which need priority attention; these include jobs, energy, sustainable cities, food security and sustainable agriculture water, oceans and disaster readiness. Rio + 20 are a joint endeavour of the entire UN system. A dedicated secretariat is responsible for coordinating and facilitating inputs to the preparatory process from UN bodies. The preparatory process is led by 11 members Bureau composed of UN Ambassadors from all regions of the world. The host country Brazil is leading the logistical preparation on the ground. Sustainable development meets the needs of the present without compromising the ability of future generations to meet their own needs. The long term global development and sustainable development consists of three pillars: economic development, social development and environmental protection.

Climate changes in Himalaya and UNDP

Recent climate changes have had significant impact on high-mountain glacial environment in Himalaya, India. Rapid melting of snow/ice and heavy rainfall has resulted into dam and lakes on June 16 and 17, 2013. Recognising climate changes and their impacts in the Himalayan Mountains, a mountain environment group of India founded a pan-Himalayan initiative called "Climate Himalaya" to divest its efforts in climate adaptation and Sustainable Mountain Development. It is evident that the Hindu Kush Himalayan region is in a state of crisis, affecting the population vulnerable under various climatic and environmental changes.

The floods have killed hundreds of people in India's Himalayan region of Uttarakhand and left tens of thousands in need of aid and rehabilitation. The disaster which was triggered by heavy pre-monsoon rains on June 15 and 16, 2013 has been dubbed a Himalayan Tsunami by local media. Entire villages are reportedly buried by landslides, while many roads, bridges and building have been swept away. There were 1,00,000 visiting pilgrims for famous and holy temple Kedarnath, Badrinath, Gangotri and Yamunotri, when the flash floods and landslides hit. Melting of glaciers coupled with monsoon rains, triggering overflow of rivers, were responsible for the Uttarakhand tragedy, a top official of the Ministry of Earth sciences said, "The flood was not only due to the rains but also because of the melting of snow. Rains came almost two weeks early in the state. During that period, winter snow was already there. Now when the rain came, snow melted and flowed down along with the rains which increased the volume of water in the rivers significantly".

Role of UNDP (United Nation Development Programme)
The UNDP, United Nations Development Programme, says that early recovery is a multi-dimensional process of recovery that begins in humanitarian settings. It is an integrated and coordinated approach, using humanitarian mechanism to gradually turn the dividends of humanitarian action into sustainable crisis recovery, resilience building and development opportunities.

When a crisis strikes, UNDP works to help ensure that the humanitarian response to the emergency to provide clean water, sanitation, food and shelter, is a long-term development work. This approach to humanitarian work, called 'early recovery' is integrated into the work of all humanitarian actors and helps orient the entire humanitarian response to contribute to rebuilding communities, creating an environment for recovery, building the capacities of local communities and institutions and integrating risk reduction into programme interventions.

The UNDP leads early recovery and has inherited responsibilities from the Inter- Agency Standing Committee (IASC), the body responsible for inter-agency cooperation in the humanitarian system. UNDP hosts the Cluster Working Group on Early Recovery (CWGER) whose roles include promoting and clarifying early recovery as a concept and ensuring it is being adopted in humanitarian response in affected countries. The humanitarian response emphasises the importance of building community capacity and to strengthen individuals and communities resilience to future disasters; and reduce dependence on relief.

UNDP is not only promoting this approach through CWGER; but as an implementing agency. For example, in the time of earthquake or flood, UNDP will work with its long term partners to ensure public services and functioning as early as possible after the crisis. They do not get help only from outside, but can rebuild their own homes and communities, improve their long term early potential to combat vulnerability to poverty and disasters.

Early recovery helps to improve coordination between humanitarian and development actors and helps saving lives, money and protects development achievements and opportunities. In this sense early recovery augments humanitarian assistance operations, supports spontaneous recovery initiatives by affected communities and establishes the foundations for resilience and long-term recovery as soon as possible after a disaster or crisis.

Bonn climate change conference - 2013
The second session of the Ad Hoc Working Group on the Durban platform for enhanced Action (ADP 2) took place between April 29 and May 3, 2013 at the world conference centre Bonn, in Bonn, Germany.

Addressing the media on the final day of the Bonn climate change conference, UNFCCC Executive Secretary Christiana Figures summed up what has been a very productive week, both in advancing the negotiations towards a universal climate agreement and in raising ambition to combat climate change now and adapt to its effects. ADP Co-Chairs special event was held for stakeholders, focussing on central elements and design aspects of the 2015 agreement, as well as ways to catalyse action and build a practical results-based approach to increase pre- 2020 ambition. At a round table on reducing emissions governments had a focused view of commitments for the 2015 agreement, in particular how the agreement could be more ambitious emission reduction action. They showed strong interest in exploring systems to clarify and review their adequacy with achieving the 20C target. At the round tables, the adaptation and implementation of climate action were discussed. There was a clear tendency to see adaptation and mitigation as "two sides of same coin". A discussion began on finance in the context of the 2015 agreement with emphasising the need for support to enable mitigation and adaptation in a transparent manner. There were discussions on adaptation, on finance, technology and capacity building support as well as transparency of action and support. A transparency system is essential to quantify mitigation efforts and support to keep track of progress towards the 20 goal.

A workshop on opportunities for mitigation and adaptation related to land use was held. Experts present their views on managing land use and forests and scientific perspective on carbon conservation and sequestration-as well as an overview of experience on the ground in the area of land use and climate change. The discussions were on implementation of national climate change strategies and mainstreaming of climate change considerations in policies and issues such as addressing sustainable natural resources management, territorial planning, forest monitoring and agricultural production. Food security, sustainable livelihoods, economic and productivity gains, bio-diversity conservation and poverty represent a major driving force behind national action on climate change. Participants gave their views on the importance of international cooperation and involvement of international organisations such as FAO, UN- REDD programme and REDD+ partnership.

Kyoto protocol on GHG emission
The Kyoto Protocol is an international agreement, an amendment and international treaty linked to the United Nations Framework Convention on Climate Change to reduce global warming. The temperature is increasing and it will be unavoidable after 150 years of industrialisation. Kyoto Protocol is legally binding on the ratifying nations and stronger than those of UNFCCC. The Kyoto Protocol agreed to reduce emissions of six greenhouse gases that contribute to global warming: carbon dioxide, methane nitrous oxide, sulphur hexafluoride, Hydro Fluorocarbons (HFCS) and Per Fluorocarbons (PFCS). The Kyoto Protocol treaty was negotiated on December 11, 1997 at the city of Kyoto, Japan and came into force on February 16, 2005.

The Kyoto Protocol is a legally binding agreement under which industrialised countries will reduce their collective emissions of greenhouse gases by 5.2 per cent compared to the year 1990. The goal is to lower overall emissions from six greenhouse gases, calculated as an average over the five- year period of 2008-12. National targets range from 8 per cent reductions for the European Union and some others to 7 per cent for the US, 6 per cent for Japan, 10 per cent for Russia and 8 per cent for Australia and 10 per cent for Iceland.[72]

The Kyoto Protocol was an agreement negotiated by many countries in December 1997 and it was opened for signature on March 16, 1998 and closed a year later. Under terms of agreement, the Kyoto Protocol did not take effect until 90 days after it was ratified by at least 55 countries involved in the UNFCCC. Another condition was that ratifying countries had to represent at least 55 per cent of the world's total carbon dioxide emissions for 1990. The first condition was met on May 23, 2002, when Iceland became the 55th country to ratify the Kyoto Protocol. When Russia ratified the agreement in November 2004, the second condition was satisfied and the Kyoto Protocol entered into force on February 16, 2005.

The Kyoto Protocol agreement in Article 2.a (ii) described for protection and enhancement of sinks and reservoirs of greenhouse gases not controlled by the Montreal Protocol under international environmental agreements; promotion of sustainable forest management practices, afforestation and reforestation.

- In Article 2.a (v), progressive reduction or phasing out the market imperfections, fiscal incentives, tax and duty exemptions and subsidies in all greenhouse gas emitting sectors that run counter to the objective of the convention and application of market instruments.
- Art 2.1.a (vi) - Encouragement of appropriate reforms in relevant sectors aimed at promoting policies and measures which limit or reduce emissions of greenhouse gases not controlled by the Montreal Protocol.
- Art 2.1.a (vii) - Measures to limit or reduce emissions of greenhouse gases not controlled by the Montreal Protocol in the transport sector.
- Art 2.1.a (viii) - Reduction of methane emissions through recovery and use in waste management as well as in the production, transport and distribution energy.
- Art 2 (2) - Reduction or limitation of emissions of greenhouse gases from aviation and marine bunker fuels, working through the International Civil Aviation Organisation and the International Maritime Organisation.
- Art 2 (3) - Annex I shall strive to implement policies and measures under this Article in such a way as to minimise adverse effects including the adverse effects of climate change, effects on international trade and social, environmental and economic impacts on other parties, especially developing country parties.
- Art 3(3) - The net changes in greenhouse gases emission by sources and removals by sinks resulting from direct human-induced land-use change and forestry activities, limited to afforestation, reforestation and deforestation since 1990.
- Art 3 (7) - In the first quantified emission limitation and reduction commitment period from 2008 to 2012, the assigned amount for each

party included in Annex I shall be equal to Annex B of its aggregate anthropogenic carbon dioxide equivalent emissions of the greenhouse gases.

- Art 4 (1) - Provided those commitments that their total combined aggregate anthropogenic carbon dioxide equivalent emissions of the greenhouse gases.
- Art 5 (1) - Methodologies for estimating anthropogenic emissions by sources and removals by sinks of all greenhouse gases not controlled by the Montreal protocol shall be those accepted by the Intergovernmental Panel on Climate Change by the conference of parties.
- Art 6 (1) - Aimed at reducing anthropogenic emissions by sources or enhancing anthropogenic removals by sinks of greenhouse gases in any sector of the economy, provided that -
  - ○ Any such project has the approval of the parties involved.
  - ○ Any such project provides a reduction in emissions by sources or an enhancement of removals by sinks.
  - ○ It does not acquire any emission reduction units if it is not in compliance with its obligations under Articles 5 and 7.
  - ○ The acquisition of emission reduction units shall be supplemental to domestic actions for the purposes of meeting commitments under Article 3.

The protocol's major feature is that -

- It has mandatory targets on greenhouse gas emissions for the world's leading economies which have accepted it and reducing their overall emissions of such gases by at least 5 per cent below existing 1990 levels in the commitment period 2008 to 2012.
- Commitments under the protocol vary from nation to nation.
- The agreement offers flexibility in how countries may meet their targets. Their emissions by increasing 'sinks' - forests, which remove carbon dioxide from the atmosphere.
- The Kyoto Protocol is a complicated agreement. The protocol not only has to be an effective against a complicated worldwide problem- it also has to be politically acceptable.
- Some mechanisms of the protocol had enough support that they were set up in advance of the protocol's entry into force. For example, the clean development mechanism through which industrialised countries can partly meet their binding emissions targets through 'credits' earned by greenhouse gas reducing projects in developing countries. There are three Kyoto mechanisms which are:

Emissions trading

- The Kyoto Protocol limits on emissions by the world's major economies, a prescribed number of emission units. Individual industrialised countries will have mandatory emissions targets they must meet.
- The protocol allows countries that have emissions units to spare, but to sell this excess capacity to countries that are over their targets. This so-called, 'carbon market'. Carbon dioxide is the most widely produced greenhouse gas and emissions of other greenhouse gases will be recorded and counted in terms of their, 'carbon dioxide equivalents' is both flexible and realistic.
- Emissions units will be involved in trades and sales. Countries will get credit for reducing greenhouse gas totals by planting or expanding forests, for carrying out 'joint implementation projects' with other developed countries and the protocol's clean development mechanism, which involves funding activities to reduce emissions by developing nations.

The Clean development mechanism

- The Kyoto Protocol does not limit on the greenhouse gas emissions of developing nations. The greenhouse gas emissions of developing countries are growing in China and India.
- The mechanism has drawn extensive interest from rich and poor countries and steps have been taken to put it into operation even before the protocol takes effect. It is cost-effective and gives flexibility to industrialised countries trying to meet their targets.
- The mechanism is meant to work bottom-up-to proceed from individual proposals to approval by donor and recipient governments to the allocation of "certified emissions reduction" credits.
- The clean development mechanism is overseen by an Executive Board. To be certified, by the clean development mechanism Executive Board, a project must be approved by all involved parties to reduce emissions, and promise reductions that would be additional to any that would otherwise occur.

Joint implementation

- Joint implementation is a programme of mutual help for countries with emission's targets under the Kyoto Protocol that allows industrialised countries to meet part of their required cuts in greenhouse-gas emissions by paying for projects that reduce emissions in other industrialised countries.

- The operation of the joint implementation mechanism is similar to that of the 'clean development mechanism'. Joint implementation projects, industrialised countries must meet requirements under the protocol for accurate inventories of greenhouse gas emissions.

Under the protocol, countries' actual emissions have to be monitored and precise records have to be kept of the trades carried out. Parties must keep a national registry to track and record transactions under the mechanisms. The Secretariat keeps an independent transaction log to verify that transactions are consistent with the rules of protocol and expert review teams have been set up to ensure compliance. The Kyoto Protocol like the convention is also designed to assist countries in adapting to the inevitable effects of climate change and facilitates the development of techniques that can help increase resilience to climate change impacts. The adaptation fund was established to finance adaptation projects and programmes in developing countries that are parties to the Kyoto Protocol. The Kyoto Protocol is generally seen as an important first step towards a truly global emission reduction regime that will stabilise greenhouse gas concentrations. As a result of the protocol, governments have already put and are continuing to put in place legislation and policies to meet their commitments; a carbon markets has been created and more and more businesses are making the investment decisions needed for a climate-friendly future. The protocol provides much of the essential architecture for any new international agreement or set of agreements on climate change.

Mitigation

The United Nations agreement for climate change among nations is in two strategies - one is mitigation and another is adaptation. Those nations mitigate greenhouse gas emissions to stabilise atmospheric concentrations of these gases at a safe level. These strategies help vulnerable nations to adapt the adverse consequences of global warming.

Global warming will continue until greenhouse gases in the atmosphere stabilise, it will be zero if net global annual emissions of these gases decline. The global emissions are rising rapidly in majority of nations. A major focus of the UN's climate change security strategy must be to facilitate emissions abatement in both developed and developing nations. The global climatic strategies have been carried out largely in the context of the 1992 UN Framework Convention on Climate Change and its 1997 Kyoto Protocol. Developed Nations largely ignored the political commitment they made under the convention to return their emissions to

1990 levels by 2000. Even if the Kyoto treaty goes into force, it will cover only 25 per cent of global emissions and not those of the United States and China, the world's two largest national emitters.[73]

Since 1992 Earth Summit, the environment has lost ground politically, sub- merged under the broader sustainable development agenda. To speed mitigation efforts, the Secretary General must raise the visibility of climate change and play a more active role in overcoming obstacles to emissions mitigation. One complication is that while developed nations should take the lead in reducing emissions, but in developing nations emission abatement could be more cost effective.

Most often, climate change mitigation scenarios involve reductions in the concentrations of greenhouse gases, either by reducing their sources or by increasing their sinks. The UN defines mitigation in the context of climate change, as a human intervention to reduce the sources or enhance the sinks of greenhouse gases. Examples, include using fossil fuels more efficiently for industrial processes or electricity generation, switching to renewable energy (solar energy or wind power), improving the insulation of buildings and expanding forests and other 'sinks' to remove greater amounts of carbon dioxide from the atmosphere. The main international treaty on climate change is the United Nations Framework Convention on Climate Change.[74] In 2010, parties to the UNFCCC agreed that future global warming should be limited to below $2^{0}C$ to the pre- industrial level.[75]

The climate change mitigation is the stabilisation of greenhouse gas concentrations in the atmosphere. The UNFCCC has the ultimate objective of preventing dangerous anthropogenic interference of the climate system. In Article 2 of the convention it has been stated that Greenhouse gas concentrations are stabilised in the atmosphere at a level where ecosystems can adapt naturally to climate change, food production is not threatened and economic development can proceed in a sustainable fashion.[76] The most important GHG emitted by human activities is carbon dioxide (CO_2) to the atmosphere for faster than natural processes can remove it.[77]

UN system's work on climate change mitigation and predict the following grounds for reduction

Reducing emissions
Climate change is the need to reduce emissions. The six main greenhouse gas emissions are projected to rise by 25 - 90 per cent by 2030 compared to 2000. Both developed and developing nations must take measurable, reportable and verifiable mitigation action. In 2010, governments agreed

that emissions need to be reduced, so that global temperature increases are limited below 2⁰C.

Reduction commitments
Most developed countries have announced mid-term-target reductions for 2020, but most of these targets are far short of the IPCC range, which would be necessary to limit temperature increase to 2⁰C. Global emission need by 2015 to reach a reduction of 50 per cent by 2050 to avoid worst effects of climate change.

Avoiding emissions
Mitigation refers to the policies and measures designed to reduce greenhouse gas emissions. Measures can include reducing demand for emissions-intensive goods and services, increasing the use of low-carbon technologies and renewable energy. By this way, the level of greenhouse gases in the atmosphere can be slowed and ultimately stabilised.

Absorbing emissions
Another way to mitigate the impacts of climate change is by enhancing 'sinks'- reservoirs that absorb CO_2, living existing forests and planting new trees are two examples of how this can be achieved. Launched in September 2008, the UN - REDD focusses on these mitigation aspects.

The role of forests
Deforestation is responsible for up to 20 per cent of greenhouse gas emissions worldwide. Most of the forest land is cleared for agricultural use. Forests are carbon sinks which able to permanently absorb about one-tenth of global CO_2 emissions into biomass, soil and forest products.

Different emission sources
Over the last three decades, all greenhouse gas emissions increased by 1.6 per cent per year with CO_2 emissions from fossil fuels use at 1.9 per cent per year. The largest growth in greenhouse gas emissions has come from energy supply and road transport.

Clean energy investments
Huge investments will be needed to increase energy capacity in developing countries over the next 10 - 20 years and these investments can be directed to lower-emission energy sources, such as renewable energy.

Greenhouse gases on the rise

Without additional action by governments, emission of the six main greenhouse gases is there - carbon dioxide, methane, nitrous oxide, sulphur hexafluoride, PFCs and HFCS. Between 1970 and 2004, emissions of these gases increased by 70 per cent.

Government intervention can help

By adopting stronger climate change policies, governments could slow and reverse these emission trends and ultimately stabilise the level of greenhouse gases in the atmosphere.

Avoiding the worst climate change impacts

Mitigation efforts over next two to three decades will determine to a large extent the long term global mean temperature increase. Properly designed climate change policies can be part and parcel of sustainable development and the IPCC's findings confirm that sustainable development paths can reduce greenhouse gas emissions and reduce vulnerability to climate change.

Governments and policies

Governments are successfully using a wide range of policies and measures to address climate change, including regulations and standards, taxes and charges, voluntary agreements, subsidies, financial incentives, research and development programs and information instruments. The most effective policy mix will vary from country to country.

Policies to guide investment

Government policies and private sector investment decisions are critical for the more than US $20 trillion that is expected to be invested in energy infrastructure from the present time till 2030 that will have long term impacts on greenhouse gas emissions.

Removing barriers to innovation

For policies to be effective, governments need to pay special attention to identify and remove barriers for innovation. These can include market prices that do not incorporate externalities such as pollution, misplaced incentives, vested interests, lack of effective regulatory agencies and imperfect information. Direct and indirect subsidies for fossil fuel use and agriculture remain common practice, although coals have declined over the past decade, particularly in industrialised countries.

Estimating economic impacts
Economists use model to estimate the economic impacts to reduce emissions. Studies indicate that there is substantial economic potential for the mitigation of global greenhouse gases emissions over the coming decades. Economic models produce lower cost estimates when they use baselines with slowly rising emissions and when they allow technological change to accelerate as carbon prices rise. Costs are also reduced when the Kyoto Protocol's flexibility mechanisms are more fully implemented. Comparing the social cost of carbon estimates with the carbon prices for different levels of mitigation shows that the social cost of carbon comparable to and possibly higher than, carbon prices for even most scenarios assessed by the IPCC.

Economic benefits
Climate policies bring many benefits include technological innovation, tax reform, increased employment, improved energy security and health benefits from reduced pollution. Projections of future greenhouse gas emissions are highly uncertain.[78] In the absence of policies to mitigate climate change, GHG emissions could rise significantly over the 21st century.[79] GHG concentrations are unlikely to stabilise this century without major policy changes.[79]

Methods should be applied to mitigate climate change and GHGs:-

Fuel switching
A 2011 study by noted climate research scientist, Tom Wigley found that CO_2 emissions from fossil fuel may be reduced by using natural gas rather than coal to produce energy and also found that additional methane (CH_4) from leakage adds to the radiative forcing of the climate system, the reduction in CO_2 forcing that accompanies the transition from coal to gas. The methane leakage from coal mining; changes in radiative forcing due to changes in the emissions of sulphur dioxide in the efficiency of electricity production between coal- and- gas fired power generation.

Carbon capture and storage
Carbon capture and storage (CCS) is a method to mitigate climate change by capturing CO_2 from power plants and storing it away safely instead of releasing it into the atmosphere. The IPCC says CCS could contribute between 10 per cent and 55 per cent worldwide carbon-mitigation efforts over the next 90 years. CCS could potentially capture about 90 per cent of all the carbon emitted by the plants.[80] In 2011, the total CO_2 storage

capacity of all 14 projects in operation or under construction is over 33 million tons a year. This is broadly equivalent to preventing the emissions from more than six million cars from entering the atmosphere each year.[81]

Energy efficiency and conservation
'Energy efficiency' is the goal to reduce the amount of energy required to provide products and services. Energy conservation is broader than energy efficiency in that it encompasses using less energy to achieve a lesser energy service, for example, through behavioural change as well as encompassing energy efficiency. Reducing energy use is seen as a key solution to the problem of reducing greenhouse gas emissions. According to the International Energy Agency, improved energy efficiency in buildings, industrial processes and transportation could reduce the world's energy needs in 2050 by one-third and help control global emissions of greenhouse gases.[82]

Sustainable transport
Modern energy efficient technologies, such as plug-in hybrid electric vehicles and development of new technologies such as hydrogen cars, may reduce the consumption of petroleum and emissions of carbon dioxide. The air transport and truck transport to electric rail transport would reduce emissions significantly.[83]

Urban planning and building design
Urban planning also has an effect on energy use. Between 1982 and 1997 the amount of land consumed for urban development in the United States increased by 47 per cent while the nation's population grew by only 17 per cent.[84] The 'smart growth' practices include compact community development, multiple transportation choices, mixed land uses and practices to conserve green space. These offer environmental, economic and quality of life benefits and they also serve to reduce energy usage and greenhouse gas emissions.

New building can be constructed using passive solar building design, low-energy building or zero-energy building techniques, using renewable heat sources. Existing building can be made more efficient through the use of insulation, high efficiency appliances, double or tripled-glazed gas-filled windows, external window shades, building orientation and sitting to reduce the amount of greenhouse gases emitted.

Reforestation and avoided deforestation
20 per cent of total greenhouse gas emissions were from deforestation in 2007. Stern review found eight countries responsible for 70 per cent of global deforestation emissions. Creating negative carbon dioxide emissions remove carbon from the atmosphere. For example, direct air capture, bio-char, bio-energy with carbon capture and storage and enhanced weathering technologies. These processes are sometimes considered as variations of sinks or mitigation.[85]

Carbon dioxide removal and solar radiation management
Carbon dioxide removal has been proposed as a method of reducing the amount of radiative forcing. Geological storage of carbon may make carbon dioxide air capture viable commercially.

The main purpose of solar radiation management seeks to reflect sunlight and thus reduce global warming. The ability of stratospheric sulphate aerosols to create a global dimming effect has made for use in geoengineering projects.[86]

Population control
Various organisations promote population control as a means for mitigating global warming through family planning, reproductive health care, public education and improving access of women to education and economic opportunities.

Politics of global warming
Many countries, both developing and developed are aiming to use cleaner technologies.[87] Use of these technologies aids mitigation and could result in reductions in CO_2 emissions. Policies include targets for emission reductions, increased use of renewable energy and increased energy efficiency.

The main current international agreement on combating climate change is the Kyoto Protocol which came into force on February 16, 2005. The Kyoto Protocol is an amendment to the United Nations Framework Convention on Climate Change (UNFCCC) that all countries reduce their emissions of carbon dioxide and find other greenhouse gases. Actions to mitigate climate change are sometimes based on the goal on achieving a particular temperature target. The target has been suggested to limit the future increase in global mean temperature to below 2°C.

Developing countries need particular support of financial and technical to mitigate carbon emissions achieving Kyoto Protocol's

Clean Development Mechanism (CDM). The UNFCCC Copenhagen climate conference was the Copenhagen Accord, in which developed countries promised to provide US $30 million from 2010-2012 of new and additional resources.[88] The overseas Development Institute have found four main understandings to mitigate climate change, they are:

- Climate finance classified as aid, but the above 0.7 per cent Official Development Assistance (ODA) target.
- Increase on previous year's Official Development Assistance (ODA) spent on climate change mitigation.
- Rising ODA levels that include climate change finance but where it is limited to a specific percentage.
- Increase in climate finance not connected to ODA.

There is a conflict between the OECD (Organisation of Economic Co- operation and Development) states about budget deficit cuts, the need to help developing countries adapt to develop sustainably and the need to ensure that funding does not come from cutting aid to other important Millennium Development Goals.

According to the most stringent scenarios of the IPCC, a long term goal in line with the latest science would include:

- A peak in emissions in the next 10-15 years.
- And a decline of 50 per cent over 2000 levels by 2050.
- This would stabilise emissions at round 450 parts per million CO_2 eq in the atmosphere and correspond to a 2-2.4^{0}C rise in temperature.

Commitments under the UNFCCC

The UNFCCC sets an overall framework for international efforts to tackle the challenge of climate change. Parties to the convention agreed to a number of commitments to address climate change:-

- To develop and periodically submit national reports containing information on the greenhouse gas emissions of that party and describing the steps it has taken and plans to take to implement the convention.
- To put in place national programs and measures to control emissions and to adapt to the impacts of climate change.
- To promote the development and use of climate-friendly technologies and the sustainable management of forests and other ecosystems.
- Developing countries (Non- Annex I countries), did not make commitments to reduce or limit greenhouse gas emissions at the time when the convention was being negotiated. The convention applies the principle of "common but differentiated responsibilities".

As a result industrialised countries (Annex - I Parties) under the convention have additional commitments:-

- Undertake policies and measures with the specific aim of returning their greenhouse gas emissions to 1990 levels by 2000.
- Provide more frequent and more detailed national reports and must separately provide yearly reports on their national greenhouse gas emissions.
- Promote and facilitate the transfer of climate friendly technologies to developing countries and to countries with economies in transition.

Future mitigation action in a post - 2012 framework
- Deep emissions cut by industrialised countries are needed and these countries must continue to take the lead in mitigation, given their historic responsibilities and economic capabilities.
- A future climate change regime will require further engagement of developing countries, in particular those whose emissions already, or will in the near future, significantly contribute to atmospheric concentrations. This will be important project economic growth and energy demand in developing countries.
- Developing countries may need to limit their emissions while safeguarding economic growth and poverty eradication.
- The UNFCCC acknowledges the need to protect forests as part of efforts to climate change.
- Tropical deforestation was excluded from the Kyoto Protocol due to controversies surrounding sovereignty, scientific uncertainty and implications for efforts to reduce fossil fuel emissions.
- Discussions on reducing emissions from deforestation in developing countries are now under way within the UNFCCC process, at the initiative of developing countries.
- According to the Intergovernmental Panel on Climate Change (IPCC), there is significant mitigation potential, including with the increased use of clean technologies and improved energy efficiency for all sectors mitigation costs in 2030 would not exceed 3 per cent of global GDP. Available mitigation options can yield multiple societal and environmental benefits.
- The carbon market has great potential for cost-effective mitigation, but needs long-term policy certainty in demand beyond 2012 to continue to deliver.

Cancun conference mitigation action plan - 2010

Developed country Emissions Reduction Target

During 2010 many countries submitted their existing plans for controlling greenhouse gas emissions to the climate change secretariat. Industrial countries presented their plans in the shape of economy-wide targets to reduce emissions, mainly up to 2020. All industrialised countries submitted economy-wide emission reduction targets. The implementation of these targets will begin in 2011. Every two years, the industrialised countries will report of progress and submit detailed annual inventories of greenhouse gas emissions reduction target.

Further specific decisions under the kyoto protocol

The Kyoto Protocol's Clean Development Mechanism (CDM) was strengthened to drive major investments and technology into environmentally sound and sustainable emission reduction projects in developing world. This will be done by means of a loan scheme to encourage clean development mechanism project activities in countries.

Governments agreed to allow carbon capture and storage projects in the CDM provided that a range of technical issues and safety requirements are resolved and fulfilled. Governments agreed that in second commitment period, the Kyoto Protocol's emissions trading and project-based mechanisms, which encourage clean technology investment from industrialised countries into developing countries. The agreement reached in Cancun on land use, land use change and forestry (LULUCF) called for the submission of reference levels for forest management. Kyoto negotiations have been looking at how countries include forest management in their greenhouse gas accounts that forests naturally absorb carbon dioxide.

Decisions addressing developing country mitigation plans

During 2010, many developing countries submitted their plans to limit the growth of their emissions, with appropriate and adequate support from industrialised countries in the form of technology cooperation, finance and help in capacity- building. These plans are known as NAMA (National Appropriate Mitigation Action). The Cancun decisions now provide a formal international registry for these plans to strengthen the ways and means, those NAMAs where countries require international support in the form of technology, finance or capacity building.

Developing countries will provide information on the action for which they are seeking support; whereas industrialised countries will provide information on support for these actions. Developing countries will also increase reporting of progress towards their mitigation objectives, although in a differentiated way to that of industrialised countries.

Reduction of emissions through stronger actions of forests
Governments also agreed to launch concrete action on forests in developing nations, which will increase going forward. The full financing options for the implementation of such Mitigation Actions in the forest area will be addressed during 2011.

Cost-effective means to achieve mitigation goals
In the course of 2011, Governments will also continue work towards establishing one or more new market-based mechanisms to both enhance and promote the cost-effectiveness of mitigation actions. The establishment of such a mechanism will be considered in Durban.

Addressing economic and social consequences of response measures
In some cases the implementation of actions that reduce emissions could result in negative economic or social consequences for other countries. As a result, government decided to convene a forum in 2011 to further discuss this and to establish a work programme to address such consequences.

Adaptation
While mitigation tackles the causes of climate change, adaptation focusses on dealing with climate impacts. Adaptation refers to the adoption of policies and practices to prepare for the effects of climate change.

Adaptation options are available in various sectors such as Water: Expanded rain water harvesting, water storage and conservation.

Agriculture: Adjustment of planting dates and crop variety, crop relocation. Infrastructure: Including coastal zones, creation of marshlands as a buffer against sea level rise and flooding.

Energy: Use of renewable sources, energy efficiency.

Developing and least developed countries are most vulnerable to the impacts of climate change, yet most in need of capacity - energy services, infrastructure and agricultural technologies - to adapt to it. Small Island developing states, in particular face increased risks of sea-level rise, beach erosion and severe weather events which impact economically important sectors like tourism and fishing.

According to the IPCC, future vulnerability depends, not only climate change but also on the type of development that is pursued. Sustainable development can reduce vulnerability and to be successful, adaptation should be implemented in the context of national and international sustainable development plans.

Adaptation to the adverse effects of climate change is vital in order to reduce the impacts of climate change that are happening now and increase resilience to future impacts. The UNFCCC on adaptation issues that are under the various convention bodies including:

- The Cancun adaptation framework, which resulted from negotiations on enhanced action on adaptation and is consisted of five clusters: implementation, support, institutions, principles and stakeholder engagement.
- Nairobi work program on impacts, vulnerability and adaptation to climate change development and transfer of technologies, research and systematic observation under the Subsidiary Body for Scientific and Technological Advice (SBSTA).
- Issues related to National Adaptation Programs of Action (NAPAs), and supporting adaptation through finance, technology and capacity-building under the subsidiary Body for Implementation (SBI).

Successful adaptation not only depends on governments but also on the active and sustained engagement of stakeholders (Nairobi work programme), including national, regional, multilateral and international organisations, the public and private sectors, civil society and other relevant stakeholders. The main adaptation issues are given below:

Bali action plan
The Bali Action Plan adopted at COP 13 in Bali, December 2007, identified adaptation as one of the key building blocks required for a strengthened future response to climate change to enable the full, effective and sustained implementation of the convention through long-term cooperative action, now up to and beyond 2012.

Cancun adaptation framework
At the Cancun climate change conference in December 2010, parties established the Cancun Adaptation Framework (CAF) with the objective of enhancing action on adaptation. This conference established Adaptation Framework, which will strengthen action on adaptation in developing countries through international cooperation. It will support better planning and implementation of adaptation measures through increased financial and technical support, and through strengthening and establishing regional

centres and networks. The Cancun Adaptation Framework will also boost research, assessments and technology cooperation on adaptation, as well as strengthen education and public awareness.

The Cancun Adaptation Framework also established an Adaptation Committee to promote the implementation of stronger action on adaptation by providing technical support and guidance to countries, strengthening knowledge-sharing and promoting synergy between a range of stakeholders. The conference also established a process for Least Developed Countries (LDCs) and other interested developing countries to formulate and implement National Adaptation Plans (NAPs) to identify and address their medium and long-term adaptation needs.

Durban, COP-17-2011
At the Durban climate change conference in November/December 2011, parties advanced the implementation of Cancun Adaptation Framework (CAF) by agreeing on:

Modalities, procedures and composition of the adaptation committee
Promote the implementation of enhanced action on adaptation under the convention through the following functions:
- Providing technical support and guidance to the parties.
- Sharing of relevant information, knowledge, experience and good practices.
- Promoting synergy and strengthening engagement with national, regional and international organisations, centres and networks.
- Providing information and recommendations, drawing on adaptation good practices, for consideration by the COP when providing guidance for implementation of adaptation actions including finance, technology and capacity building;
- Considering information communicated by parties on their monitoring and review of adaptation actions, support provided and received.

Activities to be undertaken under the work program on loss and damage
The COP established a work program in order to consider approaches to address loss and damage associated with climate change impacts in developing countries that are particularly vulnerable to the adverse effects of climate change, as part of the Cancun Adaptation Framework (decision 1/CP.16, Paragraphs 25-29). Addressing loss and damage associated with the adverse effects of climate change requires comprehensive, inclusive and strategic responses and the parties agree that the convention has

the important and fundamental role to enable coherent and synergistic approaches.

Modalities and guidelines for the national adaptation plans
The National Adaptation Plan (NAP) process was established under the Cancun Adaptation Framework (CAF). It enables parties to formulate and implement NAPs as a means of identifying medium-and-long-term adaptation needs and developing and implementing strategies and programs to address those needs. It is continuous, progressive and iterative process which follows a country-driven, gender-sensitive, participatory and fully transparent approach.

Doha, COP-18-2012
At the COP 18 in Doha, Qatar, parties continued the adaptation implementation decisions are:
* The technical guidelines for the development of national adaptation plans.
* The continuity of the work program on loss and damage including the establishment of institutional arrangements at COP-19.
* A three year work plan for the adaptation committee.

Nairobi work programme
The Nairobi Work Programme (NWP) is undertaken under the Subsidiary Body for Scientific and Technological Advice (SBSTA) of the UNFCCC. Its objective is to assist all parties, in particular developing countries, including the least developed countries and Small Island developed states to:
* Improve their understanding and assessment of impacts, vulnerability and adaptation to climate change and
* Make informed decisions on practical adaptation actions and measures to respond to climate change on a sound scientific, technical and socio-economic basis, taking into account current and future climate change and variability.

The NWP is implemented by parties, inter-governmental and non-governmental organisations, the private sector, communities and other stakeholders. The SBSTA encourages active engagement of adaptation stakeholders in the implementation of the NWP under mandated programme activities and work areas. The NWP disseminates knowledge and information on adaptation and highlights the work of partners as widely as possible through a variety of knowledge products and publications. Organisations, institutions and private sector companies at all levels and in a wide range of sectors can become engaged with the programme.

National adaptation programmes of actions
NAPAs provide a process for the Least Developed countries (LDC) to identify priority activities that respond to their urgent and immediate needs with regard to adaptation to climate change. NAPAs works on the limited ability of the LDCs to adapt to the adverse effects of climate change.

In May 2013, 49 countries had completed and submitted their NAPAs to the secretariat. The main content of the NAPA document is a list of ranked priority adaptation activities/projects, as well as short profiles of each activity, designed to facilitate the development of project proposals for implementation of the NAPA. Priority sectors/areas addressed in the NAPAs are agriculture and food security, water resources, coastal zones, early warming and disaster management. Most LDCs are in the process of implementing their NAPAs. NAPAs are meant to provide LDCs with an opportunity to identify there, "urgent and immediate needs" for adapting to climate change.[89]

UN chief hails role of science in combating climate change
On June 8, 2013 UN Chief Ban Ki-Moon said that:
* Science plays a key role in finding new ways to combat climate change and governments must use scientific data to mobilise resources and take action against the global threat.
* Global warming and greenhouse gas emissions are a significant cause of climate change and has risen on the global agenda.
* Ban pointed out that in recent years climate change has risen to the top of the UN's agenda as an urgent priority that is affecting all countries in increasingly extreme ways, from flooding in Asia to tornadoes in the United States and drought in the Sahel.
* Ban said that these type of people and institutions need to analyse the trends, sort out the data and produce useful information for forming policy.
* Next year, Ban will convene a high level meeting of world leaders to mobilise political will for a binding climate change agreement by 2015 and he emphasised that the continuous contribution of the scientific community will be crucial for decision-making in this regard.
* Ban said that over the next two years leaders at the highest level from governments, business, finance, science and civil society would be

engaged and harness the full strength of the UN system to catalyse ambitious action to reduce emissions and strengthen climate resilience.
- He also encouraged scientists to build collaborations that transcend borders and that help build capacity for research and decision-making in developing countries since they are often the most vulnerable to climate change and do not have the ability to generate and apply relevant climate information.
- To successfully reduce the risks posed by climate change, everyone will need full engagement including the scientists and meteorologists.

On February 11, 2013, Ban Ki-Moon warns against the gathering threat of climate change at the Council on Foreign Relations, focussing on the crisis in Syria and the threat of climate catastrophe. He noted that science has 'long sounded the alarm' on the gathering threat of climate change, warning against projected increases of extreme weather events, development reversals, forced displacements, and tensions over resources.

Ban Ki-Moon highlighted that few leaders understand the need to make climate change the central piece of security, economic and financial management. He called for investing in mitigation. He underscored the potential of the renewable industry to lead the transition towards a low-carbon future, as outlined in his Sustainable Energy for All initiative.

Ban welcomed US President Barack Obama's 'new resolve' to address climate change and give it high political priority. He called for a stronger sense of collective responsibility and on governments to keep their promise to reach a global, legally binding climate change agreement by 2015.

The future United Nations System

United Nations must adapt to changing international conditions and take the lead in the development of policies and coordination of action to meet the challenges of the future. The UN of the next century would still remain an organisation of nation-states responding to its members concerns and needs. By the mid-21st century, the nature of statehood and assumptions about national sovereignty will have evolved in response to global needs and to an enlarged sense of world community. The communications revolution will have created a greater awareness of the inter-connection of human society.

To provide states and people with security from violence and disorder, the present approaches will be made systematic. The separate functions of an enhanced UN system - peace-keeping and conflict resolution, human rights, social and economic development, peace building - will be better integrated to achieve 'human security' for all people. The Security Council will have to be expanded by a more representative structure and by the willingness of governments to contribute regularly to the effectiveness of UN security operations. The veto power will have been modified.

By next century the Security Council will have developed better methods of dealing with civil and ethnic conflicts within states' borders, benefitting from early warning and analysis that will permit the more effective exercise of preventive diplomacy. The United Nations will have a capacity for rapid deployment immediately upon the decision of the Security Council; Regional organisations will also play a more effective role in partnership with the UN. New methods of response combining military, police and civilian elements will have been devised. The United Nations security system of the future will have learned the successes and failures of peace-keeping missions of the early 1990s. The provision of the requisite forces and adequate financing will have been put on a systematic basis.

Human rights become a more significant feature of UN action. The human rights agenda would intrude upon national Sovereignty will have given way to recognition that societies are better protected and their prospects for sustained human development improved, by systems of representative governments whereby the rights of individuals and group are assured.

In an improved international system, member states will have fully recognised the need to link their actions more effectively in the economic and social spheres. The United Nations and its many agencies will have learned to work together and supervising the economic and social activities of the UN system will have been recognised. The enhanced UN system will have begun to harmonise trade practices, technological cooperation and the monetary policies of member states and international institutions. This activity will have the support of national, economic and financial agencies as well as the private sector as contributing to their long-term advantage. The social agenda of the UN would also be tackled through a much improved co-ordination of agencies as well as through active cooperation with the non-governmental and private sector. The necessity of achieving sustainable development for all will be recognised as involving the natural environment and improvement in the status of women worldwide.

The international civil service of the highest integrity, competence and commitment, this means that members states will need to respect the relevant charter provisions and act in support of them. It will require improvements in personnel policy for recruitment, training and promotion as well as in gender equity. The quality of leadership will be decisive and governments must exercise leadership that reflects their commitment to reform.

The United Nations is an association of sovereign states and must respond to the governments of the world. Any plans for improving international machinery must take account of the growing activism of civil society. The non- governmental groups to participate in the evolving United Nations system will require much imaginative work if the world organisation is to evolve with the times. Thus, the United Nations will be recognised by all as the best hope of humankind in confronting the dangers and opportunities of the years ahead

References

1. *UNFCCC (2006a): dialogue on long-term cooperative action to address climate change by enhancing implementation of the convention, at https://unfccc.int/meetings/ bonn_may_2006/ meeting/6327.php.*

2. *Vienna, August 27-31, 2007, at the 4th workshop under the "Dialogue on long-term cooperative action to address climate change by enhancing implementation of the convention".*

3. *IPCC, (2001a). Climate change 2001: Impacts, adaptation and vulnerability. (Contribution of Working Group II to the Third Assessment Report of the IPCC) Cambridge University Press, and IPCC, Cambridge, available online at: http://www. grida.no/climate/ipcc-tar/wg2/ index.htm.*

   IPCC, (2001b). Summary for policy makers, climate change 2001, impacts, adaptation and vulnerability. (A report of Working Group II of the IPCC) Cambridge University Press and IPCC, Cambridge, available online at: http://www.ipcc.ch/pub/un/syreng/ wg2spm.pdf.

4. *The report defined people affected by natural disasters as those who for a time either lost their home, animals, crops, livelihoods or health as a result of a natural disaster; see UN/ ISDR (2003).*

5. *Doyle (2003); Haines and Patz (2004); WHO (2002) and WHO (2003); Doyle, Alistair (September 30, 2003). "16,000 die yearly from global warming", Reuter's news service, available online at: http://www.planetark.com/dailynewsstory.cfm/ newsid/22420/story.htm. Haines, A. and J.A. Patz. (2004). "Health effects of climate change". Journal of the American Medical Association, 29 (9), pp. 99-103. World Health Organisation (WHO) (2002). The world health report 2002. Geneva WHO, available at http://www.who.int/whr/2002/ chapter4/en/index7.html. WHO (2003). Climate change and human health: risk and responses. Geneva: WHO, available online at http://www.who.int/globalchange/climate/en/ccScREEN. pdf.*

6. *Since 1990, aid has helped to crop failure and poverty leave people vulnerable to starvation in Somalia. More than two people per 10,000 die each due to production failure and malnutrition Failure, available at http://www.oxfam.org.au/2011/07/ famine-in-somalia-what-it-means/*

7. *Concept of Human Security (CHS), 2003:4. In the middle of the 1990s the Concept of Human Security is introduced as a reflection of general by passing the state to offer*

the ultimate argument for just war theory that enhance human freedoms and human fulfillment has the potential to enrich International Human Security, available online at http://annales.umcs. lublin.pl/tt_p.php%3Frok%3D...

8. M. Koskenniemi, 'The place of low in collective security', Michigan Journal of International law 1996-17, p. 456.

9. A. Orakhenshivili, collective security, Oxford University Press, Oxford, 2011, p. 2

10. Commission on Human Security 2003. Human Security now, Final Report, CHS, New York, 2003.

11. P. Hough, understanding global security, Routledge, London, 2004, pp. 2-21.

12. Von Tigerstrom 2007, supra note 54, pp. 211-212.

13. H. Nasu, 'Operationalising the "Responsibility to Protect" and Conflict Prevention: Dilemmas of Civilian Protection in Armed Conflict', Journal of Conflict and Security Law-2009-14, pp. 209-241.

14. CHS: 2003:6. Concept of gender, its historical development and related issues such as patriarchy, state security without human security and vice-versa, available at http://dialnet. unirioja.es/descarga/articulo/4055768/1.pdf.

15. Based on the UNDP Human Development Report of 1994, Oxford University Press, Oxford, 1994.

16. CHS: 2003:2. The concept of 'human security', developed in the 1994 human development. This essay adopts the narrow definition of Human Security has become a source of threats to its own people.

17. CHS: 2003:10. Useful tools for applying the human security concept, including a step-by-step strategy for resilience to difficult situations, available online at http://hdr.undp.org/en/media/ HS_Handbook_2009.

18. G. Gentili, 'European Court Of Human Rights: An Absolute Ban On Deportation Of Foreign Citizens To Countries Where Torture Or Ill-Treatment Is A Genuine Risk', International Journal Of Constitutional Law 2010-8, pp. 311-322; L. Skoglund, 'Diplomatic Assurances Against Torture- An Effective Strategy? A Review of Jurisprudence and Examination of The Arguments, Nordic Journal Of International Law, 77, 2008, pp.319-364.

19. P. Mathew, Resolution 1373- A Call To Pre- Empt Asylum Seekers? (Or "Osama, The Asylum Seeker") In: J. Mc Adam (Ed.) Forced Migration, Human Rights and Security, Portland, Hart, OR, 2008, pp. 19-61.

20. K, Wellens, 2003- The UN Security Council And New Threats To The Peace: Back To The Future', Journal Of Conflict And Security Law 2003-8, pp. 15-70; I. Osterdahl, Threat To The Peace: The Interpretation By The Security Council Of Article 39 Of The UN Charter, Uppsala: Isustus Forlag, 1998, pp. 85-88; R. Cryer, 'The Security Council and Article 39: A Threat To Coherence? Journal of Armed Conflict Law 1996-1, pp. 161-195; P.H. Kooijmans, 'The Enlargement Of The Concept "Threat To The Peace", In: R.-J. Dupuy (Ed.) The Development Of The Role Of The Security Council, (Dordrecht: Martinus Nijhoff, 1993), pp. 111-121.

21. UN Doc S/ PV. 4087, January 10, 2000.

22. D.P. Fidler, The UN and The Responsibility to Practice Public Health, Journal of International Law and International Relations, 2005-2, pp. 58-59; L. Elliott, 'Imaginative Adaptations: A Possible Environmental Role for the UN Security Council Contemporary Security Policy 2003-24(2), pp.47-68; C. Tickell, 'The Inevitability of Environmental Security', In: G. Prins (Ed.) Threats without Enemies: Facing

Environmental Insecurity, Earthscan, London, 1993, p. 23.

23. *Paul Collier, V.L Elliot, Havard Hegre, Anke Hoeffler, Nicholas Sambanis and Marta Reynal- Querol. Breaking the conflict trap: Civil war and Development policy, Oxford University Press, 2003.*

24. *World Bank, World Development Report 2011: Conflict, Security and Development, Washington, 2011.*

25. *Paul Collier, Wars, guns and votes; Harper Collins 2009, and World Bank, 2011.*

26. *General Assembly Resolution A/RES/55/2.*

27. *Report of Secretary- General, in larger freedom: towards development, security and human rights for all, New York, 2005.*

28. *General Assembly Resolution A/RES/60/1, 2005, world summit outcome, New York, 2005, para.9.*

29. *OECD/DAC Gender Equality Tipsheets, "Conflict, Peace Building, Disarmament, Security". See www.oecd.org/dac/gender.*

30. *Resolving conflict in Solomon Islands; "The women for peace Approach," Alice pollard. Development bulletin, November 2000, at http://devnet.anu.edu.au/db53. html.*

31. *Isis international, 2013. First international conference on "Ancient Indian Wisdom and Modern World" March 29-31, 2013.*

32. *United Nations Security Council June 11, 2007.*

33. *GPF (Global Policy Forum), founded in 1993, is an organised seeking to promote accountability of international organisations such as the United Nations to promote citizen participation on peace in international studies for scholars, policymakers, activists etc., available at http:// www.globalpolicy.org/*

34. *IPCC (2007). Climate change 2007: The Physical Science Basis. Summary for Policy Makers. Contribution of Working Group I to the Fourth Assessment Report of the Intergovernmental Panel On Climate Change, February 5.*

35. *Ibid. 36. Ibid. 37. Ibid.*

38. *G8 (2008), G8 Summit Declaration on Environment and Climate Change, Hokkaido, Tokyo Summit, Sapporo, July 8, p. 39. IPCC- 2007.*

40. *International Scientific Steering Committee, 2005, Avoiding Dangerous Climate Change: International Symposium on the Stabilisation of Greenhouse Gas Concentrations, Exeter, February 1-3.*

41. *G8 (2008).*

42. *G5 (2008). G5 Statement Issued by Brazil, China, India, Mexico and South Africa, Hokkaido Tokyo Summit, Sapporo, July 8.*

43. *Major economies leaders (2008), Declaration of leaders meeting of major economies on energy security and climate change, Tokyo, Hokkaido, July 9.*

44. *ASEAN et al., (2007). Cebu Declaration on East Asian Energy Security.*

45. *Milanovic B (2005). Global income equalities. Presentation to the council on Foreign Relations, New York, December 13.*

46. *World Bank (2006 b). Energy Poverty Issues and G8 Actions: Discussion Paper, Moscow- Washington, World Bank.*

47. *Prescott- Allen, R. The Wellbeing of Nations: A Country-By- Country Index of Quality of Life and The Environment, Island Press, Washington, 2001.*

48. *Sen A. Development as Freedom, Anchor Books, New York, 1999.*

49. *The model was used to describe the impacts of trade liberalisation, but it is also*

applicable to economic growth.

50. *OECD Framework is meant to describe the effects of trade, but can also serve to describe the effects of economic growth.*

51. *Stern (2006: chapter 25). Stern Review on the Economics of Climate Change. London, Her Majesty's Treasury.*

52. *See world summit on sustainable development (2002 Para, 9), World Bank (2006b), IEA (2004, Chapter 10) and UNDP (2005). World Summit on Sustainable Development (2002). Plan of Implementation of the World Summit on Sustainable Development, New York, United Nations Department of Economics-and-Social-Affairs. World Bank (2006b). Energy Poverty Issues and G8 Actions: Discussion Paper, Moscow- Washington, World Bank. IEA (2004). World Energy Outlook 2004. Paris, International Energy Agency. UNDP (2005). Energising the Millennium Developments Goals: A Guide to Energy's Role in Reducing Poverty, New York, United Nations Development Programme.*

53. *Bradley R and Baumert K, (Eds.) Growing in the Greenhouse: Protecting the Climate by Putting Development First, World Resources Institute, Washington, 2005.*

54. *See IEA (2008c); Energy Technology Perspectives 2008. Paris, International Energy Agency. Bacon R and Matter A (2005). The Vulnerability of African Countries to Oil Price Shocks: Major Factors and Policy Options. The Case of Oil Importing Countries. Energy Sector Management Assistance Program (ESMAP) Report. 308 (05).*

55. *The IEA (2008b). In Support of the G8 Plan of Actions. Energy, Efficiency Policy Recommendations. Paris, International Energy Agency.*

56. *Malhotra P and Rehman I (2004). Fire without Smoke. Delhi, the Energy Research Institute. 57. UNEP Riso's CDM Pipeline of projects (www.cdycdm.org).*

58. *There is a great deal of overlap between advancing development goals sustainably.*

59. *United Nations General Assembly (2000); United Nations Millennium Declaration, Resolution 55/2, September 8, United Nations (2002), Report of the International Conference on Financing for Development, A/CONF.198.11.*

60. *IEA (2008a), World Energy Outlook 2008, Paris, International Energy Agency.*

61. *Baumert K et al., Navigating the Numbers: Greenhouse Gas Data and International Climate Policy, World Resources Institute, Washington, 2005.*

62. *UNFCCC, (1997). Proposal Elements of a Protocol to the United Nations Framework Convention On Climate Change, Presented By Brazil In Response to the Berlin Mandate, Paper No.1 of the Note by the Secretariat: Implementation of the Berlin Mandate: Additional Proposals from Parties, Addendum. FCCCC/AGBM/1997/ Misc.1/Add.3.*

63. *UNFCCC (1992: Article 3, 4), United Nations Framework Convention on Climate Change, FCCC/INFORMAL/84.*

64. *IISD Reporting Services- Upcoming meetings'. Lisd. Ca. retrieved April 8, 2010.*

65. *Tim Flannery (November 2009), Copenhagen and beyond: conference Bound, The monthly, archieved from the original on May 6, 2010, retrieved April 8, 2010.*

66. *'Closing the Gigaton Gap', Mother Jones, November 24, 2010, retrieved January 29, 2012.*

67. *"Dates and Venues of future sessions" (PDF) retrieved 2010-10-13.*

68. *Calendar, UN Frame work convention on climate change, United Nations, retrieved December 8, 2011.*

69. *Harvey. Fiona, Vidal, John (December 11, 2011). "Global climate change treaty in sight after Durban breaks through", The Guardian (London), retrieved December 11, 2011.*

70. *Black, Richard (December 11, 2011), "Climate talks end with late deal", BBC News, retrieved December 11, 2011.*

71. *Harvey, Fiona; Vidal, John (December 11, 2011). "Durban deal will not avert catastrophic climate change say scientists", The Guardian (London), retrieved December 11, 2011. 72. United Nations Environmental Program - Kyoto Protocol official site.*

73. *Russian President Vladimir Putin signed the Kyoto Protocol on November 5, 2004, clearing the way for the international treaty to take effect in February 2005. 74. UNFCCC (March 5, 2013), Introduction to the convention UNFCCC.*

75. *UNFCCC, Conference of the parties (COP) (March 15, 2011). 30th Meeting of the CDM Small Scale Working Group.*

76. *Roger, H-H, et al., (2007), Ultimate objective of the UNFCCC.*

77. Meehl, G.A., T.F. Stocker, W.D. Cllins, P. Friedlingstein, A.T. Gaye. (2007). Global Climate Projections. In: Climate Change 2007: The Physical Science Basis.

78. *Fisher, B.s et al., ch 3: 2007. Issues related to mitigation in the long term context. IPCC, 2007*

79. *Rogner H-H et al., Introduction total GHG emissions in IPCC, 2007.*

80. *Robinson, Simon (2010-01-22); How to reduce carbon emissions: capture and store it?*

81. *Global status of CCS Report: 2011 is the institute's flagship publication and consolidates the current understanding of the level and nature of global carbon capture storage (ccs), available at http://www.globalccsinstitute.com/publications/global-status-ccs-2011.*

82. *Sophie Hebden (2006-06-22), invest in clean technology says IEA report.*

83. *Lowe, Marcia D. (1994, April). Back on track: The global Rail Revival, Retrieved 2007-02-15. 84. Fulton, William, Pendall, Rolf, Nguyen, Mai: Harrison, Alicia (2001). "Who sprawls most? How growth patterns differ across U.S".*

85. *OECD Environment outlook to 2050, climate change chapter, pre-release version, OECD, 2011. "IEA Technology Roadmap carbon capture and storage 2009".*

86. *Launder B. and J.M.T Thompson (2008), Global and Arctic climate engineering: numerical model studies.*

87. *World Bank, 2010: World Development Report 2010, Development and Climate Change.*

88. *Jessica Brown, Neil Bird and Liane Schalatek (2010) climate finance additionality: emerging definitions and their implications overseas Development Institute.*

89. *"UNFCCC Conference of parties, seventh meeting". And the fifteenth sessions of the subsidiary bodies, October 29-November 9, Marrakesh, Morocco, available online at http://unfccc.int/ cop7/*

8. CONCLUSIONS

Climate change and security
Climate change undermines human security in the present day and will increasingly do so in future. Climate change is a security problem for some states, cultures and communities through its likely impacts on ecosystems and their inhabitants and through its indirect effects on development and political stability. It reduces people's access to natural resources that are important to sustain their livelihoods. Climate change is also likely to undermine the capacity of states to provide the opportunities and services that help people to sustain their livelihoods and which help to maintain and build peace. These direct and indirect impacts of climate change on human security increase the risk of violent conflict.

Climate change poses risk to human in security principally through its potentially negative effects on people's well-being. There is need for more research on the ways it may undermine human security in understanding the people's vulnerability which is still sufficiently uncertain for the purposes of designing effective adaptation strategies. There is a need for systematic, comparative and cross-scale research to enhance understanding of the connections between climate change, human security and violence. This includes understanding the ways in which it may affect environmental changes and their capacities to avoid or adapt their livelihoods that can be sustained and their needs and values can continue to be satisfied. It also requires understanding of the ways in which people may respond if climate change undermines livelihoods and outcome might be an increased propensity for people to engage in violence as an alternative livelihood strategy. It enhanced understanding of climate insecurity and also the ways in which climate change challenges states' capacity to protect livelihoods and maintain peace if it exists. Such research on climate insecurity can reduce uncertainty about the human dimensions of climate impacts and enhance knowledge of potential adaptation strategies to avoid human insecurity as well as an increased risk of violent conflict.

Environmental security has exposed the inadequacy of militarised practices of security, the nature of sovereignty in the face of environmental change and to elevate environmental problems from the level of 'low politics' to 'high politics'. So, those states would commit as much energy and resources to address environmental problems as the security problems. However, priority to environmental security increased resources and energy to enhancing environmental security. So, understanding climate change as a security issue risks making it a military rather than a foreign policy problem and a sovereignty issue rather than a global commons problem.

A critical and ambiguous concept in the UNFCCC is its reference to 'dangerous' levels of climate change. It is on vulnerability, security encapsulates danger much better than concepts such as sustainability, vulnerability or adaptation and it offers a 'framework' in which danger can be recast as wide spread risks to welfare and sovereignty. Security can also serve as an integrative concept which links local (Human security), national (national security) and global (international security) levels of environmental change and response. It also integrates mitigation and adaptation as both are essential to security from climate risks. It requires understanding not just these social-ecological interaction in places, but also the many economic, political, cultural and social interactions between different places and the ways these might be altered by climate change. It requires understanding different groups' capacities to adapt to change and the limits of those capacities. The multi-level, cross scale and longitudinal approach enables understanding of the political economy of climate insecurity and management of them. Such research can reduce uncertainty about the human dimensions of climate impacts, enhance knowledge of potential adaptation strategies and contribute to dialogue about the levels of greenhouse gas emissions that may be considered to be dangerous.

Definitions for security and climate change need to be constructed in such a way that policy makers have incentives to pursue mitigation and adaptation, not merely to focus on GHG emissions as the only suitable goal and outcome of the current Paris Climate Change agreement. Integration of energy and environment as key concepts for both environmental systems and possible solutions for cooperation is the essence of security thinking.

Climate change and water security
To help develop practical adaptation and mitigation strategies for agricultural water management in developing countries, the following recommendations for immediate action can be made:
• Ensure better prediction of the impacts on agricultural systems. This can be carried out using the typology and decision analysis.

- Provide assistance in developing and applying downscaling techniques to better analyse agro-climatic future and in the process, build local capacity in modelling and climate adaptation.
- The investment needs for different solutions, which takes into account long- term embodied and operational energy use. These tasks are required for all agricultural impact and adaptation studies, and sit at a higher strategic level than work irrigation and water use in agriculture.

Above these three activities, it is necessary to expand the density, detail and frequency of monitoring of climatic hydrologic systems in order to confirm the evolution of trends and modelled predictions, refine the assessment of impacts and manage adaptive strategies accordingly. Improved information on the nature and dynamics of key production systems is also required including: higher resolution and more detailed mapping and management of soils; ground water mapping and monitoring of water use; adaptation of cropping systems and practical forecasting of drought and flood. It will be necessary to evaluate, document and disseminate good practice at farm, system and strategic levels as it emerges.

Particular attention should be paid to identifying and promoting effective 'no regrets' activities for adaptation and mitigation. A global picture of agricultural impacts can be assembled from regional and national studies that work at an appropriate level of details. There is a strong argument for global studies to be built from the bottom up in the future in order to perform key crop sectors as cereals.

Many countries will still be pre-occupied with livelihoods and food security, so it will be important that well-targeted and coordinated work continues to implement sustainable development. This will require agricultural and water services to make strong and open partnerships with key environmental groups, ranging from international and local organisations to line agencies and environmental departments.

Climate change impacts will be global and development assistance should not overlook highly impacted communities such as small island nations. Many donors and organisations have declared a strong commitment to continued development and will increasingly develop through a climatic sensitive lens. Solid and appropriate advice will be needed in the development and management of water resources for agriculture and in the establishment of climate resilient food production systems.

Short and medium-term recommendations: Short term until 2030: Existing agricultural water management systems at national and basin level need to be analysed with respect to AR4 specifically and this will entail:-

- Monitoring the relative contribution of rain-fed and irrigated production to global food balances to determine the long term sensitivity of food production systems to climate change.
- Elaborating vulnerability mapping such as the joint FAO/IIASA initiative that includes the food insecurity, poverty and environmental global GIS database.
- Determining the operational room to across river basin systems on the basis of updated assessments of the partition between surface and ground water sources of supply with the aim to improve the data for carrying out meaningful sensitivity analyses.
- Building in as much operational flexibility as possible into local irrigation- scheme-level water management strategies in anticipation of both increased demand and the need to adjust operational supply.

FAO can expect requests from member countries to improve the understanding of climate change impacts and adaptation strategies.

Medium term until 2050: Investment plans and operational adjustments will need to be prepared to address national and sub-regional and regional issues. These plans and adjustments will comprise:-

- Large surface irrigation systems fed by glaciers and snow melt;
- Ground water systems in arid and semi-arid areas, where rainfall will decrease and become more variable;
- Upstream watersheds, where a combination of irrigated agriculture, rain fed agriculture, pasture and forestry is practised;
- Large deltas, which may be partly submerged by sea-level rise, increasingly prone to flood and storm and cyclone damage and intrusion through surface and ground water respectively;
- Seasonal storage systems in the monsoon regions where the proportion of storage yields will decline but peak flood flows are likely to increase;
- Supplemental irrigation areas, where the consequences of irregular rainfall are mitigated by short-term interventions to capture and store more soil moisture or runoff.

The development of adaptive capacity to reduce adverse impacts of global change in rural areas of developing countries requires analyses at various spatial scales and an understanding of the linkages across the various scales. At the farm level, households adjust to global change by changing farm practices. These local actions in turn, influence climate and global change. At the basin level, basin authorities influence both land and water allocation and carry out purposeful adaptations to global change. Purposeful adaptation can be either tactical, in response to climate or other global changes or strategic in anticipation of future global change. At the national level, governments and authorities influence ecosystem services and human well-being. They also carry out purposeful adaptations, including changes in price, trade and investment policies to anticipate or respond to global change. At the regional level, organisations and institutions have the potential to mitigate global change impacts through changes in trading regimes, and the development of regional transportation and communications infrastructure. Important global factors also affect water and food security at the local level, such as world food trade and competition for water generated by the world economy.

Increasing water availability and increasing the reliability of water in agriculture, i.e. through irrigation is one of the preferred options to increase productivity and contribute to poverty reduction. However, as a result of predicted climate change, semi-arid and sub-humid tropical areas that would greatly benefit from increased irrigation may see water availability changing temporarily and spatially, not only declining but also unfavourably distributed over the growing season.

Positive examples include many carbon sequestration practices involving increased crop cover, including agro-forestry and use of improved rotation systems that are more resilient to climate variability, thus providing good adaptation in view of increased pressure on water and soil resources. Short- term plans to address food insecurity provide access to water resources or encourage economic growth must be placed in the context of future climate change to ensure that short-term activities in a particular area do not increase vulnerability to climate change in the long term. Policy attention is needed in

- Developing long-term water policies and related strategies, taking into account country specific legal, institutional, economic, social, physical and environmental conditions. Policies and strategies will also need to integrate the different sectors depending on water-rain-fed and irrigated agriculture, livestock, fisheries, forestry, nature and biodiversity protection, manufacturing and industry and municipal water use. Water policies need to address such issues as upstream - downstream competition over water resources and equitable allocation of water across regions and generations.
- Increasing water productivity by promoting efficient irrigation and drainage systems.

- Improved watershed and resource management, integrating the different natural resources water, soil, flora and fauna through the promotion of integrated water resources management processes.
- Enhancing water availability through better use of ground water storage, enhancing ground water recharge where increasing surface water storage. Given the current economic situation of many water-stressed countries, however, managing demand is equally important: reducing water consumption and improving water use efficiency.
- Institutional and governance reforms that balance demand and supply across sectors and that mainstream climate change adaptation.
- Enhancing stakeholders' participation in water development and climate change adaptation.
- Improve information and early warning systems to provide land and water users with timely and adequate information and knowledge about availability and suitability of resources to promote sustainable agriculture and prevent further environmental degradation. Information exchange and dialogue between the agriculture, water and climate communities is vital, not only at national levels but also at trans-boundary river basin levels.
- Human resource, capacity and skills development of policy makers and end users to help them deal with new challenges.
- Increase investments in agriculture and rural development. The 2003 Maputo declarations called for African governments to target 10 per cent of their national budget to the agricultural and rural development sector. This is clearly justified that the environmental, economic and social importance of agriculture in Sub-Saharan Africa (SSA), the anticipated impacts of climate change on agriculture especially in semi-arid and sub-humid areas and the role of agriculture has to play in climate change adaptation and mitigation.

Adaptation measures in water management are often under-represented in national plans or in international investment portfolios. Therefore, significant investments and policy shifts are needed. These should be guided by the following principles:-

- Mainstream adaptations within the broader development context;
- Strengthen governance and improve water management;
- Improve and share knowledge and information on climate and adaptation measures and invest in data collection;
- Build long-term resilience through stronger institutions and invest in infrastructure and in well-functioning ecosystem;

- Invest in cost-effective and adaptive water management as well as technology transfer;
- Increased national budgetary allocations and innovative funding mechanisms for adaptation in water management.

Application of these principles would require joint efforts and local to global collaboration among sectoral, multi-sectoral as well as multi-disciplinary institutions. Responding to the challenges of climate change impacts on water resources also requires the development of deliberate and context specific adaptation strategies. Countries are urged to improve and consolidate their water resources management systems and to identity and implement 'no regrets' strategies, which has positive development outcomes that are resilient to climate change.

UN policy on water

- Establish water as a human right and focus on improving water quality and sanitation to save lives.
- Integrate UN work on water across agencies by building on the UN-declared international water decade (2005-2015).
  - Survey UN water activities across agencies to identify successful programs and areas of integration;
  - Establish a Global Fund of water to provide for funding and coordinating the UN's water activities;
  - Create a forum to identify and articulate the needs of stakeholders in the Global South for transboundary water management, dispute resolution and conflict transformation;
  - Components of the UN system working on water policy must move beyond more technical management questions and instead assess water and development issues within the broader context of peace and security.
- Develop an integrated, systematic program of preventive water diplomacy based on World Bank and Global Environmental Facility Frameworks.
- Facilitate development of institutional frameworks for dialogue on water issues at the basin level to encourage cooperation among parties, establish a coordinating mechanism.
- Support institutional frameworks and investments in appropriate and strategic water projects. Recruit and train facilitators in hydrology, international law and conflict prevention.
- Establish international standards for gathering and analysing

hydrological data and develop a database that can be accepted by all stakeholders.

Climate change and food security

FAO can play a decisive lead role through its GEA efforts in highlighting and promoting sustainable rural development and climate change action in the food and agriculture sectors. Such efforts can support and develop UNFCCC focus on climate change adaptation and mitigation to provide new momentum by moving beyond carbon credits for food and agriculture activities, by recognising their intrinsic climate, ecosystem services and community benefits. The overriding goal is to generate sufficient financial flows necessary to address climate change and rural development in LDCs by 2020. Climate change projects use in both regulatory and voluntary markets, identifying pilot activities with joint adaptation and mitigation benefits as well as positive implications for food security, ecosystem resilience and rural development. The efforts should be directed to develop new methodologies for those agricultural activities of relevance to LDCs that can already enter carbon markets, range of ecosystem, development and social services they also provide. The FAO GEA efforts should focus on developing strong partnerships with key public and private players in order to strengthen the role of food and agriculture activities within future climate policy agreements, focussing on both regulatory carbon markets as well as on new markets based on payment for the range of ecosystem and social services provided by agriculture.

Over the 21st century, the world will need to produce significantly more food in order to deliver a basic, but adequate diet to everyone. The amount of food required will be greater in diets and the trends in the existing management regimes of food systems continue. Concurrent efforts are needed to establish a sustainable global food system with climate-resilient agriculture production systems, efficient use of resources, improved marketing and distribution of infrastructure, low-waste supply chains and healthy diets. Agricultural production must be accompanied by concerted action to reduce greenhouse gas emissions from agriculture to avoid climate change and threats to long- term viability of global agriculture.

The recent assessment reports on global food security, the Commission of Sustainable Agriculture and Climate Change has identified critical points and purposes of the following evidence based actions to deliver long-term benefits to communities in all countries:-

- Integrate food security and sustainable agriculture into global and national policies:
  - Establish a work programme on mitigation and adaptation in agriculture in accordance with the principles and provisions of UNFCCC based on article 2, as a first step to inclusion of agriculture in the mainstream of international climate change policy.
  - Make sustainable, climate- friendly agriculture central to green growth and Rio+ 20 Earth Summit.
  - Finance 'early action' to drive change in agricultural production systems towards increasing resilience to weather variability while contributing significantly to mitigating climate change. This includes supporting national climate risk assessments, developing mitigation and adaptation strategies and programme implementation.
  - Develop common platforms at global, regional and national levels for policy action related to climate change, agriculture, and crisis response and food security at global, regional and national levels. These include country level coalitions for food security and building resilience, particularly in countries most vulnerable to climate shocks.
- Significantly, raise the level of global investment in sustainable agriculture and food systems in the next decade:
  - Implement and strengthen the existing G8 programmes and commitments to sustainable agriculture and food security including long-term commitments for financial and technical assistance in food production and to empower small hold farmers.
  - Enable UNFCCC Fast Start Funding, major development banks and other global finance mechanisms to prioritise sustainable agriculture programmes that deliver food security, improved livelihoods, resilience to climate change and environmental co-benefits. Such programmes should emphasise on improving infrastructure and land rehabilitation.
  - Adjust national research and development budgets, and build integrated scientific capacity to reflect the significance of sustainable agriculture in economic growth, poverty reduction and long-term environmental sustainability and focus on key food security issues.
  - Increase knowledge of best practices and access to innovation by supporting revitalised extension services, technology transfer to communities, low- to high- income countries and women farmers.

- Sustainably intense agricultural production while reducing greenhouse gas emissions and other negative environmental impacts of agriculture:
  - Develop, facilitate and reward multi-benefit farming systems that enable more productive and resilient livelihoods and ecosystems with emphasis on closing yield gaps and improving nutrition.
  - Introduce strategies for minimising ecosystem degradation and rehabilitating degraded environments, with emphasis on community- designed programmes.
  - Empower marginalised food producers particularly women to increase the range of appropriate crops by strengthening land and water rights, increasing access to markets, finance and insurance and enhancing local capacity.
  - Identify and modify subsidies such as for water and electricity that provide incentives for farmers to continue agricultural practices that deplete water supplies or destroy native ecosystems.
  - Economic incentives for sustainable agriculture with strengthening governance of land tenure to prevent further loss of forests, wetlands and grass lands.
- Develop specific programmes and policies to assist population and sectors that are most vulnerable to climate changes and food insecurity:
  - Develop funds that respond to climate shocks, such as 'Index-linked funds' that provide rapid relief when extreme weather events affect communities through public-private partnerships, based on agreed principles.
  - Strengthening market databases, promoting open and responsive trade systems, establishing early warning systems and allowing tax-free export and import for humanitarian assistance.
  - Create and support safety nets and other programmes to help vulnerable population in all countries become food secure.
  - Establish emergency food reserves and financing capacity that can deliver rapid humanitarian responses to vulnerable population threatened by flood crises.
  - Create and support platforms for harmonising and coordinating global donor programmes, polices and activities, paying particular attention to systematically integrating climate change risk management, adaptation and mitigation co-benefits and improved local nutritional outcomes.

- Food access and consumption patterns to ensure basic nutritional needs are met and to foster healthy and sustainable eating patterns worldwide:
  - Address chronic under-nutrition and hunger by harmonising development policy and coordinating regional programmes to improve livelihoods and access to services among food insecure rural and urban communities.
  - Promote positive, changes in the variety and quantity of diets through innovative education campaigns, marketing practices of retailers and processors with public health and environmental goals.
  - Promote and support and evaluate food security, nutrition and health, practices and technologies across supply chains, agricultural productivity and efficiency, resource use and environmental impacts and food system costs and benefits.
- Reduce loss and waste in food systems, targeting infrastructure, farming practices, processing, distribution and household habits:
  - In all sustainable agriculture, development programmes include research and investment components focussing on reducing waste, from production to consumption by improving harvest management and food storage and transport.
  - Develop integrated policies and programmes that reduce waste in food supply chains such as economic innovation to enable low income producers to store food during periods of excess supply and to separate and reduce food waste.
  - Promote dialogue and convene working partnerships across food supply chains to ensure and reduce waste are effective and efficient.

    Create comprehensive, shared, integrated information systems that encompass human and ecological dimensions:
  - Sustain and increase investment in regular monitoring, in land use, food production, climate, the environment, human health and well-being worldwide.
  - Support improved transparency and access to information in global food markets and invest in interlinked information systems with common protocols that build on existing institutions.
  - Develop, validate and implement the decision support systems that integrate biophysical and socio-economic information and enable policy- makers in trade, agriculture, nutritional security and environmental consequences.

Climate change is making it urgent for more widespread and significant changes in farming practices to increase productivity and use of natural resources more efficiently and sustainably. The CCAFS and FAO incorporate lessons learned from innovative action research approaches and from development work in the field of gender and climate change. These materials will help to inform practitioners how gender can be mainstreamed into development activities and action research on climate change adaptation and mitigation.

Recommendations

Many opportunities exist to enhance the role of the food and agriculture sectors in supporting climate change responses with strong links to sustainable development. A lead role of FAO through its GEA programme could help to ensure that land-based project activities are increasingly mainstreamed in future climate agreements, guaranteeing that strong sustainability components and greening of the economy through agriculture is promoted and enhanced, especially in LDCs. Land-based solutions leading to natural and managed ecosystems that are more resilient to future climate lead to less GHG emissions or increased carbon storage.

Implementing these strategies represent 'good practice' agriculture including the adoption of traditional and less intense cultivation practices than carrying capacity of the natural systems and positive ecosystem services and community benefits to carbon credits.

The following actions are suggested for developing a strong and useful role of the FAO GEA until 2020 and beyond 2020.

- Policy Action resulting in inclusion of more explicit references to food and agriculture in future climate agreements;
- Design and develop enhanced sustainability criteria for land-based adaptation and mitigation projects, for use both in regulatory and voluntary markets, focussing on food security, ecosystem resilience and rural development opportunities;
- Design and develop new methodologies for land-based adaptation and mitigation projects, both regulatory and voluntary markets, targeting for the CDM, a range of new projects that can generate permanent carbon credits for regulatory markets- that is reductions in non-CO_2 GHG rather than carbon sequestration; while exploring potential for new markets, focussing on payments for ecosystem and social services beyond carbon;
- Develop and maintain for Measurement, Reporting and Verification (MRV) rules facilitating the promotion of land based activities. MRV

systems could be promoted as a means to enhance participation by communities in LDCs;

- Access Climate Green Funding for new agriculture activities that exhibit both adaptation and mitigation components, targeting joint NAMA and AF international funds;
- Lobby for substantial climate funding beyond carbon, building on strengthened sustainability criteria, based on ecosystem services and community benefits associated with GEA sponsored projects; water, biodiversity, soil conservation, organic practices can be valued.

While new ideas for funding of land-based projects are essential to expand opportunities for a sustainable and green agriculture, significant efforts should focus on ensuring that land-based projects can be expanded regulatory markets post 2020 in order to secure the large financial flows needed to fight climate change and achieve sustainability in LDCs. GEA FAO efforts should strive to:-

- Enhance partnership with key public and private players towards reaching a new climate agreement after 2012, one that allowed more food and agriculture project activities in flexible mechanisms, supported by high- quality sustainability targets and range of ecosystem service indicators developed by FAO.
- Design alternatives to utilise non-permanent land-based credits in regulatory markets. For instance, explore requirements that future and trade compliance buyers be required to maintain in their portfolios. This would be equivalent to establishing a tax on the price of future regulated emission reductions in order to pay for food and agriculture projects.
- Promote a full-fledged market in ecosystem and social services beyond carbon for agriculture projects activities and programmes with strong climate components.

Recommendations for the UNFCCC relating in agriculture to climate change

- Promote bio-diversity climate resilient small scale agriculture based on agro- ecological principles;
- Support appropriate technology development and transfer that enhance sustainability of food production systems;
- Include safeguards which protect bio-diversity equitable access to resources by rural people, food security, the right to food, the rights of indigenous people and local population as well as the welfare of farm animals while promoting poverty reduction and climate adaptation;

- Explore opportunities to sustainably reduce emissions from the agricultural sector;
- Reduce emissions from the conversion of other land to agriculture;
- For developing countries the priorities should be on agriculture sustainability, climate resilience, and food security, and parties must provide resources for promoting bio-diversity, resilient small-scale agriculture and appropriate technology development and transfer;
- Developing countries must progress toward full and comprehensive accounting of the emissions associated with their agricultural activities including bio-energy production and use.

Policy recommendations for agriculture

UNFCCC at Doha, Qatar, 2012 (Nov 26-Dec 7)
Cop 18 at Doha, Qatar discusses with how to feed a growing population in the face of climate change and how to reduce GHGs within the agricultural sector, further discussions are needed within the UNFCCC to promote agricultural systems and approaches that can contribute to adaptation, mitigation and food security solutions at local, national and international levels. The programme of work or other further work under the UNFCCC on agriculture and climate change should explore at minimum three priority areas:-

Achieving synergies between mitigation and adaptation efforts in agricultural systems:
The UNFCCC should explore the way in which mitigation and adaptation benefits can be derived from the careful design and management of agricultural systems and surrounding landscapes. Climate change mitigation and adaptation actions related to agriculture should be integrated into strategies to help the sector adopt
land use options. Many of the agricultural systems or management practices that help reduce GHG emissions from agriculture or enhance carbon storage within agricultural land or in the surrounding landscape also enhance the overall adaptive capacity of agricultural lands, making them more resilient to climate change. Many adaptation strategies seek to sustain agricultural productivity in the face of climate change and also enhance overall carbon stocks in the soil and vegetation, contributing to climate change mitigation. There is a lot of potential to improve food security, climate adaptation and climate mitigation both in and outside of agricultural landscapes.

Prioritising needs of the most vulnerable social groups and ecosystems:
Engaging vulnerable and marginalised groups, who are often the intended beneficiaries of adaptation measures, can help to achieve sustainable agriculture goals and multiple benefits such as enhanced food security and improved livelihoods for those who are in the most precarious conditions. For this reason, interventions to enhance the adaptive capacity of agricultural systems and agricultural communities should be targeted to reach women, small-scale producers, the rural poor, people at the agriculture- forest frontier and other vulnerable groups.

Bio-diversity and ecosystems are highly impacted by agriculture; both are significant contributors to the sustainability of agricultural productivity. Measures should target work in areas in which bio-diversity and ecosystem services are critical to agricultural systems. The UNFCCC can prioritise the needs of the most vulnerable groups and ecosystems by focussing on participatory processes, financing and market access that specifically include marginalised actors such as women and small-scale producers; support for local and traditional knowledge through recovery, documentation, respect for indigenous rights; integration of water resource, ecosystem and bio-diversity considerations, and poverty eradication.

Promoting integrated, landscape level approaches to climate change and food security:
The UNFCCC should explore the ways in which landscape scale approaches can contribute to climate solutions and also enhance food security. Integrated landscape approaches take into account the social, economic and environmental trade-offs and synergies of different agricultural systems, management practices and landscape configurations and consider the multiple products and benefits that agricultural landscapes provide. Improving agricultural productivity or enhancing the adaptive capacity of crop, an integrated landscape approach manages agricultural plots, farms and landscapes to both climate change mitigation and adaptation providing the benefits to meet food security and livelihood needs. There is now ample literature on integrated landscape management which can serve as a guide for UNFCCC approaches to work in agricultural landscapes. UNFCCC's discussion that the developing countries should have clear signals from developed countries that financial, technology transfer and capacity building resources are available to address agriculture and climate change adaptation and mitigation to ensure the fulfilment of food security and resilience goals.

Climate change and health

The environmental and social impacts of climate change will mostly act to the detriment of human safety, health and survival. At the same time, a low carbon economy would make a contribution to addressing many of the priority health conditions that are currently overburdening health systems.

In most countries, the health sector has been slow to recognise the serious implications of climate change for population health and for the healthcare system. Individuals as free agents are responsible for their own health related actions, impedes recognition that a population's way of living and its collective environmental conditions are the prime determinants of the rates and patterns of health and disease within that population. Therefore, the health sector must widen its field of vision in order to play a substantive role in the policy discourse to link effective sectors of government and to assist the source of risk to the health of present and future generations. Climate change causes increase in poor health, child deaths, hardship, displacement and conflict and if it causes geopolitical instability and international tensions, then the security of global health and the health of global security will be at great risk.

Climate change will have significant health impacts, both domestically and globally. While all of changes associated with the process are not predetermined, the actions taken today certainly help to shape environment in the decades to come. However, aggressive mitigation actions can significantly blunt the worst of the expected exposures. Climate changes and health issues transcend national borders and climate change health impacts in other countries are likely to affect health. Famine, drought, extreme weather events and regional conflicts - all likely consequences of climate change - are some of factors that increase the incidence and severity of disease as well as contributing to other adverse health impacts to address climate change-related decision making at local, regional, national and global levels.

There are significant research needs to help direct adaptation activities and inform mitigation choices going forward. Such needs include integrating climate science with health science; integrating environmental, public health, and marine and wildlife surveillance; applying climate and metrological observations to public health issues; and down-scaling long term climate models to estimate human exposure risks and burden of disease. Several themes emerged during the creation of these documents including systems and complexity, risk communication and public health education, co-benefits of mitigation and adaptation strategies and urgency and scope. These are discussed below:

System and complexity

Public health agencies can develop evidence-based prevention strategies and the health care community can pursue secondary controls and respond to health incidents when prevention is not effective. Research on these links should focus not only on the direct health effects of climate change but also on the complex relationships between different exposure pathways and health risks.

The fundamental importance of ecosystems in climate change impacts and significant roles that environmental factors play in human health, climate change and health research should also focus on location and environmental conditions. A huge amount of diverse information will be needed at all governance levels such as local, regional, national and global. There is also a need to identity common elements of strategies that may be generalised from one community to another. Research is needed on how to identify common features of locales that will help identify them as having similar responses to climate change, how to determine and develop optimal strategies for interventions and how to develop and implement communication tools that will effectively help communities respond to their particular situation.

Risk communication and public health education

Knowledge is one of the most strategic tools in reducing health problems in any environment and to understand what is harmful and how to avoid such harm. Knowledge of the health impacts associated with climate change will have limited value without effective communication and education strategies to increase public awareness and understanding of the specific risks involved and the complexity of the issues. Communication with particularly vulnerable individuals and populations as well as with health care professionals and public health officials tasked with protecting communities is of immense significance. Such warning systems might be more effective through multiple channels. Public health professionals need to be highly vigilant for opportunities to increase the range and impact of early warning systems on vulnerable populations.

Mitigation and adaptation

Mitigation strategies need to balance the economic costs of emission reductions with the costs of environmental degradation. Many mitigation and adaptation strategies reach across health endpoints and may be both beneficial and problematic for a wide array of diseases. In the transportation sector, reduce CO_2 emissions which will reduce the effects of climate change on human health. Thus, the cost of emission reduction should include reduced health cost due to effective mitigation. Basic research on health and environmental factors and implementation to develop new models are needed to allow for a careful use of health. Health impacts, both positive and negative and possible mitigation strategies prior to widespread implementation need to be taken into account.

Adaptation efforts may have important health co-benefits that need to be identified. Adaptation efforts need to focus on urban design. There is considerable need for study of broader adaptation issues that will affect many different groups. Health affects of adaptation practices in food, water and chemical use. Regional water shortages will cause a greater intensity of recycling and reuse of water with concomitant increase in risks of human exposure to water borne pollutants and pathogens. Many new ideas and options are being developed. As with all emerging technologies, it is important to examine their effects on health, both positive and negative, so that the best options for society can be identified and adopted.

Scope and urgency
The necessary research on climate change, health impacts, the health effects of mitigation and development of appropriate adaptation strategies will not occur spontaneously. To be successful, research program needs to be integrated, focussed, inter-disciplinary, supported and sustainable to adjust to new information. The effort must also be multinational, multiagency and multidisciplinary bringing strengths' together. This research will require capacity building in a number of areas, especially in climate sciences and disease and ecosystem surveillance necessary to support the health sciences.

Climate change threatens many of the natural and built systems that protect and preserve nation's health. Climate change could impact on public health systems if they are not appropriately strengthened. Researching into the vulnerability of these systems will be critical to identifying areas mostly in need of attention, avoiding mistake, limiting human suffering and ultimately saving lives.

Gender health
Policies to promote mitigation activities that have strong co-benefits in health and other development needs provide a potential, political bridge across the 'development gap', between rich and poor countries. Adaptation strategies need to take into account women's and men's relative and different capacities, power, social resilience, vulnerabilities and sources,

because gender norms, roles and relations can either enable or constrain adaptive capacities.

IPCC acknowledges that disasters affect men and women differently on number of levels including economically, socially, and psychologically. There are few researches on sex and gender differences in vulnerability to and impacts of climate change especially health-related impacts, the gender mainstreaming in climate change response activities, sustainable and equitable development, a clear focus on adaptation and mitigation, a strong commitment to resources and empowerment of individuals to build their own resilience.

Equity and social justice cannot be achieved without recognising the differences in vulnerability and strengths of women and men, and various factors that contribute to vulnerability. General sensitive research is needed to better understand the health impacts of climate change in general, and extreme events, in particular. There is need for the development of gender-responsive and accessible health services that reach the poorest population to particular health needs of women and men throughout their entire life-cycle.

Recommendation

Steps can be taken to lessen climate change and reduce its impacts on health and health of future generations. Some of these steps can yield benefits for health, environment, economy and society at the same time. The government should support adaptation to and mitigation of climate change to create healthier, more sustainable communities. The goals of the NIEHS climate change and human health program efforts to:-

- Provide research on human health impacts related to climate change and adaptation.
- Raise awareness and create new partnerships to advance key areas of health research and knowledge development on human health effects of climate change.
- Serve as an authoritative source of information on human health effects of climate change for NIEHS stakeholders including the public.
- Represent NIEHS science in climate change research and policy activities at the NIH, HHS, federal government and international levels.

Human induced climate change is an emerging threat that rightly commands widespread policy and public attention. Along with other rapid changes associated with global population and economic growth, climate change strains existing weak points in health protection systems for reconsideration of public health priorities. The most effective responses are likely to be strengthening of the key functions of environmental management, surveillance and response to safeguard health from natural disasters and changes in infectious disease patterns.

Climate change, natural disasters and role of UN

This IPCC report has shown that all key drivers - climate change, poorly planned development, poverty and environmental degradation - influence the risk of weather event becoming a disaster. Thus, these factors need to be managed collectively. In the coming decades, disaster losses are expected to continue to rise due to the increasing exposure of populations, assets and environmental degradation compounded by climate change.

In many places, local trends in average temperature and precipitation due to climate change are already being observed and reliable projections for the future can improve planning decisions. Key impacts of climate change will be due to changes in climate variability and weather extremes. Organisations working on disaster risk reduction and development will need to establish new partners such as national meteorological offices or global centres on climate research. Some methods and tools for disaster risk assessment may need to be adjusted to address better hazard trends.

The poor and most vulnerable will be the most directly affected. The close interactions between poverty, development trends and climate change are likely to pose significant challenges to the global objectives of ending poverty and promoting shared prosperity by 2030. The actions will be needed to provide the poor with the resources, information and knowledge required to become more resilient. UNHCR's policy to tackle the effects of climate change will be reflected in three distinct areas: operational management; protection strategies; and advocacy.

A new approach is needed to tackle risk management into work on climate change and the introduction of climate change into natural hazards and developing planning. The key concepts in that new approach should be capacity-building and resilience. It may be necessary to instigate a process of managed retreats from those areas that will become unusable, involving re- location to areas that offer security and opportunity. To deal with such serious matters, national decision-making will require strong, sustainable and accepted institutional structures and a population and civil society educated in the issues and alternatives. LDCs are unlikely to have the capacity or resources to set the focus on relief, where MDCs has its focus on risk reduction and prevention.

IOM calls for integrating migration - related programmes into comprehensive action for the benefit of vulnerable countries and communities affected by the impact of climate change, environmental degradation and other factors of vulnerability such as poverty. It recommended:

- Raise policy and public awareness of the importance of DRR and CCA in addressing environmental migration and the need for increased consideration of the issues of human mobility in DRR and CCA strategies conducive to sustainable development. Interaction can be facilitated at the national and regional levels through state-appointed focal points for the Hyogo Framework for action and climate change.

- Support further research into developing multi-hazard assessment of the cross-cutting issues of mobility, gender equality, health and security. That would require further investigation which include, inter alia, the role of remittances in recovery or the impact of migration on urbanisation.

- Develop policy coherence at the national and international levels by mainstreaming DRR and CCA into migration management policies and practice; this requires strengthening linkages with other relevant policy domains, such as development, environment and humanitarian portfolios.

- Minimise forced displacement with adequate resources to meet the growing challenge of climate change including measures to ensure adequate assistance and protection for people on the move as a result of environmental factors.

- The role of migration as part of DRR and CCA is to develop temporary and circular labour migration schemes with environmentally vulnerable communities to strengthen the developmental benefits of such migration.

- Social protection holds poor for safety and future weather extremes and to tackle increasing levels of risk and vulnerability. There is a need to further develop on evidences based on how to effectively combine social protection measures to mitigate vulnerability to climate change in different contexts. This could include:-

- Capturing further lessons from existing case studies to support learning in other countries.

- Combining the long-term study of poverty impacts and social responses to climate change with trends and projections for future climate hazards.

- Building evidence on the economic costs and benefits of different social protection measures for climate change adaptation.
- Generating evidence of the cost effectiveness of social protection measures relative to alternative interventions.
- Taking a longer term perspective for social protection initiatives that takes into account the changing nature of shocks and stresses.
- Developing climate risk assessments for use of the social protection programme design and implementation.
- Developing practical guidance on the design and implementation of appropriate adaptation methods, particularly women, children and the elderly.
- Supporting civil society to help the poor build voice to demand access to social protection instruments.
- Reviewing existing adaptation funding guidelines and criteria to identify opportunities to integrate appropriate social protection responses.
- Strengthening synergies and linkages between academics and practitioners from across the three disciplines to strengthen understanding, co-ordination and good practices.

Disaster and climate resilience requires cost effective in long run. Policies to promote ecosystem, safer building practices and strengthened early warning have all proven effective in saving lives and assets. Using a learning-by-doing approach will help develop expertise and knowledge, support local solutions and provide incentives for capacity maintenance and expertise retention.

Many countries lack the incentives to mainstream climate and disaster risks into economic planning and investment decisions. Political cycles favour short-term development decisions and government employees often have little incentive to participate in inter-sectoral committees to address problems. Thus, climate and disaster resilience get reflected in strategic sectoral programs and budgets as part of the core work program of participating stakeholders.

Disasters provide opportunities to build political will to integrate resilience measures into recovery and development. Adequate, practicable and long-term financing is needed to bring about transformative change. Some countries have been able to identify disaster and climate risks, plan and prioritise investments and combine funding from different sources to optimise implementation. This report focussed on the experience of the World Bank in climate and disaster resilient development, many other partners-including international organisations, national experts, civil society partners, and multi-lateral and bilateral donors- have also worked extensively in this field. The loss and damage agenda could help promote closer integration of the actions needed to manage this risk.

UNESCO, UNHCR and IOM played vital role and focussed on the goal of diminishing vulnerability, particularly amongst the poor and enhanced co- ordination between climate resilience and disaster risk management. A new approach is needed to underpin the incorporation of risk management into work on climate change and the introduction of climate change into natural hazards and development planning. The approach needed the capablility of dealing with the long-term transformations that climate change may bring and ways in which people respond, both at the national, regional and local level. The key concepts in that new approach should be capacity-building and resilience.

Adaptation to climate change may involve some very difficult political choices. For example, long-term changes to land use are likely to be required such as affecting agriculture and forestry, the use of coasts, river resources and settlement patterns and infrastructure.

There are proactive approaches to the long-term challenges that accelerated climate change patterns. But LDCs are unlikely to have the capacity or resources to respond similarly. Risk management in MDCs has its focus on risk reduction and prevention. For LDCs, the focus has generally been on relief. This difference reflects economic disparity. Risk management cannot address under-lying causes of poverty, but it can help to build those structures that will enable a greater degree of self-help. The mechanisms, resources and capacity do exist. The challenge finds the means of developing closer linkages between UNFCCC and UN/ISDR.

Disaster Risk Reduction, Management, and adaptation must remain central to the discussions and progress must be made to effectively and fairly address the increased risks. The disaster risk reduction can be implemented immediately under the guidance of the Hyogo Framework, to provide adaptive capacity, to increase resilience to future threats and to reduce the existing unacceptable and growing levels of disaster risk.

Finally, climate and disaster risk management need to be integrated much more closely with development planning and targeted poverty alleviation programs.

Climate change induced displacement and UN
Legal and policy responses have to involve a combination of strategies. Physical adaptation needs to be financed and developed and migration

options including opportunities for economic, family and educational migration, need to be accepted as a rational and normal adaptation strategies, while movement can be a sign of vulnerability, it can also be a means to achieve security and attain human rights, especially when it can be planned. Solutions need to be developed within a human rights framework such as the fundamental principles of humanity, human dignity, human rights and international cooperation.

Localised or regional responses may be better able to respond to the particular needs of the affected population in determining who should move, when, in what fashion and with what outcome. The slow onset of some climate change impacts, such as rising sea levels, provides a rare opportunity to plan for responses rather than relying on a remedial instrument in the case of spontaneous flight. Crucially, policy responses to climate related movement must not operate in a vacuum. The complement policies relating to development, housing, family planning and the carrying capacity of particular environments.

The policies are needed to conduct regional assessments of environmental vulnerability; produce integrated assessments of climate risks and regional impacts; prioritise development and assistance policies to increase adaptative capacity for those are most vulnerable; prepare contingencies for future migration.

Climate sensitive development policies

A first important area where policy-makers can respond to climate change is in the form of climate sensitive development policies- i.e. pro-poor climate change adaptation policies that build local resilience and adaptive capacity, reducing the need for the poor to migrate away from affected areas. These should include new policies to build specific adaptive capacity among affected populations in areas such as African Sahel as well as the integration of climate change concerns into existing policies to ensure that programmes do not further undermine the resilience of the poor when faced with climate change.

How developed countries adapt to climate change will also have significant impacts on the nature and extent of international migration flows. However, IPCC report on the societal impacts of climate change- is also that policy based on trying to prevent migration resulting from climate change is doomed to failure. Migration can have both positive and negative effects for the individuals who move and for the areas they move from and to. Policies need to be developed which support those who will migrate in the future as a result of climate change.

Policies to support migrants

Policies aimed at migrants and migrations that are linked to climate change might range across a number of areas, including:-

Incorporation of migration into National Adaptation Programmes of Action (NAPA) which are designed to help less developed countries identify and rank priorities for adaptation to climate change (Brown, 2007, 26).

Incorporation of both migration and climate change into national development plans.

- Policies aimed at ensuring the social protection of more vulnerable or poorer migrants. There is likely to be some increase in international migration, particularly across borders within Africa and South Asia, this might include measures to improve the social benefits across borders; however, more important is ensuring the protection of poorer migrants who move from rural to urban or rural to rural areas within developing countries.

- Policies aimed at those moving to slum areas of large cities. Indeed, the provision of basic housing, safe water, basic health and education, as well as employment in large cities remain some of the principal challenges for the 21st century, the challenges to increase in the context of climate related migration.

- Policies aimed at defusing tensions where migration exacerbated by climate change may involve the crossing of a sensitive border. A case in point is the border between Bangladesh and India where the construction of 3.6m high fence is the current policy response to ongoing tensions between the two countries.

- Policies need to expand the definition of a 'refugee' to include environmental or climate change related factors.

- Policies must be framed to support the relocation of affected populations. The most contentious of all possible policy options, since relocation could be seen as a forced migration, one that pro-poor policies should be seeking to avoid.

Research into migration and climate change

Given the absence of local and regional data, a final area in which policy makers must take action is to support further research to understand the specific causes and consequences of migration associated with climate change.

Donors and national governments must play a key role. International collaboration lies at the policy responses to mitigate the migratory impacts of environmental change. Policy initiatives to accomplish these objectives require high level coordinated dialogue between governments, inter-governmental and non- government agencies.

The donors, national governments in collaboration with UN organisations such as UNHCR, UNEP, UNDP, IOM; the World Bank and INGOs should:-

- Promote high level dialogue in order to develop, strengthen and harmonise international understanding of concepts, knowledge base, vocabulary and experience related to the multiple cause-effect links between environmental degradation, socio-economic impacts and environmentally-induced forced migration;
- Promote fruitful collaboration among the environmental and social sciences, including the development of common terminology, statistical methods, indicators and databases;
- Analysts and practitioner should be engaged in disasters, climate change and development planning;
- Generate, collate and disseminate reliable data on the number of people migrating as a result of environmental change;
- Promote the development of more sophisticated typologies of environmentally- induced migration;
- Enhance knowledge of livelihood, resilience, successful adaptation, preparedness and coping strategies used by local populations to mitigate the impacts of environmental change and its potential to induced conflict or forced displacement;
- Support research which will enhance understanding of the relationship between environmental change and conflict;
- Advocate clarification of international institutional responsibilities for promoting and coordinating policy responses to environmental change and forced migration;
- Develop a comprehensive, accepted and concrete definition of environmental forced migrants but without risk to any erosion of current international refugee law;
- Recognise that use of incorrect terminology give governments grounds to disregard advocacy on behalf of the environmentally displaced;
- Promote the development of adequate and appropriate protection instruments to safeguard the rights, needs and human security of environmentally displaced populations;
- Encourage governments to sign up and adhere to the Guiding Principles for Internal Displacement and to recognise their applicability to the protection needs of those displaced as a result of climate change within country borders;

- Adopt proactive development policy responses to the potential migratory impacts of climate change which stress coping capacities, adaptation and sustainability and strengthen the incorporation of resilience strategies in programmes and projects;
- Recognise that sustainable adaptation measures must be locally and regionally place specific; that will contribute to both vulnerability and poverty reduction;
- Promote policy responses which mainstream the participation of local partners and community-focussed approaches to adaptation and enhancing resilience;
- Promote the development of appropriate funding regimes to support protection and assistance mechanism;
- Promote the integration of environmental policies and responses in relief, recovery and development programmes in situations of conflict and forced displacement;
- Offer greater support to national disaster preparedness and responses agencies;
- Urge developing countries to integrate the impacts and responses to climate change into Poverty Reduction Strategy and Conflict Reduction Strategies;
- Develop principles and practices for development policies, projects and programmes and require donors and development agencies urgently to adopt them.

Role of United Nations in Meeting the Threats of Climate Change
Climate change will trigger profound global change, and these changes could pose genuine risks to international peace and security. Managing these changes will require well-conceived actions within the UN system, while climate change could contribute to armed conflict and violent. Preventing large-scale humanitarian catastrophes from climate related droughts, floods, crop failures, mass migrations and exceptionally severe weather remains the most significant policy challenge.

The UN needs to improve substantially the effectiveness of international efforts to mitigate emissions. The UN secretary-general incorporates climate change in the creation of a new senior-level office that would be charged with building political support for addressing all global environmental challenges that promote sustainable development.

Climate change is already occurring and will continue for decades, the UN must place equal emphasis on helping nations adapt to global warming. The UN system needs to work harder to prevent and respond to humanitarian crisis, which will increasingly be fuelled by climate factors. Climate change plays a role in disasters which will be difficult to predict from traditional humanitarian crises. Many of the UN's existing disasters efforts' need to be strengthened by shifting emphasis from disaster response to prevention and by integrating awareness of the consequences of climate change into this work programmes.

UN system must also launch a new effort aimed at dealing more directly with the security risks associated with humanitarian disasters in weak and totalitarian states, where climate change is most likely to trigger regional insecurity. The UN needs to develop powerful analytic tools such as fully coordinated vulnerability index, preventative diplomacy through the creation of senior-level disaster prevention coordinator and multilateral intervention and the development of new norms and institutional arrangements regarding use of force.

Land degradation, climate change and water quality - these are the environmental issues that threaten security. The high-level panel's report should recommend that the Security Council mainstream environmental issues into its security operations with environmental conflict experts and sharing conflict- related environmental data and analysis across UN agencies. The high-level panel should recommend that the UN coordinate international efforts that bind environment and conflict. By protecting the earth, the UN Security Council can help preserve the peace.

A two part strategy is needed to deal with the inevitable adverse effects of climate change. First the UN should strengthen those programs that handle disaster and humanitarian crises and that are already beginning to take climate change into account. Second, the UN should create a new effort focussed on predicting, preventing and handling climate change - related disasters in weak states and those with repressive governments.

Strengthening ongoing disaster work
Shift priority from relief to prevention:-

There is too little awareness of the priority of disaster-risk reduction among countries with responsive decision makers. The UN's Inter Agency Task Force on Disaster Reduction (IATF/DR) and the Inter Agency Secretariat of the International Strategy for Disaster Reduction (UN/ISDR) are existing frameworks in which early warning systems and vulnerability assessment are embedded.

Integrate disaster and climate planning:-

The UN needs to integrate the consequences of climate change more fully into its security, natural disaster prevention and humanitarian response activities. Climate change bodies are in danger of reinventing the wheel on disaster prevention and response. The existing network of disaster experts should be more fully integrated into the IPCC reporting process to avoid this potential problem.

New strategy needed for vulnerable states

Improve early warning systems and vulnerability indices: - The UN system needs for predicting which states and regions are most vulnerable to severe security threats related to climate change. The UN's Humanitarian Early Warning System is an internal UN tool to identify countries in pre-crisis situations and the UN's office for the coordination of humanitarian affairs (OCHA) is an external system that focusses on natural disasters and complex emergencies. At the regional and country levels, OCHA has an Integrated Regional Information Network, primarily for sub-Saharan Africa. In agriculture, FAO has the Global Information and Early Warning System on Food and Agriculture UN Development Programme and UN Environment Programme have developed a Disaster Risk Index.

Preventative diplomacy

UN has identified high-risk countries; it should develop contingency plans for the consequences of climate change. The contingency plans include for providing shelter, nutrition, medicines and policies. UN officials should also share information concerning disaster prevention with relief agencies such as the UN High Commissioner for Refugees, the international Red Cross and the broader NGO relief community.

The international community needs to revisit norms and institutional arrangements concerning the use of force in response to disasters with respect to terrorism and weapons of mass destruction. The UN should be facilitating this dialogue including potential climate induced catastrophes in its programs for post conflict reconstruction. The UN should take steps on accounts of Global Security against climate change as:-

- The UN Security Council should take a leadership role by making environmental security a priority and moving environmental issues from the technical to the security domain and should report annually to the security council on emerging environmental threats to security;
- Coordinate international efforts on environmental security, especially within the UN system creating a new post of High Commissioner for Environment or sustainable development establishing a high level policy forum;

- Form a global think tank for environmental security, fostering data sharing, technology transfers and institutional learning across international organisations;
- Utilise environmental cooperation as a practical pathway to building confidence and peace. Develop a method to quantify peace dividends to justify the additional costs of investing in environmental projects in conflict zones;
- Improve management, leadership and coordination among UN agencies by conducting regular rotations of management staff among environment, development and security agencies;
- Create a 'fusion centre' to coordinate early-warning intelligence across agencies. Integrate environmental considerations into the production of the Security Council's confidential 'watch list' of countries at risk;
- Identify conflicts between the trade, human rights and environmental regimes: modify trade rules that encourage unsustainable development, integrate climate change and development planning-use renewable energy to meet growing demand in the developing world and provide carbon-free energy;
- Carry environmental security through to the project and program level;
- Create incentives to shift expenditures from conventional security to human security by creating a UN fund to match reductions in military spending with money devoted to sustainable development or human security;
- The Secretary-General should elevate mitigation by making it a matter of personal diplomacy and advocating it directly to world leaders;
- Increase the public awareness of climate change and reduce the public vulnerability to hazards through education;
- Undertake an internal review of how climate change affects UN's mission and an 'accountability examination' to assess the impact of UN policies on climate change; and
- Seek the participation of the private sector and encourage technological investment in climate change mitigation.

Recommendations

UNFCCC decision 23/cp18 made at Doha in 2012 on Gender Balance there are two sets of recommendations were made as follows:-

The first recommendation was to the UNFCCC secretariat
- The second set of recommendations to the national government.

Recommendations to UNFCCC secretariat
- Strategies for climate change interventions, at all levels must take different impacts of climate change on men and women. The fact that women are impacted more severely than men by climate change needs to be acknowledged in all planning and implementation stages of climate response initiatives at all levels;
- Parties ensure increased women participation in climate change at national levels;
- There is need to make targeted effort to build the capacity of women to effectively participate in the UNFCCC processes;
- The UNFCCC should facilitate availability of funds to support the participation of women delegates to the UNFCCC negotiations and processes through initiatives such as the Women Delegates Fund (WDF) administered by Women's Environment and Development Organisation (WEDO) through the Global Gender and Climate Alliance (GGCA);
- Sustained training and capacity building on gender mainstreaming in climate change both at the national and international levels;
- Strengthening appreciation for gender in climate change through awareness creation of the important role that gender mainstreaming plays in managing climate change; documenting effects of climate change on different groups of women and men; and highlighting the best practices on gender and climate change;
- Promote/strengthen research on gender and climate change;
- Monitoring and reporting on gender representation in the COPs and taking measures to encourage conformity with the gender balance requirement among the parties.

Recommendations to national governments
- The government should create a national fund to support women delegates to various UNFCCC processes;
- Adopt and comply with affirmative action of at least 30 per cent representation;
- Capacity building of delegates before COPs to facilitate effective participation;
- Formulation of gender policy and strategy on climate change. The capture of indigenous knowledge in policy formulation and implementation must be ensured;
- Create awareness among the policy makers at national and country levels;

- Promote research on gender to establish evidence based on gender and climate change and highlights of the best practices on addressing gender.

For the United Nations to confront successfully the challenges and opportunities of the future, new institutional mechanisms are needed to prevent intra-state and inter-state conflicts and to promote sustainable economic and social progress. The new Economic Council, a new Social Council and Security council, all three are working together on behalf of human security and sustainable development. The UN's economic and social agendas, the Economic Council and the Social Council would coordinate policy and programs through a Global Alliance for sustainable development providing security from violence.

Security from violence
- Expand the membership of the Security Council not more than 23 members taking into consideration the principles of participation and equity in a universal organisation;
- Limit the use of veto related to chapter VII or to other decisions entailing the use of military personnel;
- Articulate a clear mandate for all peace keeping, peace enforcement, collective security and custodian operations;
- Establish an ad hoc military authority for each enforcement operation which is directly under UN command;
- Establish a UN Rapid Reaction Force capable of immediate deployment upon the decision of the Security Council;
- Establish a joint General Assembly-Security Council Working Groups to promote progress towards global disarmaments at the regional levels;
- Establish a Security Assessment Staff and Agencies to support the efforts of the Secretary-General and the three councils to protect and promote human security.

Promoting economic betterment
- Economic council is the principal organ of the United Nations system. The Economic Council would integrate the work of all UN agencies and international institutions, programmes and offices engaged in economic issues. It would promote the monetary and trade policies of member states and encourage international cooperation on transfers of technology and resources;

- The 23 members should be chosen on rotation by the General Assembly, taking into consideration geographic representation, population and a balance between national economies of different size;
- The Economic Council should have a Standing Advisory Committee, composed of distinguished and talented individuals drawn from various disciplines and professional fields and from the private sector;
- Representative of UN agencies, financial institutions and non-state actors would provide input to the Economic Council's deliberations.

Protecting the social issues
- Establish a Social Council as a principal organ of the United Nations system. The Social Council should integrate UN activities relating to issues of social development such as the protection of environment, education, health care, population and migration; the promotion of human rights and freedom of cultural expression;
- The Social Council should have a Standing Advisory Board to distinguish individuals drawn from various disciplines, professional fields and other groups involved in social and human rights policies;
- Representatives of UN specialised agencies, regional organisations and non-state actors would provide the policy implementation;
- The Social Council should strengthen and rationalise the Centre for Human Rights.

Leadership, organisation and resources
- The organisation needs to eliminate by transforming, rationalising or abolishing certain units. The three council arrangements recommended would provide such an opportunity. The Secretary-General should convene a committee of permanent representatives and senior staff to consider reforms;
- The financial crisis of the organisation needs to be addressed by securing the commitment of member states to pay their assessments; re-evaluating the assessment formula and making more efficient use of existing resources.

With the emergence of non-conventional security threats more particularly climate change, the United Nations being freed of the cold war moorings and long consignment to understanding of security in Neo-realist paradigm, during these years has played an active and performative role with a missing zeal to address, combat and wipe out the impending dangers of climate change with the whole hearted support and participation from the richest to the poorest countries of the world.

BIBLIOGRAPHY

- *1951 Refugee Convention, the mandate includes: victims of manmade disasters and persons of concern of the High Commissioner: see ECOSOC resolution 2011 (LXI) of August 2, 1976, available at, http://www.unhcr.org/refworld/docid/3ae69ef418.html.tabid=79.*
- *1998 guiding principles on internal displacement, available at: http://www2.ohchr.org/english/issues/idp/standards.htm.*
- *A.Orakhenshivili, collective security, oxford; oxford university press, 2011, p. 2.*
- *According to the Vienna declaration and programme of action (world conference on human rights) 1993, article 5: all human rights are universal, indivisible and interdependent and interrelated.*
- *Adapted from: FAO (2008). Climate change and food security: A framework documents.*
- *ADB et al., "poverty and climate change: reducing the vulnerability of the poor through adaptation", VARG multi development agency paper, United Nations Development Projects (UNDP) United Nations, New York, 2003.*
- *Adger et al., climate change impacts adaptation and vulnerability, 2007.*
- *Adger, N. and N. Brooks, country level risk measures of climate-related natural disasters and implications for adaptation to climate change. Working paper 2b. Tyndall centre for climate change research, Norwich, 2003. http://www.tyndall.ac.uk/publications/working-papers/wp 26.pdf.*
- *Adger, W. & Kelly, 1999, Social Vulnerability to Climate Change and the Architecture of Entitlements' Mitigation and Adaptation Strategies for Global Change.*
- *Adger, W. Institutional Adaptation to Environmental Risk under the Transition in Vietnam', Annals of the Association of American Geographers, vol. 90, no. 4, 2000, pp. 738-758.*
- *Ahmed et al., Climate volatility on poverty vulnerability in developing countries. Environmental Research Letters, vol. 4, no. 3, 2009. http://iopscience.iop.org/1748-9326/4/3/034004/fulltext/*
- *Ainsworth and long-2005(2005) what have we learned from 15 years of tree-air CO2 enrichment (FACE)? A mela- Analysis of the responses of Photosynthesis, canopy properties and plant production to rising COS.In: New phytol, 1652005, PP.35/-372.*
- *Al- Ahmed A. (2009, December 3). Jeddah flood deaths shame Saudi royals. The Guardian, retrieved from http://www.guardian.co.uk*
- *Albritto DL, Meiro-Filho LG, Technical summary, in climate change, 2001.*
- *Alcamo, J., D. Van Vuuren, C. Ringler, W. Cramer, T. Masui, J. Alder and K. Schulze. Changes in nature's balance sheet: model - based estimates of future worldwide ecosystem services. Ecology and Society, vol. 10, no. 2, 2005, p. 19.*

- *Alexander, D. (2002a) 'from civil defence to civil protection-and back again 'Disaster Prevention and Management.*
- *Alexander, D. (2002b) Principles of Emergency Planning and management. Terra publishing Harpenden.*
- *Aloisi, S. 'Senegal mulls "Green Wall" to stop Desert Advance'. Reuters August 1, 2005. http://forests.org/articles/redear.asp?linked=44784.*
- *Aniello C, Morgan k, Busbey A, Newland L, Mapping micro urban heat islands, 1995.*
- *Arnell NW, Climate change and global water resources. Global environmental change- Human and policy dimensions 2004, vol. 14, pp. 31-52.*
- *Arnold JEM, 2008, managing ecosystems to enhance the food security of the rural poor.*
- *ASEAN et al., (2007). Cebu Declaration on East Asian Energy Security.*
- *At the 13th COP, Bali December 2007. It provides the roadmap toward a new international climate change agreement to be concluded by 2009, and that will ultimately lead to post-2012 international agreement on climate change.*
- *Badjeck et al., 2010: Impacts of climate variability and change on fishery- based livelihoods.*
- *Baechler 1999a. Environmental degradation in the south as a cause of armed conflict, in Carius, A., and Lietzmann, K. (eds.), Environmental Change and Security: A European Perspective (Berlin: Springer-Verlag).*
- *Baechler G., 1999b. Environmental degradation and violent conflict: hypotheses, research agendas and theory building, in Suliman, M. (ed.), Ecology, Politics and Violent Conflict (London and New York: Zed Books).*
- *Barker T, Bashmakov I, Bernstein L, Bogner JE, Bosch PR, Dave R, Davidson OR, Fisher BS, Gupta S, Halsnaes K et al: Technical Summary. In Climate Change, 2007: Mitigation Contribution of Working Group III to the Fourth Assessment Report of the Inter-governmental Panel on Climate Change. Edited by Metz B, Davidson OR, Bosch PR, Dave R, Meyer LA (Cambridge, United Kingdom/New York, NY, USA: Cambridge University Press, 2007).*
- *Barker T, Bashmakov I, Bernstein L, Bogner JE, Bosch PR, Dave R, Davidson OR, Fischer BS, Gupta S, Halsnaes K et al., Technical Summary: In Climate Change 2007: Mitigation, Contribution of Working Group III to the Fourth Assessment Report of the Inter-governmental Panel on Climate Change. Edited by Metz B, Davidson OR, Bosch PR, Dave R, Meyer LA (Cambridge, UK/ New York, NY, USA: Cambridge University Press, 2007).*
- *Barnett T.P., J.C. Adam, D.P. Lettenmaier, Potential impacts of a warning climate on water availability in snow dominated regions, Reviews, Nature Vol. 438:17, pp. 303-308.*
- *Barnett, J & Adger, W. N. Climate dangers and atoll countries, Climatic Change, 2003.*
- *Barnett, J. (2006), Climate Change, Insecurity and Justice. In W.N. Adger, J. Paavola, M.J. Mace, & S. Huq (eds.), Fairness In Adaptation to Climate Change, Cambridge (MA: MIT Press, 2006).*
- *Barnett, J. Security and climate change. Global Environmental Change, 2003.*
- *Barnett, j., 2001b. The Meaning of Environmental Security: Ecological Politics and In the New Security Era (London and New York: Zed Books).*
- *Barnett, J., and Dovers, S., Environmental security, sustainability and policy.*

Pacifica review, 2001.

- *Barnett, J., Global Warming and the Security of Atoll-Countries, 2002.*
- *Batchelor C., A. Singh, M.S. Rama Rao and J. Butterworth. Mitigating the potential unintended impacts of water harvesting, 2005, at http://www.nri.org/projects/ WSSIWRM/Reports/ water% 20 harvesting % 20 impacts.pdf.*
- *Bates, B.C, Kundzewicz, Z. W., Wu, S. and Palutikof, J. P. (Eds.). 'Climate Change and Water'. Technical Paper of the Intergovernmental Panel on Climate Change (Geneva: IPCC Secretariat, 2008).*
- *Baumert K et al., Navigating the Numbers: Greenhouse Gas Data and International Climate Policy. Washington, World Resources Institute, 2005.*
- *Beddington JR, Asaduzzaman M, Clark ME, Fernandez, Bremauntz A, Guillou MD, Howlett DJB, John MM, Lin E, Mamo T, Negra C, Nobre CA, Schools RJ, Van Bo N, Wakhungu J. 2012. What Next for Agriculture after Durban?*
- *Berkes, F, J. Colding and C. Folke (eds.) Navigating social-ecological systems: building resilience for complexity and change(Cambridge, U.K: Cambridge University Press, 2003).*
- *Berkes, F, J. colding, and C. Folke (cds). Navigating social - ecological systems: building resilience for complexity and change (Cambridge, UK: Cambridge University Press, 2003).*
- *Berstein, P., et al., Effects of restrictions on international permit trading: the MS-MRT model, in weyant, J., (ed.), the costs of the Kyoto Protocol: a multi-model evaluation, The Energy Journal, 1999.*
- *Bhatt, M. Corporate Social Responsibility and Disaster Reduction: local overview of Gujarat, case study for corporate social Responsibility and Disaster Reduction. A Global Overview, DFID-funded study conducted by the Benfield Grieg hazard research centre, University College London, 2002. http://www.benfieldhrc.org/ siteroot/disaster-studies/csr/csr-gujurat.pdf.*
- *Black, Richard (December 11, 2011). "Climate talks end with late deal". BBC News, retrieved December 11, 2011.*
- *Boano, C, zetter, R, and Morris, T, (2008) Environmentally displaced people understanding the Linkages between environmental change, Livelihoods and forced migration, Refugee study center policy Brief NO.1 (RC : Oxford), p.4.*
- *Boyce, J., et al., Power distribution, the environment and public health: a state-level analysis. Ecological Economics, 1999.*
- *Bradley R and Baumert K, (eds.) (2005). Growing in the Greenhouse: Protecting the Climate by Putting Development First. Washington, World Resources Institute.*
- *Brody A. et al., Gender and climate change, 2008.*
- *Brown et al., A review of paired catchment studies for determining changes in water yield resulting from alterations in vegetation, Journal of Hydrology, vol. 310, no. 1-4, 2005, pp. 28-61.*
- *Brown, N. Climate, ecology and international security. Survival, 1989.*
- *Brown, N., Climate, ecology and international security, 1989.*
- *Brown,O. 2008, migration and climate change in international organisation for migration (IOM) Research series no.31 (IOM,Geneva,2008)*
- *Bruinsma J, ed., World agriculture: Towards 2015/ 2030, A Food and Agriculture Organisation Perspective (London: Earthscan, 2003).*
- *Bruinsma J. (ed.), The Resource Outlook To 2050: By how much do land, water*

and crop yields need to increase by 2050? Expert meeting on how to feed the world in 2050 Food and Agriculture Organisation of the United Nations, Economic and Social Development Department, Rome, 2009.

- C. Tickell, 'The Inevitability of Environmental Security', In: G. Prins (Ed.) Threats without Enemies: Facing Environmental Insecurity (London: Earth Scan, 1993), p. 23.

- CA. 2007. Water for Food, Water for Life. The Comprehensive Assessment of Water Management in Agriculture D. Molden (Ed.) London, Earthscan and Colombo, International Water Management Institute.

- Callow, R., Mac Donald, A., Nicol, A. And Robins, N. (na) 'Ground water security and drought in Africa - linking water availability access and demand' (unpublished manuscript).

- Campbell - Lendrum, D et al., Environmental burden of series, 2007.

- Cannon T., Gender and Climate hazards in Bangladesh, 2002.

- Cascio, J, Environmental refugee (2005) world changing, change. Your thinking, Time Magazine at http://www. World changing.com/archives/003618.html.Cited 2009, August 10.

- Casey Brown and Upmanu Lall, "Economic Development: The Role of Variability and a Framework for Resilience, Natural Resources Forum, 2006.

- Cash D. Climate change adaptation report. Beston Executive office of energy and Environmental Affairs and Adaptation Advisory Committee, 2011.

- CCAFS: Proposal for CGIAR Research program 7: Agriculture and food security, Copenhagen, 2011.

- CESCR (Committee on Economic, Social and Cultural Rights (1999): General Comment No- 12. The Right to adequate food (Art-11), Geneva.

- Chant S., Gender, cities and the Millennium Development Goals in the global south, 2007.

- Chronic Poverty Research Center (CPRC), University of Manchester.

- Climate Adaptation Working Group, 2009. Shaping Climate Resilient Development: A Framework for Decision Making.

- Collier, P. and S. Dercon. African Agriculture in 50 Years: Small holders in a rapidly changing world? Expert meeting on how to feed the world in 2050 Food and Agriculture Organisation of the United Nations, Economic and Social Development. Rome, 2009.

- Commission on Human Security 2003. Human Security now, Final Report, New York: CHS.2003.

- Confalonieri et al., climate change impacts, adaptation and vulnerabilities, 2007.

- COP 9 (Conference of Parties 9) 2002. 'The special climate change fund (SCCF), decision 5/ cp-9'. COP 9, Milan. http://unfccc.int/cooperation-and-support/funding/ special-climate-change-fund/items/2602.php.

- Costello A, et al., Managing the health effects of climate change, 2009.

- Council directive 2004/83/EC of April 29, 2004 on minimum standards for the qualification and status of third country Nationals or state less persons as refugees or as persons who otherwise need International protection and the content of the protection granted (2004) OJL 304/12.

- Cutter et al., social vulnerability to environmental hazards, 2003.

- D.P.Fidler, The UN and The Responsibility to Practice Public Health, Journal of

International Law and International Relations, 2005-2, pp. 58-59;

- *Daszak P, Cunningham AA, Hyatt AD, Actatrop, Zool, Feb 23; Review anthropogenic environmental change and the emergence of infectious diseases in wild life.*
- *David grey and Claudia W. Sadoff: water security for growth and development water policy 2007*
- *Davidson OR. 2011- Strategies to mitigate climate change in a sustainable development framework.*
- *Davidson, D.J/ Williamson T. /Parkins, J.R., Understanding Climate Change risk and vulnerability in northern forest-based communities In: Canadian Journal of Forest Research, vol. 33, no. II, 2003, pp. 2252-2261.*
- *de Janvry and Sadoulet, E., Subsistence farming as a safety net for food-price shocks, Development in practice, vol. 21, no. 4-5, 2011, pp. 449-456.*
- *Defries R, Rosenweig C, Towards a whole-landscape approach for sustainable land use in the tropics, 2010.*
- *Dell et al., Temperature and income: Reconciling new cross-sectional and panel estimates. American Economic Review, vol. 99, no. 2, 2009, pp. 198-204.*
- *Devereaux, S. and Sabates-Wheeler R., Transformative social protection. IDS working paper 232, 2004, retrieved from www.ntd.co.uk/idsbookshop/details. asp?id=844. The order has been changed to reflect the need to focus on preventive and promotional measures needed when these fail.*
- *DFID (2004b) key sheet 06. Adaptation to climate change: making development disaster- proof. DFID, London.*
- *"Disaster Risk Reduction Strategies and Risk Management Practice: Critical Elements for Adaptation to Climate Change" Submission to the UNFCCC Adhoc Working Group on Long Term Cooperative Action by The Informal Taskforce on climate change of the Inter-Agency Standing Committee and The International Strategy for Disaster Reduction November 11, 2008.*
- *Dovers, S. Institutions for sustainability, Tela, 7 (Melbourne: the Australian Conservation Foundation, 2001).*
- *Doyle, Alistair (2003, September 30). "16,000 die yearly from global warming". Reuters news Service, available online at http://www.planetark.com/dailynewsstory. cfm/newsid/22420/story. htm.*
- *Drager N, Fidler DP, foreign policy, trade and health, 2007.*
- *Dye, P. and D. Versfeld. Managing the hydrological impacts of South African plantation forests: An overview, Forest Ecology and Management, vol. 251, no. 1-2, 2007, pp. 121-128.*
- *Easterling WE, Aggarwal PK, Batima P, Brander KM, Erda L, Howden SM, Kirilenko A, Morton J, Soussana JF, Schmidhuber J, Tubiello FM (2007) "Food, Fibre and Forest Products". In Climate Change 2007: Impacts, Adaptation and Vulnerability, Contribution of Working Group II to the Fourth Assessment Report of the Intergovernmental Panel on Climate Change, Palutikof, P.J. van der Linden and C.E. Hanson. Cambridge, UK, Cambridge University Press.*
- *Easterling WE, Aggarwal PK, Batima P, Brander KM, Erda L, Howden SM, Kirilenko A, Morton J, Soussana J-F, Schmidhuber J et al., Food, Fibre and Forest Products. In Climate Change, 2007: Impacts, Adaptation and Vulnerability. Contribution of Working Group II to the Fourth Assessment Report of the Inter-governmental Panel on Climate Change. Edited by Parry ML, Canziani of, Palutikof JP, Van der Linden*

PJ, and Hanson CE (Cambridge, UK: Cambridge University Press, 2007), pp. 273-313, IPCC.

- *Eckstein, H., & Gurr, T. Patterns of authority: A structural basis for political inquiry (New York: Wiley, 1975).*
- *Eckstein, H., and Gurr, T., Patterns of Authority: A Structural Basis for Political Inquiry (New York: Wiley, 1975).*
- *Edward page, Department of politics and international studies, University of Warwick, SGIR 7th pan-European, International Relation Conference Stockholm, September 9-11, 2010.*
- *Edwards, M. Security implications of a worst-case scenario of climate change in the South-west Pacific, Australian Geographer, 1999.*
- *Ehrlich, P. & Ehrlich, A., Population growth and environmental security, Georgia Review, 1991.*
- *El-Hinnawi, E., Environmental Refugees (Nairobi: UNEP, 1985).*
- *Elverland, S,"20 Million climate displaced in 2008,"Norwegian Refugee council, June 8, 2009. at: http://www.nrc.no/? did= 9407544.*
- *Emanuel, K., Divine Wind: The History and Science of Hurricanes. Oxford university press, New York, NY, 2005.*
- *Esty et al., State failure task force: phase II findings, Environmental Change and Security Project Report, 5, 1999, pp. 49-72.*
- *European commission 2008 - the report goes on to identify the Arctic, Latin America, Africa, Central Asia as most at risk from climate induced violent conflict .*
- *Ever Land, S., 20 Million Climate Displaced in 2008; Norwegian Refugee Council, June 8, 2009, available at: http://www.nrc.no/?did=9407544.*
- *Ezzati M, Lopez A, Murray C, (eds.), Comparative quantification of Health risks, Geneva, World Health Organisation, 2004.*
- *FAO (1996), Rome Declaration and World Food Summit Plan of Action, retrieved from www. fao.org/docrep/003/x8346E/X8346e02.htm#p1-10.*
- *FAO (2002) World Agriculture: Towards 2015/2030, an FAO Perspective (Rome/ London, FAO/ Earthscan Publishers, 2002).*
- *FAO (2003) World Agriculture: Towards 2015/2030 (Rome: FAO, 2002).*
- *FAO (2006) SOCO; The demand for the products of irrigated agriculture in Sub-Saharan Africa. FAO Water Report 31. Rome.*
- *FAO and CIFOR. Forests and floods drawing in fiction or thriving on facts? 2005.*
- *FAO and WFP (2009): The state of food insecurity in the world: Economic crises -impacts and Lessons Learned.*
- *FAO, (2008) GM food safety assessment: tools for trainers, In press, Expert Meeting on Global Perspectives on Fuel and Food Security: Technical Report, February 18-20, 2008. Rome.*
- *FAO, 2000. Global Agro - Ecological Zones. Version 1.0. FAO Land and Water Digital Media Series 11. Rome. CDROM.*
- *FAO, 2004a. Carbon sequestration in dry land soils, World soil resources reports, No. 102. Rome, available at ftp://ftp.fao.org/agl/agll/docs/wsrr102.pdf.*
- *FAO, 2007a. Adaptation to climate change in agriculture, forestry and fisheries: perspective, framework and priorities. Report of the FAO Interdepartmental Working Group on Climate Change. Rome.*
- *FAO, 2011a: These numbers would imply that labour productivity for women is*

much higher than for men with little evidence to support that.

- *FAO, The state of food insecurity in the world. Food and Agriculture organisation, 2002.*
- *FAO.2010.AQUASTATdatabase, at http://www.fao.org/nr/water/aquastat/main/index.stm.*
- *Figures presented in the study monitoring disaster displacement in the context of climate change, IDMC/ OCHA, 2009, available at : www.internaldisplacement. org/8025708F004CFA06/ (httppublications)/451D224B41C04246C12576390031 FF63? Open Document.*
- *Fischer G, Shah M, Van Velthuizen H, Climate Change and Agricultural Vulnerability. A Special Report Prepared as a Contribution to the World Summit on Sustainable Development (International Institute for Applied Systems Analysis, Laxenburg, Austria, 2002).*
- *Fischer G., M. Shah, F. Tubiello and H. Van Velthuizen. 2005. Socio- Economic and Climate Impacts on Agriculture; an Integrated Assessment, 1990-2080. Phil. Trans. R. Soc. B (360) 2067-283; Shah et al., 2008.*
- *Fischer, G., F.N. Tubiello, H. Van Velthuizen and D.A. Wiberg. 2007. Climate change impacts on irrigation water requirements: Effects of mitigation, 1990-2080. Technological forecasting and social change 74(2007) 1083-1107; Nelson, G.,M. Rosegrant, J. Koo, R. Robertson, T. Sulser, T. Zhu, C. Ringler, S. Msangi, A. Palazzo, M. Batka, M. Magalhaes, R. Valmonte-Santos, M. Ewing and Le. D. 2009. Climate Change impact on agriculture and costs of adaptation. IFDRI. Washington, DC.*
- *Fischer, G., Shah M., Van Velthuizen, H. and Nachtergaele, F.O. Global Agro-Ecological Assessment for Agriculture in the 21st Century, 2001.*
- *Flor Cruz, J (2009, May 13) China Marks earthquake anniversary C N N, retrieved from wwww.cnn.com*
- *Foley J.A, Ramankutty,N.Brauman, K.A, Cassidy,E.S, Gerber, J.S, Johnston, M., Mueller, N.D, O'comell C., ,Ray,D.K West, P.C Balzer, C., Bennet,E.M, Carpenter, S.R., Hill, J., Monfreda, C., Polas Ky, S., Tilman,D and Zaks, D.P.M. (2011), Solutions for a cultivated planet. http://dx.doi. org/10.1038/nature 10452.*
- *Food and Agriculture organisation. The State of Food Insecurity in the World 2001 (Food and Agriculture Organisation, Rome, 2002).*
- *Foresight (2010). Synthesis report C11: ending hunger. Foresight project on global food and farming futures. The government office for science, London, retrieved from www.bis.gov.uk/assets/bispartners/foresight/docs/food-and-farming/synthesis/11-631-c11-ending-hunger.pdf.*
- *Foresight, the Future of Food and Farming: Executive Summary. The Government Office for Science, London, 2011, retrieved from www.bis.gov.uk/assets/bispartners/ foresight/docs/food-and-farming/11-547-future-of-food-and-farming-summary.pdf.*
- *Frank et al., The future of public health in Canada, 2011.*
- *Friels, Bowenk, Campbell Lendrum D, Frumkin H, Mc Michael AJ, Rasanathan k, Annu Rev, Public Health, 2011.*
- *From the International Covenant of Economic, Social and Cultural Rights (1966).*
- *Frumkin et al., Review climate change: the public health response, 2008.*
- *Fuel for life; household energy and health, Geneva: WHO, 2006.*
- *G. Gentili, 'European Court Of Human Rights: An Absolute Ban On Deportation Of Foreign Citizens To Countries Where Torture Or Ill-Treatment Is A Genuine*

Risk', *International Journal Of Constitutional Law 2010. L. Skoglund, 'Diplomatic Assurances against Torture- An Effective Strategy? A Review of Jurisprudence and Examination of the Arguments, Nordic Journal of International Law 2008-77, pp. 319-364.*

- G5 (2008). *G5 Statement Issued by Brazil, China, India, Mexico and South Africa. Hokkaido Toyko Summit. Sapporo. July 8.*
- G8 (2008). *G8 Summit Declaration on Environment and Climate Change. Hokkaido Tokyo Summit. Sapporo. July 8.*
- Gleditsch, N., (Eds.), *Environmental Conflict (Boulder: Westview Press).*
- Gleditsch, N., *Armed conflict and the environment, in Diehl, 2001.*
- Gleik, P., *Effects of climate change on shared fresh water resources, in Mintzer, I., (ed.), Confronting Climate Change (Cambridge: Cambridge University Press, 1992).*
- Global Governance project (2012). *Forum on climate Refugees< Retrieved on May 5, 2012.*
- Global humanitarian forum 2009, *The Anatomy of a silent Ciris, Human impact report climate change, Geneva.*
- Global status of CCS Report: *2011 is the institute's flagship publication and consolidates the current understanding of the level and nature of global carbon capture storage (ccs). Available on, http://www.globalccsinstitute.com/publications/ global-status-ccs-2011.*
- Godfray HCJ Pretty J, Thomas SM, Warham EJ, Beddington JR: *Linking policy on climate and Food Science 2011.*
- Goodhand J. *'Enduring Disorder and Persistent Poverty: A Review of Linkages between War and Chronic Poverty', World Development, 2003.*
- Goodin, R. *The theory of institutional design (Cambridge: Cambridge University Press, 1996).*
- Gore, A (presenter), Guggenheim, D. (Director) *An inconvenient truth: A global warming (Motion picture). United States paramount pictures, 2006.*
- Gough, M. *'Human Security: The Individual in the Security Question-The Case of Bosnia', Contemporary Security Policy, 2002.*
- Gough, M. *Human security: the individual in the security question-the case of Bosnia. Contemporary security policy, no. 23, 2002, pp. 145-191.*
- GPF (Global Policy Forum), *founded in 1993, is an organised seeking to promote accountability of international organisations such as the united nations to promote citizen participation on peace in international studies for scholars, policymakers, activists etc., available at http:// www.globalpolicy.org/*
- Greenhouse gas 'Sinks': *IPCC Fourth Assessment Report Working Group II, Glossary of Terms: http://195.70.10.65/pdf/glossary/ar4-wg2.pdf*
- Greenhouse gas abatement strategies for animal husbandry Agriculture, *Ecosystems and Environment.*
- Guiding principles on internal Displacement on *http://www2.ohchr.org/english/ issues/idp/ standards.htm.*
- Gulbler Dj, Meltzer M, Adv virus Res, *Review impact of Dengue fever on the developing world, 1999.*
- H. Nasu, *'Operationalising the "Responsibility to Protect" and Conflict Prevention: Dilemmas of Civilian Protection in Armed Conflict', Journal of Conflict and Security*

Law-2009.

- *Haines A, Patz JA, JAMA, Review Health effects of climate change, 2004.*
- *Haines A. et al., public health benefits of strategies to reduce GHG emissions; overview and implications for policy makers, 2009.*
- *Haines, A. and J.A. Patz., "Health effects of climate change". Journal of the American Medical Association, vol. 29, no. 9, 2004, pp. 99-103.*
- *Hallegatte, S., A cost effective solution to reduce disaster losses in developing countries: Hydro- Meteorological Services, Early Warning and Evacuation. Policy Research Working Paper 6058, Sustainable Development Network, World Bank, Washington, DC, 2012.*
- *Hallegatte, S., Corfee - Morlot, J., Green C., and Nicholls, R. J. 2013. Future Flood Losses in Major Coastal Cities. Nature Climate Change, doi: 10-1038.*
- *Harakunarak and Aksornkoae. Life-saving belts: post-tsunami re-assessment of mangrove ecosystem values and management in Thailand, 2005.*
- *Harmer A. and J. Macrae (eds.) 2004. Beyond the continuum. The changing role of aid policy in protracted crises. Humanitarian Policy Group (HPG) Research Report 18. Overseas Development Institute (ODI), London. http://www.odi.org.uk/hpg/papers/HPGrepoat.18.pdf.*
- *Harvey M, Pilgrims, food policy: The new competition for land: food, energy and climate change, 2011.*
- *Harvey, Fiona; Vidal, John (December 11, 2011). "Durban deal will not avert catastrophic climate change say scientists". The Guardian (London), retrieved December 11, 2011.*
- *Harvey. Fiona, Vidal, John (December 11, 2011). "Global climate change treaty in sight after Durban breaks through". The Guardian (London), retrieved December 11, 2011.*
- *Hauge, W., and Ellingsen, T. Causal path ways to conflict. In P. Diehl, & N. Gleditsch (eds.), Environmental Conflict (Boulder: West View Press, 2001), pp. 36-570.*
- *Hegerl GC, Zwiers FW, Braconnot P. et al., understanding and attributing climate change, 2007.*
- *Helberg, R., S. Jorgenson and P. Seigal (2008b) "climate change: challenges for social protection in Africa", World Bank, Washington, D.C. http://ssrn.com/abstract=1174774.*
- *Heltberg, R., S. Jorgenson and P. Seigal (2008a). "Addressing human vulnerability to climate change: Towards a 'no regrets' Approach, World Bank, Washington DC, at http://ssrn.com/ abstract=1158177*
- *Hertel TW, Rosch SD, 2010: Climate change, Agriculture and poverty. Appl Econ Perspect Policy 2010, 32:355-385 doi:10.1093/aepp/ppq016.*
- *Hewitt, K. Regions of Risk: A Geographical Introduction to Disaster. Longman, Harlow, 1996.*
- *Hoegh- Guldberg et al., Pacific in Peril: Biological, Economic and Social Impacts of Climate Change on Pacific Coral Reefs (Amsterdam: Greenpeace, 2000).*
- *Hofer, T. and B. Messerli, Floods in Bangladesh, History, Dynamics and Re-thinking the role of the Himalayas, UNU Tokyo, pp. 468; Calder; Brown et al., 2005: A review of paired catchment studies for determining changes in water yield resulting from alterations in vegetation. Journal of Hydrology, vol. 310, no. 1-4, 2006, pp. 28-61.*

- *Hofimann, C.A., L. Roberts, J. Shoham and P. Harvey, Measuring the impact of Humanitarian aid. A review of current practice. HPG Research Report 17. ODI, London, 2004.*
- *Homer- Dixon, t. and Percival, v., Environmental Scarcity and Violent Conflict: Briefing Book. American Association for the Advancement of Science, Toronto, 1996.*
- *Homer-Dixon, T., On the threshold: environmental changes as causes of acute conflict. International security, 1991.*
- *Howden, et al., Adapting agriculture to climate change, 2007, Available at: http:// pubs.giss. nasa.gov/abs/ho03300x.html.*
- *Hussain, I. Pro-Poor Intervention Strategies in Irrigated Agriculture in Asia-Poverty in Irrigated Agriculture: Issues, Lessons, Options and Guidelines, Final Synthesis Report Submitted to the Asian Development Bank, International Water Management Institute (IWMI), Colombo, Sri Lanka, 2005.*
- *Hydrological, meteorological and climatological disasters include storm, flood, wetmass movement, extreme temperature, drought and wildfire. Data downloaded from EM - DAT database, centre for research on the Epidemiology of Disasters (CRED), http://www.emdat.be/*
- *Hyogo Framework for Action 2005 - 2015: Building the resilience of nations and communities to disasters (HFA), see part B, paragraph 4(i)C: http://www.unisdr. org/hfa/hfa.htm.*
- *I. Osterdahl, Threat to the Peace: The Interpretation By The Security Council Of Article 39 Of The UN Charter, Uppsala: Isustus Forlag, 1998.*
- *IAASTD, 2008. Agriculture at a crossroads: The synthesis Report Science and Technology. Washinton, DC, USA: International Assessment of Agricultural knowledge, Science and Technology for Development. http://www.agassessement. org.*
- *IASC Operational Guidelines on the protection of persons affected by natural disasters and the related pilot manual, Brookings-Bern Project on International Displacement, March 2008, available at: http://www.unhcr.org/refworld/ docid/49a2b8f72.html.*
- *ICESCR (Art 2, Para-1): International Covenant on Economic, Social and Cultural Rights. New York, 1999.*
- *ICIMOD, Inventory of Glaciers, Glacial Lakes and Identification of Potential Glacial Lake Outburst Flood (GIFs) Affected by Global Warming in the Mountains of Himalayan Region (Kathmandu: ICIMOD, 2007). DVD/CD-ROM.*
- *IEA (2004). World Energy Outlook 2004. Paris, International Energy Agency.*
- *IEA (2008a). World Energy Outlook 2008. Paris, International Energy Agency.*
- *IFAD, Rural poverty report: new realities, new opportunities for tomorrow's generation (Rome: International Fund for Agricultural development, 2011).*
- *IFRC. World disaster report, Geneva. International Federation of Red Cross and Red Crescent societies, 2007.*
- *IISD Reporting Services- Upcoming meetings'. Lisd. Ca, retrieved April 8, 2010.*
- *ILO, 2007. Chapter 4, Employment by sector, In key indicators of the labour market (KILM), 5th edition. Available at: www.ilo.org/public/english/employment/strat/ kilm/download/kilm04.pdf.*
- *In UN general assembly, 45 session, the report of secretary general, November 8,*

1990, p.q
- *Ingram J. Ericksen P, Liverman D (eds.): Food Security and Global Environmental Change (London, UK: Earthscan, 2010).*
- *INRA/CIRAD. Agrimonde: Scenario and Challenges for Feeding the World in 2050 (Versailles: Editions Quae, 2011).*
- *Inter Agency standing committee's Informal Task Force on climate change: - http:// www. humanitarianinfo.org/iasc/page LOader, aspx? page=content-news-news details & news id = 134.*
- *Inter-governmental Oceanographic Commission. Tsunami risk assessment and mitigation for the Indian Ocean. Knowing your tsunami risk and what to do about it. UNESCO Manuals and Guides 52, 2009.*
- *IOM Guatemala, Survey on Remittances 2008 and Environment, Working Notebooks on Migration, No. 26, 2008.*
- *IPCC (Inter-governmental Panel on Climate Change) 2001b. Summary for Policymakers: Climate Change 2001: Impacts, Adaptation, and Vulnerability, In Climate Change 2001: Impacts, Adaptation, and Vulnerability. Contribution of Working Group II to the Third Assessment Report of the Intergovernmental Panel on Climate Change (Cambridge: Cambridge University Press).*
- *IPCC (Inter-governmental Panel on Climate Change). 2001a. Technical Summary: Climate Change 2001: Impacts, Adaptation, and Vulnerability, In Climate Change 2001: Impacts, Adaptation, and Vulnerability. Contribution of Working Group II to the Third Assessment Report of the Intergovernmental Panel on Climate Change (Cambridge: Cambridge University Press).*
- *IPCC 2008 Technical Paper VI In: Climate Change and Water. Geneva IPCC Secretariat, pp. 210.*
- *IPCC 2012. Managing the risks of extreme events and disasters to advance climate change adaptation (Cambridge: Cambridge University Press).*
- *IPCC Climate change 2007, the Fourth IPCC Assessment Report (Cambridge University Press, 2007), at http://www.ipcc.ch/*
- *IPCC fourth assessment report, working group I report: http://195.70.10.65/ipcc report/ar4-wg1.htm.*
- *IPCC fourth assessment report, working group I, glossary of terms: http://ipcc-wg1. ucar.edu/ wg1/report/ar4wg1_print_annexes.pdf.*
- *IPCC Fourth Assessment Report, Working Group I, summary for policymakers: http://195.70.10.65/pdf/assessment-report/ar4/wg1/ar4-wg1-spm.pdf.*
- *IPCC, 2007b. Climate change 2007 - mitigation of climate change. Contribution of working group III to the fourth assessment report of IPCC (Cambridge, UK. Cambridge University Press).*
- *IPCC, 2013. Climate change 2013: the Physical Science Basis. Summary for policymakers. Working Group I contribution to the IPCC Fifth Assessment-report/ ar4/wg1/ar4-wg1-spm.pdf.*
- *IPCC, 2013. Climate Change 2013: The Physical Science Basis. Summary for Policy Makers. Working Group I Contribution to the IPCC Fifth Assessment Report.*
- *IPCC, Climate Change 2001: Impacts, Adaptation, and Vulnerability (Cambridge: Cambridge University Press, 2001).*
- *IPCC, Special report on emissions scenarios (Cambridge, UK: Cambridge University Press, 2000).*

- *IRD. 2010. Tropical marine ecosystem programme. http://www.mpl.ird.fr/suds-en-ligne/ecosys/ ang-ecosys/intro 1.htm.*
- *ISDR. Global assessment report on disaster risk reduction (United Nations: Geneva, Switzerland, 2009).*
- *Isis international, 2013. First international conference on "Ancient Indian Wisdom and Modern World" March 29-31, 2013.*
- *J. Winpenny, Financing water for all, 2003.*
- *JA Patz, D Campbell - Lendrum, T Holloway, JA Foley. Impact of regional climate change on human health, Nature 2005; 438: 310-7*
- *Jarvis A, et al., 2011: An integrated adaptation and mitigation framework for developing agriculture research: synergies and trade-offs.*
- *Jessica Brown, Neil Bird and Liane Schalatek (2010) climate finance additionality: emerging definitions and their implications overseas Development Institute.*
- *K, Wellens, 2003- The UN Security Council And New Threats To The Peace: Back To The Future', Journal Of Conflict And Security Law 2003-8,*
- *Kaplan, R., The coming anarchy, Atlantic Monthly, 1994.*
- *Keen, D. 'Incentives and Disincentives for Violence', In Berdal, M. and Malone, D. (Eds.) Greed and Grievance: Economic Agendas in Civil Wars, 2000.*
- *Keim ME-preventing disasters: public health vulnerability reduction as a sustainable adaptation to climate change, 2011.*
- *Ken worthy J. Transport Energy use and Greenhouse gases in urban passenger Transport system, 2003.*
- *Kissinger et al., 2011, Copenhagen, Denmark: CCAFS from REDD+ Agriculture.*
- *Klotzli, S., The Water and Soil Crisis in Central Asia - A Source For Future Conflicts? ENCOP Occasional Paper No. 11. Ceter for Security Policy and Conflict Research, Zurich, 1994.*
- *KolmannsKog, V.O., Future floods of refugees. A comment on climate change, conflict and forced migration (2008) Norwegian Refugees council on http://www.nrc. no/arch/- img/9268480. pdf. Cited 2009, June 26.*
- *Koppe, C.S., Kovatas, G. Jendritzky, B. Menne. Heat-waves: risks and responses, World Health Organisation. Global Change and Health, series no 2. Regional office for Europe, Copenhagen, 2004.*
- *Kundzewicz ZW, Mata LJ, Arnell NW, Doll P, Jimenez B, Miller KA, Oki T, Sen Z, Shiklomanov I A (2007). Freshwater Resources and their Management. Climate Change 2007: Impacts, Adaptation and Vulnerability. Contribution of Working Group II to the Fourth Assessment Report of the Intergovernmental Panel on Climate Change, M.L.*
- *L. Elliott, 'Imaginative Adaptations: A Possible Environmental Role for the UN Security Council Contemporary Security Policy 2003-24(2), pp.47-68;*
- *Lahsen M, et al., 2010: Impacts, adaptation and vulnerability to global environmental change.*
- *Lambert, J., Refugees and the environment: The forgotten element of sustainability Brussels, U.K. European parliament, 2002.*
- *Launder B. and J.M.T Thompson (2008), Global and Arctic climate engineering: numerical model studies.*
- *Laxmi, V., O. Erenstein and R.K. Gupta, 2007. Impact of zero tillage in India's rice - wheat systems (Mexico, D.F: CIMMYT, 2007).*

- *Leary et al., For whom the bell tolls: Vulnerability in a Changing Climate. A synthesis from the AIACC project. AIACC Working Paper No. 21 (Florida: International START Secretariat, 2006).*
- *Living with Risk: A global review of disaster risk reduction initiatives, preliminary version, UNISDR, Geneva, 2002.*
- *Lobell DB et al., Climate trends and global crop production since 1980, science 333:617-620, 2011.*
- *Lonergan, Environmental degradation and population displacement. A VISO Bulletin, Issue No. 2. Global Environmental Change and Human Security Project, Vancouver, 1993.*
- *Lopez. A., The protection of environmentally displaced persons in international Law. Environmental Law, vol. 37, no. 2, 2007, pp. 365-409.*
- *Ludi and bird: Offer great understanding in distinguishing between poverty and vulnerability. Vulnerability is dependent on the nature of the hazard, 2007, available at http://climateemergencyinstitute.com/food_sec_subsaharan_mburia.html*
- *M Ezzati et al., comparative qualification of health risks, 2004.*
- *M Parry et al., global food supply and risk of hunger, 2005.*
- *M. Koskenniemi, 'The place of low in collective security', Michigan Journal of International Law 1996-17, p. 456.*
- *MA (Millennium Ecosystem Assessment). Ecosystem services and human well-being: wetlands and water synthesis, World Resources Institute, Washington, DC, 2005.*
- *Macmillan Brown Centre for Pacific Studies Working Paper, Christchurch.*
- *Magrath, P. and Tesfu, M., Meeting the needs for water and sanitation of people living with HIV/AIDS in Addis Ababa, Ethiopia (Addis Ababa: Water Aid Ethiopia, 2006).*
- *Malhotra P and Rehman I (2004). Fire without Smoke. Delhi, The Energy Research Institute.*
- *Mannetje, L.t, The role of grasslands and forests as carbon stores. Wageningen, the Netherlands, University of Wageningen, 2006.*
- *Maskrey A. (ed.) (1993) Los disasters no son naturals. La Red/ Intermediate Technology Development Group (ITDG), Lima. http://www.desenredando.org/public/libros/1993/idnsn/index.html.*
- *Mc Carthy, J., Canziani, O., Leary, N., Dokkend. & White, K. (eds.) Climate Change 2001: Impacts, Adaptation & Vulnerability. (Cambridge: Cambridge University Press, 2001).*
- *Mc Michael AJ, et al., climate change and human health: risks and responses, WHO, Geneva, 2003.*
- *Mc Michael AJ, Githeko A. Human Health, climate change, 2001.*
- *Meehl, G.A., T.F. Stocker, W.D. Cllins, P. Friedlingstein, A.T. Gaye. (2007). Global Climate Projections. In: Climate Change 2007: The Physical Science Basis.*
- *Meze - Hausken, E. 'Migration Caused by Climate Change: How Vulnerable are People in Dryland Areas?' Mitigation and Adaptation Strategies for Global Change, 2000.*
- *Milanovic B (2005). Global income equalities. Presentation to the council on Foreign Relations. New York. December 13.*
- *Millennium Ecosystem Assessment, Geneva: WHO, 2005.*

- *Millennium Ecosystems Assessment. Ecosystems and human well-being: current state and trends: finding of the condition and trends working group, 2005.*
- *Mochizuki. K. Conflict and people's insecurity: An insight from the experiences of Nigeria, 2004.*
- *Molden, D. (ed.) Water for Food, Water for Life (London: Earthscan and Colombo: Water Management Institute, 2007).*
- *Molden, D., T. Oweis, P. Steduto, P. Bindraban, M. Hanjra and J. Kijne, Improving agricultural water productivity: Between optimism and caution. Agricultural Water Management 97, 2010, pp. 528-535.*
- *Molle, F. and J. Berkoff, Cities versus Agriculture: Revisiting Intersectoral Water Transfers, Potential Gains and Conflicts. Comprehensive Assessment of Water Management in Agriculture Research Report 10, International Water Management Institute, Colombo, 2006.*
- *Monteny, G.J., Bannink, A. and Chadwick, D. 2006.*
- *Moores, Climate change, water and China's national interest, China Security 2009.*
- *Moss RH, et al., 2011: The next generation scenarios for climate change research and assessment.*
- *Myers, N. 1993 Envirinmental refugees in a global warmed world, Bioscience, 43 (ii), pp. 752- 761.*
- *Myers, N. and kent, J, Environmental exodus: An emegen crisis in the global erena. Washington, DC. Climate Institute, 1995.*
- *Myers, N., Population, environment and conflict, Environmental Conservation, 1987.*
- *Negra C, et al., 2011 from REDD+ for Agriculture (CCAFS) Copenhagen.*
- *Nellemann, C., MacDevette, M., Manders, T., Eickhout, B., Svihus, B., Prins, A., and Kaltenborn, B. (eds.) the Environmental Food Crisis. The Environment's Role in Averting Future Good Crises. A UNEP Rapid Response Assessment (Arendal: UNDP, 2009).*
- *Nelson GC et al., 2009: climate change impact on agriculture and costs of adaptation. Food policy report.*
- *Nelson GC et al., 2010: Food security, Farming and climate change to 2050.*
- *Nelson, G.C. Are Biofuels the Best Use of Sunlight? In Handbook of Bioenergy Economics and Policy, ed. Madhu Khana and David Zilberman, 10 (New York: Springer, 2009).*
- *Newman P, ken worthy J. sustainability and cities 1999.*
- *Nicholls, R, Hoozemans, F., Marchand, M., Increasing Flood Risk and Wetland Losses Due to Global Sea-Level Rise: Regional and Global Analyses, 1999.*
- *O' Brien et al., Mapping vulnerability to multiple stressors: climate change and globalisation in India. Global environmental change, 14, 2004, pp. 303-313.*
- *O' Brien, G. and P. Read (2005) "Future UK emergency management: new wine, old skin? Disaster prevention and management.*
- *O' Brien. G. and P. Read (2004). Future UK Emergency Management: From discretion to regulation-panacea or long overdue reform? Proceedings of the International Emergency Management Society 11th Annual Conference, May 18-21, 2004, Melbourne, Australia.*
- *OECD (2003) Poverty and Climate Change. Reducing Vulnerability of the Poor through Adaptation (Paris: OECD, 2003).*

- *OFDA/CRED International Disasters Database EM - DAT.*
- *Ohlsson, L. Livelihood conflicts: linking poverty and environment as causes of conflict, 2000.*
- *Oke T. R (1973): city size and urban heat island, Atmospheric Environment.*
- *OLi Brown, 'The numbers game' in forced migration review, vol-31, October 2008.*
- *On the complex nexus linking migration, climate change and environment: Assessing the evidence, IOM, 2009, available at: index.php? Main page=product-info and products-id=539.*
- *P.Hough, understanding global security (London: Routledge, 2004), pp. 2-21.*
- *Padgham, J. (ed.), Agriculture Development under a changing climate. Opportunities and challenges for adaptation, Joint department discussion paper-issue1, World Bank, Washington, 2009.*
- *Parry et al., Climate change and hunger: Responding the challenge, world food programme (Rome, Italy, 2009).*
- *Patz, JA et al., Effects of environmental change on emerging parasitic diseases, 2000.*
- *Paul Collier, V.L Elliot, Havard Hegre, Anke Hoeffler, Nicholas Sambanis and Marta Reynal- Querol. Breaking the conflict trap: civil war and Development policy, Oxford University Press, 2003.*
- *Pauly D. Watson R, Alder J, Global trends in world fisheries, 2005.*
- *Peel, M-C., G-C. Pegram and T.A. McMahon. 2004a. Run length analysis of annual precipitation and runoff. International Journal of Climatology.*
- *Peel, M-C., T.A. McMahon and B.L. Finlayson. Continental differences in the variability of annual runoff-update and reassessment. Journal of Hydrology, 2004.*
- *Peluso, N., & Harwell, N. Territory, Custom, and the Cultural Politics of Ethnic War in West Kalimantan, Indonesia. In N. Peluso, & M. Watts (eds.), Violent Environments (Boulder: Lynne Rienner, 2001), pp. 43-68.*
- *Peterson, T. C., M. P. Hoerling, P. A. Stott and S. C. Herring, eds., Explaining Extreme Events of 2012 from a Climate Perspective. Special Supplement Bulletin of the American Meteorological Society vol. 94, no. 9, 2013.*
- *Pierre, J and G.B Peters 2005 .Governing comlex societies. Palgrave McMillan*
- *Policy: - Climate change refugees seek a new international deal (2008) climate news for business, climate change corp. http corp.com/content.asp? Content ID= 5871, Cited 2009, June 26.*
- *Portes A., Social capital: its origin and applications in modern sociology, 1998.*
- *Poverty and climate change: Reducing the vulnerability of the poor through adaptation, June 2003, World Bank. Org/povcc.*
- *Prescott-Allen, R., The Wellbeing of Nations: A Country-By-Country Index of Quality of Life and the Environment (Washington, Island Press, 2001).*
- *Pretty et al., Sustainable intensification in African agriculture. International Journal of Agricultural Sustainability, vol. 9, no. 1, 2011, pp. 5-24.*
- *Pro act network. The role of Environmental Management and co-engineering in disaster risk reduction and climate change adaptation, 2008.*
- *R. Cryer, 'the Security Council and Article 39: A Threat To Coherence? Journal of Armed Conflict Law 1996-1, pp. 161-195; P.H. Kooijmans, 'The Enlargement Of The Concept "Threat To The Peace", In: R.-J. Dupuy (Ed.) The Development Of The Role Of The Security Council, Dordrecht: Martinus Nijhoff, 1993.*

- *Rahman, A., Climate Change and Violent Conflicts, In Sulimans, M., (Ed.), Ecology, Politics and Violent Conflict (London and New York: Zed Books, 1999).*
- *Raknereud, A. And Hegre, H., The hazard of war: reassessing the evidence for the democratic peace. Journal of Peace Research, vol. 34, no. 4, 1997, pp. 385-404.*
- *Ratha.D and Shaw.w - South-South migration and remittances, Washington, DC, 2007.*
- *Redclift, M., (Eds.), Human Security and the Environment: International Comparisons.*
- *Reed, S, environment and security (2007) Topics/ core issues, climate institute, on http://www. climate. org/topics/environmental security/index.html, Cited 2009, June 30.*
- *Reno, W. 'Shadow States and the Political Economy of Civil Wars', In Berdal, M. and Malone, D. (eds.) Greed and Grievance: Economic Agendas in Civil Wars, 2000.*
- *Resolving conflict in Solomon Islands; "The women for peace Approach," Alice pollard. Development bulletin, November 2000, http://devnet.anu.edu.au/db53. html.*
- *Robine JM et al., Death toll, exceeded 70,000 in Europe during the summer of 2003, Les competes Rendus/Series Biologies, 2008, vol. 331, pp. 171-78.*
- *Rockstrom, J., J. Barron and F. Fox. Water Productivity in Rainfed Agriculture: challenges and opportunities for small holder farmers in drought-prone tropical Agro-Ecosystems, 2001.*
- *Roe. D et al., 2006, International Institute for Environmental and Development.*
- *Rogers, D., and V. Tsirkunov, Weather and Climate Resilience: Effective Preparedness through National Meteorological and Hydrological Services. Direction in Development; World Bank, Washington, DC, 2013.*
- *Rotz, C.A.2004. Management to reduce nitrogen losses in animal production. Journal of Animal science 82 (e.SUPPL) E119-E137.*
- *Ruane, J., A. Sonnino, P. Steduto and C. Deane. Coping with water scarcity: What role for biotechnologies? Land and water discussion paper no. 7, FAO, Rome, 2008.*
- *Russell 'S., International Migration: Implications for the World Bank. Human Resources Development and Operations Policy Working Papers Number 54 (Washington: World Bank, 1995).*
- *Sabine, C.L. et al., 2004: The oceanic sink for anthropogenic CO2.*
- *Sarewitz D, Pielke R, and Keykhah R, Vulnerability and risk: some thoughts from a polical and policy perspective, Risk Analysis, 23, 2003, pp. 805-810.*
- *Sattherwaite, D. et al., "Adapting to climate change in urban areas: The possibilities and constraints in low-and-middle-income nations"; Human Settlements Discussion Paper Series: Climate Change and Cities1, International Institute of Environment and Development, London, 2007.*
- *Scawthorn, C. 'Emergency Water Supply and Disaster Vulnerability'. In J. Uitto and A. Biswas (Eds.) Water for Urban Areas. United Nations University Press. Tokyo, 2000, pp. 200-225.*
- *Schmidhuber, J. and F. Tubiello, Global Food Security under Climate Change. PNAS, vol. 104, no. 50, 2007, pp. 19703 - 19708.*
- *Schmidhuber, J., J. Bruinsma and G. Boedeker. Capital Requirements for Agriculture in Developing Countries to 2050. Expert meeting on how to feed the World in 2050*

Food and Agriculture Organisation of the United Nations. Economic and Social Development Department, Rome, 2009.

- *Scott, C.A., N.I. Faruqui and L. Raschid-Sally. Waste Water Use in Irrigated Agriculture Confronting the Livelihood and Environmental Realities. Oxford University Press. SBN13: 9780851998237, 2004.*

- *Scott, L, 2008, "Climate variability and climate change. Implications for chronic poverty", working paper 108; www.chronicpoverty.org/pubfiles/108.*

- *See for example the broader definitions of refugee contained in the OAU convention governing the specific aspects of refugee problems in Africa 1969, at article 1(2). http://www.unhcr.org/refworld/docid/3ae6b36018.html and the Cartagena declaration on refugees 1984, at conclusion 3, available at: http://www.unhcr.org/refworld/docid/4538838e 10.html.*

- *Seguin J, editor, Human health in a changing climate, 2008.*

- *Sen A (1999). Development as Freedom (New York: Anchor Books).*

- *Sen, A. Development as Freedom (New York: Anchor Books, 1999).*

- *Shah, M., Fischer, G., and Van Velthuizen, H., Food Security and Sustainable Agriculture. The Challenges of Climate Change in Sub - Saharan Africa (Laxenburg: International Institute for Applied Systems Analysis, 2008).*

- *Shaw, B., When are environmental issues security issues? Environmental Change and Security Project Report, 1996.*

- *Shepherd et al., The geography of poverty, disasters and climate extremes in 2030. ODI, met office Hadley Centre, RMS Publicati, Exeter, 2013.*

- *Shvidenko, A. et al., Regional certificate full carbon account. Fusion of Remotely sensed Data, on-ground Information and Ecological Modelling. Paper presented at the EGUO5 General Assembly of the European Geosciences Union, Vienna, April 22-27, 2005.*

- *Sir John Holmes, under-Secretary-General for Humanitarian Affairs and Emergency, Relief Co-ordinator, Opening Remarks at the Dubai International Humanitarian Aid and Development Conference and Exhibition 'DIHAD 2008 Conference' & April 2008, available at : http://www. reliefweb.int/rw/rwb.nsf/db900sid/YSAR-7DHL88?open document.*

- *Skoufias, E., (ed.) The poverty and welfare impacts of climate change: Quantifying the effects, identifying the adaptation strategies. World Bank, Washington, DC, 2012.*

- *Skoufias, E., (ed.) The Poverty and Welfare Impacts of Climate Change: Quantifying the Effects, Identifying the Adaptation Strategies. World Bank, Washington, DC, 2012.*

- *Smill, V. Nitrogen in crop production: an account of global floods. Global Biogeochemical cycles, vol. 13, no. 2, 1999, pp. 647-662.*

- *Smit KR, Balkrishnan K. Mitigate climate, http://www.thecommonwealth.org/files/190381/file name/4 kirksmith, 2009pdf.*

- *Smith, P., D. Martino, Z. Cai, D. Gwary, H. Janzen, H. Kumar, B. Mc Carl, S. Ogle, F. O'Mara, C. Rice, R. Schools, O. Sirotenko, M. Howden, T. Mc Allister, G. Pan, V. Romanenkov, U. Schneider, S. Towprayoon, W. Wattenbach and J. Smith. 2008. Greenhouse Gas Mitigation in Agriculture, Phil. Trans. R. Soc. B (2008) 363, pp. 789-813.*

- *Sperling, F. (ed.) Poverty and climate change: reducing the vulnerability of the* poor through adaptation. Washington, DC: AFDB, ASDB, DFID, Netherlands, EC, Germany, OECD, UNDP, UNEP and the World Bank (VARG), 2003.

- Stern et al., "Stern review on the economics of climate change",
- Stern, N. Economics of climate change: The stern review (Cambridge: Cambridge University Press, 2007).
- Stern, N., "Key Elements of A Global Deal on Climate Change London", London School of Economics and Political Science, London, 2008.
- Stockholm Plan of Action for Integrating Disaster Risks and Climate Change Impacts in Poverty Reduction: http://www.unisdr.org/eng/partner-netw/wb-isdr/docs/stockholm-plan-of-action.pdf.
- Stone et al., urban form and extreme heat events and air pollution in cities, 2010.
- Stott PA, Stone DA, Allen MR Nature, 2004. Human contribution to the European heatwaves of 2003.
- Stott, P.A., Stone, D.A. and Allen, M.R. 2003. Human Contribution to the Europe Heat Wave in 2003. Nature 432, pp. 610-614.
- Stripple, J., Climate change as a security issue, 2002.
- Sudmeier - Rieux and Ash. Environmental Guidance Note for Disaster Risk Reduction (IUCN: Gland, 2009).
- Sudmeier- rieux et al., Ecosystems, livelihoods and disasters- an integrated approach to disaster risk management. Ecosystem Management Series No.4. IUCN, 2006.
- Summery of IASC expert meeting on Migration/Displacement and climate change September 15, 2008.
- Swain, A., Conflicts over Water: The Ganges Water Dispute. Security Dialogue vol. 24, no. 4, 1993, pp. 429-439.
- Swain, A., The Environmental Trap: The Ganges River Diversion, Bangladeshi Migration and Conflicts in India, Department Of Peace and Conflict Research Uppasala University Report, Sweden, 1996.
- Swaminathan, H., Suchitra, J.Y. and Lahoti, R. KHAS: Measuring the Gender Asset Gap (Bangalore: Indian Institute of Management Bangalore, 2011).
- Swart, R. Security risks of global environmental changes. Global Environmental Change, 1996.
- The IEA (2008b). In Support of the G8 Plan of Actions. Energy, Efficiency Policy Recommendations. Paris, International Energy Agency.
- The rights under the 1966 international covenant on economic, social and cultural rights (ICESCR), http://www.unhcr.org/refworld/ docid/4538838e10.html.
- The UN Global Assessment Report on Disaster Risk Reduction is a biennial report of the United Nations coordinated and produced by the UNISDR: http://www.preventionweb.net/english/ hyogo/gar/
- The world health report 2004 changing history: WHO; 2004.
- Thomton Pk, Gerber PJ; 2010: Climate change and the growth of the livestock sector in developing countries.
- Tim Flannery (November 2009). Copenhagen and beyond: conference Bound. The monthly, archieved from the original on May 6, 2010, retrieved April 8, 2010.
- Tubiello, F. And G. Fischer. Reducing climate change impacts on agriculture: Global and regional effects of mitigation, 2000-2080 Technological Forecasting and Social Change 741030-1056, 2007.

- *UN (2003), The Human Rights Based Approach to Development Cooperation, towards a common understanding among UN agencies.*
- *UN Framework Convention on Climate Change (May 1992). Frank Biermann and Ingrid Boas, preparing for a warmer world: Towards a global governance system to protect climate refugees (2010).*
- *UN/ ISDR, 2005: World conference on Disaster Reduction Kobe, January 18-22, 2005. http:// www.unisdr.org/*
- *UN/ISDR (United Nations inter-agency secretariat of the international strategy for Disaster Reduction) living with risk. A global review of disaster reduction initiatives. Geneva, 2004.*
- *UN/ISDR, 2004: (United Nations Interagency Secretariat of the International Strategy for Disaster Reduction), A Global Review of Disaster Reduction Initiatives. Geneva.*
- *UNDP (2005). Energising the Millennium Developments Goals: A Guide to Energy's Role in Reducing Poverty. New York, United Nations Development Programme.*
- *UNDP (United Nations Development Program). Ukun rasik a'an: East Timor human development report 2002 (Dili: UNDP, 2002).*
- *UNDP (United Nations Development Programme) 2004, Reducing Disaster Risk: A challenge for Development, UNDP. Bureau for Crisis Prevention and Recovery. New York.*
- *UNDP (United Nations Development Programme) 2004a. Reducing Disaster Risk: A Challenge for Development. UNDP, Crisis Prevention and Recovery, Disaster Reduction Unit, Geneva. http://www.undp.org/bcpr/disred/english/publications/rdr.htm.*
- *UNDP (United Nations Development Programme) 2007. Human Development Report 2007/2008: Fighting Climate Change: Human Solidarity in a Divided World (Basingstoke: Palgrave).*
- *UNDP (United Nations Development Programme), Human Development Report (Geneva: United Nations, 1994).*
- *UNDP Fighting Climate Change - Human Solidarity in a Divided World. (New York: UNDP, 2008).*
- *UNEP further estimates that over US $200 billion annually are needed from now till 2050, in order to promote sustainable growth in the agriculture sectors of LCDs.*
- *UNEP Montreal Protocol, 1987.*
- *UNEP/UNISDR. Environmental and disaster risk. Emerging perspectives 2nd edition (UNISDR Secretariat: Geneva, 2008).*
- *UNER, Towards a green economy: Pathways to sustainable Development and poverty Eradication, 2011.*
- *UNFCCC (2006a): dialogue on long-term cooperative action to address climate change by enhancing implementation of the convention. https://unfccc.int/meetings/bonn_may_2006/ meeting/6327.php.*
- *UNFCCC submission paper, 'protecting the health of vulnerable people from the humanitarian consequences of climate change and climate related disasters of Ad Hoc Working Group on Long-Term Cooperative Action, June 1-12, 2009 by WHO with UNHCR and other agencies, available at, http://www.unhcr.org/refworld/docid/4a2d189e1a.html. http://www.unhcr.org/refworld/docid/479744c42.html.*
- *UNFCCC, (1997). Proposal Elements of a Protocol to the United Nations*

Framework Convention On Climate Change, Presented By Brazil In Response to the Berlin Mandate, Paper No.1 of the Note by the Secretariat: Implementation of the Berlin Mandate: Additional Proposals from Parties, Addendum. FCCCC/ AGBM/1997/Misc.1/Add.3.

- *UNFCCC, 2011. Adaptation. http://UNFCCC.int/adaptation/items/4159.php-26 Sep, 2011.*
- *UNFCCC, available at: http://www.unhcr.org/refworld/docid/3boof2770.html.*
- *UNFPA, State of World Population 2007: Unleashing the Potential of Urban Growth (New York: United Nations Population Fund, 2007).*
- *UNHRC (United Nations Human Rights Council) 2009: Report of the Office of the United Nations High Commissioner for Human Rights on the Relationship between Climate Change and Human Rights.*
- *UNICEF (United Nations Children's Fund) 2005. Emergency Field Handbook: A Guide for UNICEF staff. UNICEF, New York, NY. http://www.unicef.org/ publications/files/UNICEF-EFH-2005.pdf.*
- *UNISDR, (2009b). Risk and poverty in a changing climate. Global Assessment Report On Disaster Risk Reduction. United Nations office for Disaster Risk Reduction, Geneva.*
- *UNISDR, Risk and poverty in a changing climate: in rest today for a safer tomorrow 2009 Global Assessment Report On Disaster Risk Reduction, United Nations, Geneva, 2009, available at www. preventionweb.net/english/hyogo/gar/report/ index.php?id=1130&pid:pih:2.*
- *United Nations Population Division 2010. World population prospects: the 2010 revision. New York: United Nations Department Of Economic And Social Affairs, United Nations Population Division, at http://esa.un.org/wpp/unpp/panel-population.htm.*
- *United Nations, The Millennium Development Goals Report 2011 (New York: United Nations, 2011).*
- *USDA. The Agricultural Model Inter comparison and improvement projects, 2011.*
- *Van Dijk et al., Forest-flood relation still tenuous- comment on 'global evidence that deforestation amplifies flood risk and severity in the developing world'. Global Change Biology, 2009.*
- *Van Ire Land et al., Climate change: socio-economic impacts and violent conflict. Dutch National Research Programme on Global Air Pollution and Climate Change. Report No. 410200 006, Wageningen, 1996.*
- *Van Ireland, E., Klassen, M., Nierop, T., Van Der Wusten, H., Climate Change: Socio-Economic Impacts and Violent Conflict. Dutch National Research Programme on Global Air Pollution and Climate Change, 1996.*
- *Van wormer, K, Besthorn, F.H. & Keefe, T., Human behaviour and the Social environment Mavro Level: groups, communities and organisations (New York, NY: Oxford University Press, 2007).*
- *VARG (Vulnerability and Adaptation Resource Group) 2005. Disaster Risk Management in a changing climate: a discussion paper. http://www.unisdr.org/ wcdr/*
- *Vasquez, J., The war puzzle (Cambridge: Cambridge University Press, 1993).*
- *Verbury PH et al., 2011: changes in using land use and land cover data for global change studies.*

- *Vermeulen S.J, Aggarwal PK, Ainslie A, Angelone C, Campbell BM, Challinor AJ, Hansen JW, Igram JSI, Jarvis A, Kristjanson P, Lau C, Nelson GC, Thorton PK, Wollenberg E. 2012. Options for support to agriculture and food security under climate change. Environmental Science and Policy, 15, pp. 136-144.*
- *Vermeulen SJ, et al., 2010: Agriculture, food security and climate change.*
- *Vienna, August 27-31, 2007, at the 4th workshop under the "Dialogue on long-term cooperative action to address climate change by enhancing implementation of the convention.*
- *Wall DH, Rabbinge R, Gallopin G, et al., Implications for achieving the Millennium Development Goals, 2005.*
- *Wallensteen, P. and Sollenberg, M., Armed Conflicts, Conflicts Termination and Peace Agreements 1989-96, Journal of Peace Research, 1997.*
- *Watson, R. Presentation of the Chair of the Inter-governmental Panel on Climate Change to the Sixth Conference of Parties of the United Nations Framework Convention on Climate Change. November 13, 2000, Geneva, at http://www.ipcc. ch.*
- *Weiss R, Mc Michael A. Social and environmental risk factors in the emergence of infectious diseases, 2004.*
- *Westra, Laura, Environmental Justice and the rights of ecological refugees (London, UK: Earth Scan, 2009).*
- *WHO (2003). Climate change and human health: risk and responses. Geneva: WHO. Available online at http://www.who.int/globalchange/climate/en/ccScREEN. pdf.*
- *WHO Framework convention on Tobacco control, 2003.*
- *WHO guidelines for the safe use of waste water, excreta and grey water-3rd edition, Geneva: WHO, 2006*
- *WHO, protecting Health from climate change, 2009.*
- *WHO, World Health Report 2002: Reducing risks, promoting healthy life, WHO, Geneva, 2002.*
- *WHO, Zoll. Regional office for South East Asia; sustainable development and healthy environment; water, sanitation and healthy.*
- *Wiggins S. 'Rising Food Prices - A Global Crisis'. Briefing Paper No. 37 (London: ODI, 2008).*
- *Wisner, B. and J. Adams, Environmental Health in Emergencies and Disasters: A Practical Guide. World Health Organisation, Geneva, 2003. http://www.who.int/ watersanitationhealth/hygiene/emergencies/emergenes 2002/en/*
- *WMO (World Meteorological Organisation), WMO statement on the Status of the Global Climate in 2003. WMO - No.966. Geneva, 2004, pp. 11.*
- *Wolf J. et al., Social capital, individual responses to heat waves and climate change adaptation, 2010.*
- *Wolf, A., 'Water wars' and water reality: conflict and co-operation along international waterways, in lonergan, S., (ed.), Environmental Change, Adaptation, and Security (Dordrecht: Kluwer Academic Publishers, 1999).*
- *Wood cock et al., Public health benefits of strategies to reduce greenhouse gas emissions: urban land transport, 2009.*
- *World Bank (2006 b). Energy Poverty Issues and G8 Actions: Discussion Paper. Moscow- Washington, World Bank.*

- *World Bank (2012c). The Sendai Report: Managing Disaster Risks for a resilient future. World Bank, GFDRR and government of Japan. Washington, DC.*
- *World Bank (2012d). Turn down the Heat. A report for the World Bank by the Potsdam institute for climate impact research and climate analytics. Washington, DC.*
- *World Bank 2013b. Managing Risk for Development. World Development Report 2014. Washington, DC.*
- *World Bank, (2008): Strategic Climate Fund. June 2008. Washington, D.C. World Bank, available at http://siteresource.worldbank.org/INTCC/Resources/strategic_climate_fund_final. pdf#stratgic_climate_fund (Accessed September 15, 2008).*
- *World Bank, 2013c. Zambia strengthening climate resilience (PPCR Phase II) Project, Report Number: 73982-zm, Washington, DC.*
- *World Bank, World Development Report 2011: Conflict, Security and Development (Washington, 2011).*
- *World Bank. (2013b). Managing risk for development. World development report 2014. Washington, DC.*
- *World Bank: World Development Report 2008: Agriculture for Development.*
- *World population prospects, the 2010 revision, New York, at http://esa.un.org/UNpd/Wpp/index. htm.*
- *World Watch institute, State of the world: innovations that nourish the planet (New York: WW Norton & Company, 2011).*
- *World Water Assessment Programme, 2009: The United Nations World Water Development Report 3: Water in a Changing World (Paris, UNESCO and London, Earthscan, 2009).*
- *www.hm-treasury.gov.uk/sternreview-index.htm, HM Treasury, London and Cambridge University Press, 2006.*
- *www.migration drc. Org & C.R Parsons, R Skeldon, T.L Walmsley and L.A Winters, International migration, Economic development and policy, Washington, The World Bank 2007.*
- *Younger et al., The built environment, climate change and health opportunities for co-benefits, 2008.*
- *Zhou XN et al., American journal of tropical medicine and Hygiene, 2008, vol. 78, pp. 188-194. WCED, Our Common Future (Oxford: Oxford University Press, 1987).*